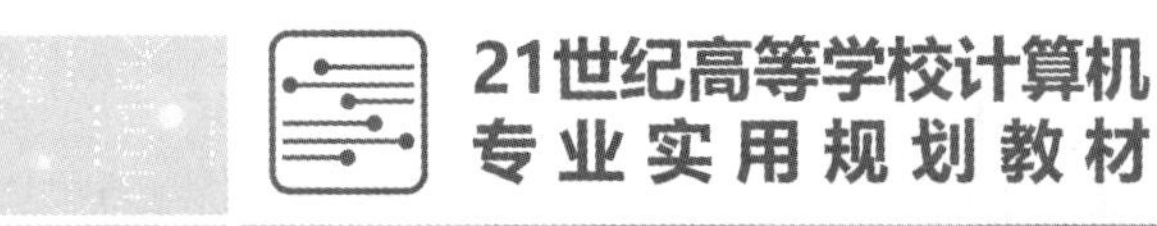

计算机专业英语

◎ 戚文静 李晓峰 刘学 李国文 编著

清華大學出版社
北京

内 容 简 介

本书从计算机科学各个领域的最新教材、专著、论文、百科等英文素材中选择各分支学科的基本概念、方法及最新发展等内容，涵盖计算机硬件、软件、数据库、网络、信息安全、人工智能、机器学习、人机交互、量子计算、大数据、云计算等领域的基础理论和应用，内容具有基础性、趣味性、广泛性和知识性。读者通过阅读和学习本书，能够掌握大量的英文专业术语、熟悉常用的语法和句式表达方法，提高读者在计算机科学相关领域的英语应用能力。同时本书精心选择的内容还可帮助读者丰富计算机学科知识、拓展科学研究视野。

本书可作为普通高等学校计算机科学与技术、网络工程、软件工程及相关专业的英语教材或计算机导论课程的双语教材，也可作为从事相关行业的科研与工程技术人员的参考用书。

图书在版编目（CIP）数据

计算机专业英语/戚文静等编著. —北京：清华大学出版社，2019（2024.8重印）
（21世纪高等学校计算机专业实用规划教材）
ISBN 978-7-302-51129-8

Ⅰ. ①计… Ⅱ. ①戚… Ⅲ. ①电子计算机–英语–高等学校–教材 Ⅳ. ①TP3

中国版本图书馆 CIP 数据核字（2018）第 202518 号

责任编辑： 黄　芝　薛　阳
封面设计： 刘　键
责任校对： 梁　毅
责任印制： 宋　林

出版发行： 清华大学出版社
网　　址：https://www.tup.com.cn, https://www.wqxuetang.com
地　　址：北京清华大学学研大厦 A 座　　邮　　编：100084
社 总 机：010-83470000　　邮　　购：010-62786544
投稿与读者服务：010-62776969，c-service@tup.tsinghua.edu.cn
质量反馈：010-62772015，zhiliang@tup.tsinghua.edu.cn
印 装 者： 三河市君旺印务有限公司
经　　销： 全国新华书店
开　　本： 185mm×260mm　　**印　张：** 22　　**字　　数：** 534 千字
版　　次： 2019 年 6 月第 1 版　　**印　　次：** 2024 年 8 月第 5 次印刷
印　　数： 5001～6000
定　　价： 59.00 元

产品编号：054041-01

出版说明

随着我国改革开放的进一步深化，高等教育也得到了快速发展，各地高校紧密结合地方经济建设发展需要，科学运用市场调节机制，加大了使用信息科学等现代科学技术提升、改造传统学科专业的投入力度，通过教育改革合理调整和配置了教育资源，优化了传统学科专业，积极为地方经济建设输送人才，为我国经济社会的快速、健康和可持续发展以及高等教育自身的改革发展做出了巨大贡献。但是，高等教育质量还需要进一步提高以适应经济社会发展的需要，不少高校的专业设置和结构不尽合理，教师队伍整体素质亟待提高，人才培养模式、教学内容和方法需要进一步转变，学生的实践能力和创新精神亟待加强。

教育部一直十分重视高等教育质量工作。2007 年 1 月，教育部下发了《关于实施高等学校本科教学质量与教学改革工程的意见》，计划实施“高等学校本科教学质量与教学改革工程（简称‘质量工程’）”，通过专业结构调整、课程教材建设、实践教学改革、教学团队建设等多项内容，进一步深化高等学校教学改革，提高人才培养的能力和水平，更好地满足经济社会发展对高素质人才的需要。在贯彻和落实教育部“质量工程”的过程中，各地高校发挥师资力量强、办学经验丰富、教学资源充裕等优势，对其特色专业及特色课程（群）加以规划、整理和总结，更新教学内容、改革课程体系，建设了一大批内容新、体系新、方法新、手段新的特色课程。在此基础上，经教育部相关教学指导委员会专家的指导和建议，清华大学出版社在多个领域精选各高校的特色课程，分别规划出版系列教材，以配合“质量工程”的实施，满足各高校教学质量和教学改革的需要。

本系列教材立足于计算机专业课程领域，以专业基础课为主、专业课为辅，横向满足高校多层次教学的需要。在规划过程中体现了如下一些基本原则和特点。

（1）反映计算机学科的最新发展，总结近年来计算机专业教学的最新成果。内容先进，充分吸收国外先进成果和理念。

（2）反映教学需要，促进教学发展。教材要适应多样化的教学需要，正确把握教学内容和课程体系的改革方向，融合先进的教学思想、方法和手段，体现科学性、先进性和系统性，强调对学生实践能力的培养，为学生知识、能力、素质协调发展创造条件。

（3）实施精品战略，突出重点，保证质量。规划教材把重点放在公共基础课和专业基础课的教材建设上；特别注意选择并安排一部分原来基础比较好的优秀教材或讲义修订再版，逐步形成精品教材；提倡并鼓励编写体现教学质量和教学改革成果的教材。

（4）主张一纲多本，合理配套。专业基础课和专业课教材配套，同一门课程有针对不同层次、面向不同应用的多本具有各自内容特点的教材。处理好教材统一性与多样化，基本教材与辅助教材、教学参考书，文字教材与软件教材的关系，实现教材系列资源配套。

（5）依靠专家，择优选用。在制定教材规划时要依靠各课程专家在调查研究本课程教材建设现状的基础上提出规划选题。在落实主编人选时，要引入竞争机制，通过申报、评审确定主题。书稿完成后要认真实行审稿程序，确保出书质量。

繁荣教材出版事业，提高教材质量的关键是教师。建立一支高水平教材编写梯队才能保证教材的编写质量和建设力度，希望有志于教材建设的教师能够加入到我们的编写队伍中来。

21世纪高等学校计算机专业实用规划教材
联系人：黄芝 huangzh@tup.tsinghua.edu.cn

前　言

“计算机英语”是计算机及相关专业的一门专业基础课程。由于计算机相关核心技术大部分源于英语国家，很多的技术文档资料、编程工具原版都以英语为主，因此较好的英文水平有利于专业和职业的长远发展；另外，计算机技术更新速度极快，如果不掌握一定的专业词汇，必然会影响对新技术的理解和消化。为了提高学生在计算机相关领域的英语应用能力，使学生在有限的学时中比较全面地掌握专业词汇、能顺利地阅读和理解专业文献，同时保证学生的学习兴趣、拓展专业知识面、避免陷于深奥的专业知识，本教材在内容的编排和选择上坚持以下三个原则。

基础性和趣味性原则：每一部分内容介绍计算机科学一个学科领域的科普性内容，避免出现较深奥和晦涩的概念及理论，易于读者阅读和理解。

广泛性和知识性原则：力求覆盖计算机科学相关领域内容，高度概括每个领域的发展历史、研究内容和发展情况，使不同层次的读者通过阅读和学习可以对该领域有全面、客观的了解。

可靠性和实效性原则：素材取自国内外最近几年计算机科学各个领域的最新教材、专著、论文、学者访谈、百科等可靠来源，使读者可以了解计算机科学与技术的最新进展。

本教材每章内容相对独立，教师可根据实际教学需要确定重点授课内容。本书第 1 章概括计算机科学的发展历史、学科分支及未来发展方向；第 2 章集合了计算机体系结构、硬件、互联网、无线网方面的基础知识；第 3 章介绍操作系统的常识，包括操作系统的定义、任务及常用操作系统的案例及对比；第 4 章介绍算法、数据结构和软件工程方面的知识；第 5 章集合了关系数据库、查询语言、信息检索和网页检索的基础概念和常识；第 6 章为人工智能相关领域的知识，包括图灵测试、知识表达和推理、机器人、计算机视觉、人工智能可能存在的风险等；第 7 章为计算图形学领域的分支和应用，包括计算机辅助设计、3D 建模、虚拟现实、数据可视化等；第 8 章为人机交互技术方面的相关知识，包含人机交互的发展、人机界面的设计原理和规则等；第 9 章为计算机安全相关知识，包括基本概念、安全措施、密码学及网络战争等内容；第 10 章介绍计算机科学与技术领域的一些最新进展，包括量子计算、深度学习、云计算和大数据方面的内容。本书不仅是一本计算机专业英语的教材，也是对计算机专业知识的一个概览，具有很强的科学性、知识性、趣味性。

本书的第 1～4 章主要由刘学副教授编写；第 5 章和第 6 章主要由李晓峰教授编写；第 7、8、10 章及附录词汇部分主要由戚文静教授编写和整理；第 9 章和第 10 章由李国文博士编写；最后由戚文静完成对本书的统稿；另外，刘宇祺、王亦佳等同学参与了本书稿的录入和校对工作，赵敬教授对本书的编写提出了很多宝贵意见。

另外，在清华大学出版社编辑的大力支持和协助下，本书得以在较短时间内出版和发行，在此对他们的工作表示诚挚的感谢。同时，感谢戚文静家人对我的理解、支持和关爱，使得本书稿能够顺利完成。

教材中难免有不足和疏漏之处，恳请广大读者提出宝贵意见和建议，以便进一步完善本教材。

戚文静

2018 年夏于映雪湖

Contents

Chapter 1 Introduction of Computer Science

1.1 History of Computer Science

The start of the modern science that we call "Computer Science" can be traced back to a long ago age. In Asia, the Chinese were becoming very involved in commerce with the Japanese, Indians, and Korcans. Businessmen needed a way to tally accounts and bills. Somehow, out of this need, the abacus was born. The abacus is the first true precursor to the adding machines and computers which would follow. For over a thousand years after the Chinese invented the abacus, not much progress was made to automate counting and mathematics. The Greeks came up with numerous mathematical formulae and theorems, but all of the newly discovered math had to be worked out by hand. Most of the tables of **integrals**, **logarithms**, and **trigonometric** values were worked out this way, their accuracy unchecked until machines could generate the tables in far less time and with more accuracy than a team of humans could ever hope to achieve.

Blaise Pascal, noted mathematician, thinker, and scientist, built the first mechanical adding machine in 1642 based on a design described by Hero of Alexandria to add up the distance a carriage travelled. The basic principle of his calculator is still used today in water meters and modern-day **odometers**. This first mechanical calculator, called the Pascaline, as shown in Fig.1-1, had several disadvantages. Although it did offer a substantial improvement over manual calculations, only Pascal himself could repair the device and it cost more than the people it replaced! In addition, the first signs of **technophobia** emerged with mathematicians fearing the loss of their jobs due to the progress.

Fig.1-1 Pascal calculating machine

Fig.1-2 Arithmometer

The **Arithmometer**, as shown in Fig.1-2, was the first mechanical calculator strong enough and reliable enough to be used daily in an office environment. This calculator could add and subtract two numbers directly and could perform long multiplications and divisions effectively by using a movable **accumulator** for the result. Patented in France by Thomas de Colmar in 1820 and manufactured from 1851 to 1915, it became the first commercially successful mechanical calculator.

While Thomas de Colmar was developing the first successful commercial calculator, Charles Babbage realized as early as 1812 that many long computations consisted of operations that were regularly repeated. He theorized that it must be possible to design a calculating machine which could do these operations automatically. He produced a **prototype** of this "**difference engine**" by 1822 and with the help of the British government and started working on the full machine in 1823. It was intended to be **steam-powered**; fully automatic, even to the printing of the resulting tables; and commanded by a fixed **instruction program**. This machine used the decimal number system and was powered by cranking a handle. The British government was interested, since producing tables was time consuming and expensive and they hoped the difference engine would make the task more economical[3].

In 1833, Babbage ceased working on the difference engine because he had a better idea. His new idea was to build an "**analytical engine**." The analytical engine was a real parallel decimal computer which would operate on words of 50 decimals and was able to store 1000 such numbers. The machine would include a number of built-in operations such as **conditional control**, which allowed the instructions for the machine to be executed in a specific order rather than in numerical order. The instructions for the machine were to be stored on **punched cards**, similar to those used on a Jacquard loom.

A step toward automated computation was the introduction of punched cards, which were first successfully used in connection with computing in 1890 by Herman Hollerith working for the US Census Bureau. He developed a device which could automatically read census information which had been punched onto card. Surprisingly, he did not get the idea from the work of Babbage, but rather from watching a train conductor punch tickets. As a result of his invention, reading errors were consequently greatly reduced, work flow was increased, and, more important, stacks of punched cards could be used as an accessible memory store of almost unlimited capacity; furthermore, different problems could be stored on different batches of cards and worked on as needed. Hollerith's **tabulator** became so successful that he started his own firm to market the device; this company eventually became International Business Machines (IBM).

Hollerith's machine though had limitations. It was strictly limited to tabulation. The punched cards could not be used to direct more complex computations. In 1941, Konrad Zuse, a German who had developed a number of calculating machines, released the first programmable computer designed to solve complex engineering equations. The machine, called the Z3, was controlled by **perforated strips** of discarded movie film. As well as being controllable by these celluloid strips, it was also the first machine to work on the **binary system**, as opposed to the more familiar **decimal system**. Binary representation was proven to be important in the future design of computers which took advantage of a multitude of two-state devices such card readers, electric circuits which could be on or off, and vacuum tubes.

By the late 1930s, punched-card machine techniques had become so well established and reliable that a large automatic digital computer, called the Harvard Mark I, was constructed, which could handle 23-decimal-place numbers and perform all four **arithmetic operations**; moreover, it had special built-in programs, or **subroutines**, to handle logarithms and trigonometric functions. Meanwhile, the British mathematician Alan Turing wrote a paper in 1936 entitled On Computable Numbers in which he described a hypothetical device, a **Turing machine**, which presaged programmable computers. The Turing machine was designed to perform logical operations and could read, write, or erase symbols written on squares of an infinite paper tape. This kind of machine came to be known as a **finite state machine** because at each step in a computation, the machine's next action was matched against a finite instruction list of possible states.

Back in America, John W. Mauchly and J. Presper Eckert at the University of Pennsylvania built giant ENIAC machine. ENIAC contained 17,468 vacuum tubes, 7,200 crystal diodes, 1,500 relays, 70,000 resistors, 10,000 capacitors and around 5 million hand-soldered joints. It weighed more than 30 short tons (27 t), was roughly 8 by 3 by 100 feet (2.4m × 0.9m × 30m), took up 1800 square feet ($167m^2$), and consumed 150kW of power[4]. This led to the rumor that whenever the computer was switched on, lights in Philadelphia dimmed. ENIAC is generally acknowledged to be the first successful high-speed electronic digital computer, it was efficient in handling the particular programs for which it had been designed and was productively used from 1946 to 1955.

In 1945, mathematician John von Neumann contributed a new understanding of how practical fast computers should be organized and built; these ideas, often referred to as the **stored-program technique**, became fundamental for future generations of high-speed digital computers and were universally adopted. The primary advance was the provision of a special type of machine instruction called conditional control transfer, which permitted the program sequence to be interrupted and reinitiated at any point, and by storing all instruction programs, instructions could be arithmetically modified in the same way as data. As a result, frequently used subroutines did not have to be reprogrammed for each new problem but could be kept intact in "libraries" and read into memory when needed. The computer control served as an errand runner for the overall process. The first-generation stored-program computers required considerable maintenance, attained perhaps 70% to 80% reliable operation, and were used for 8 to 12 years. Typically, they were programmed directly in **machine language**, although by the mid-1950s progress had been made in several aspects of advanced programming. This group of machines included EDVAC and UNIVAC, the first commercially available computers.

BASIC (Beginners All-purpose Symbolic Instruction Code) had originally been developed in 1963 by Thomas Kurtz and John Kemeny, it was designed to provide an interactive, easy method for upcoming computer scientists to program computers. By this time, a number of other specialized and general-purpose languages had been developed. A surprising number of today's popular languages have actually been around since the 1950s. FORTRAN, developed by a team of IBM programmers, was one of the first high level languages, in which the programmer does not have to deal with the machine code of 0s and 1s. It was designed to express scientific and mathematical formulas. COBOL was developed in 1960 by a joint committee. It was designed to produce applications for the business world and had the novice approach of separating the data descriptions from the actual program. In the late 1960s, a Swiss computer scientist, Niklaus Wirth, released Pascal, which forced programmers to program in a structured, logical fashion and pay close attention to the different types of data in use.

Operating systems are the **interface** between the user and the computer. Windows is one of the numerous **graphical user interfaces** around that allows the user to manipulate their environment using a mouse and icons. Other examples of Graphical User Interfaces (GUIs) include X-Windows, which runs on UNIX® machines, or Mac OS X, which is the operating system of the Macintosh. (1-1) An **application** is any program that a computer runs that enables you to get things done. This includes things like word processors for creating text, graphics packages for drawing pictures, and communication packages for moving data around the globe.

The Web was developed at CERN (European Organization for Nuclear Research) in Switzerland during 1980s. As a new form of communicating text and graphics across the Internet, it makes use of the **hypertext markup language** (HTML) as a way to describe the attributes of the text and the placement of graphics, sounds, or even movie clips. Since it was first introduced, the number of users has blossomed and the number of sites containing information and searchable archives has been growing at an unprecedented rate.

1.2 Areas of Computer Science

Computer science can be divided into a variety of theoretical and practical disciplines. Some fields, such as **computational complexity** theory, are highly abstract, whilst fields such as computer graphics emphasize real-world applications.

1.2.1 Theoretical Computer Science

1. Theory of Computation

According to Peter J. Denning, the fundamental question underlying computer science is, "What can be efficiently automated?" The study of the theory of computation is focused on answering fundamental questions about what can be computed and what amount of resources are required to perform those computations. In an effort to answer the first question, **computability theory** examines which computational problems are solvable on various theoretical models of computation. The second question is addressed by computational complexity theory, which studies the time and space costs associated with different approaches to solving a multitude of computational problems. The famous "P=NP?" problem, one of the Millennium Prize Problems①, is an open problem in the theory of computation.

2. Algorithms and Data Structures

In mathematics and computer science, an algorithm is a step-by-step procedure for calculations. Algorithms are used for calculation, data processing, and automated reasoning. More precisely, an algorithm is an effective method expressed as a finite list[1] of **well-defined instructions** for calculating a function. (1-2) Starting from an initial state and initial input (perhaps empty), the instructions describe a computation that, when executed, proceeds through a finite number of well-defined successive states, eventually producing "output" and terminating at a final ending state. The transition from one state to the next is not necessarily **deterministic**; some algorithms, known as randomized algorithms, incorporate random input.

Data structure is a particular way of storing and organizing data in a computer so that it can be used efficiently. Data structures provide a means to manage large amounts of data efficiently, such as large **databases** and internet indexing services. Usually, efficient data structures are a key to designing efficient algorithms. Some formal design methods and programming languages emphasize data structures, rather than algorithms, as the key organizing factor in software design. **Storing and retrieving** can be carried out on data stored in both main memory and in secondary memory.

3. Programming Language Theory

Programming language theory is a branch of computer science that deals with the design, implementation, analysis, characterization, and classification of programming languages and their individual features. It falls within the discipline of computer science, both depending on and affecting mathematics, software engineering and linguistics. It is an active research area, with numerous dedicated academic journals.

4. Information and Coding Theory

Information theory is related to the **quantification of information**. This was developed by Claude E. Shannon to find fundamental limits on **signal processing** operations such as compressing data and on reliably storing and communicating data. Coding theory is the study of the properties of codes (systems for converting information from one form to another) and their fitness for a specific application. Codes are used for data **compression**, **cryptography**, error detection and correction, and more recently also for network coding. Codes are studied for the purpose of designing efficient and reliable data transmission methods.

① 千禧年大奖问题，又称世界七大数学难题，是美国克雷数学研究所(Clay Mathematics Institute,CMI)于 2000 年 5 月 24 日公布的 7 个数学猜想。

1.2.2 Applied Computer Science

1. Computer Architecture and Engineering

Computer architecture, or digital computer organization, is the conceptual design and fundamental operational structure of a computer system. It focuses largely on the way by which the central processing unit performs internally and accesses addresses in memory. The field often involves disciplines of computer engineering and electrical engineering, selecting and interconnecting hardware components to create computers that meet functional, performance, and cost goals.

2. Artificial Intelligence

This branch of computer science aims to or is required to synthesize goal-orientated processes such as problem-solving, decision-making, environmental adaptation, learning and communication which are found in humans and animals. From its origins in **cybernetics** and in the Dartmouth Conference (1956), artificial intelligence (AI) research has been necessarily **cross-disciplinary**, drawing on areas of expertise such as applied mathematics, symbolic logic, semiotics, electrical engineering, philosophy of mind, neurophysiology, and social intelligence. (1-3) The starting-point in the late 1940s was Alan Turing's question "Can computers think?", and the question remains effectively unanswered although the "Turing Test" is still used to assess computer output on the scale of human intelligence. But the automation of evaluative and predictive tasks has been increasingly successful as a substitute for human monitoring and intervention in domains of computer application involving complex real-world data.

3. Computer Graphics and Visualization

Computer graphics is the study of digital visual contents, and involves syntheses and manipulations of image data. The study is connected to many other fields in computer science, including **computer vision**, **image processing**, and **computational geometry**, and is heavily applied in the fields of special effects and video games.

4. Computer Vision

Computer vision is a field that includes methods for acquiring, processing, analyzing, and understanding images and, in general, high-dimensional data from the real world in order to produce numerical or symbolic information, e.g., in the forms of decisions. As a scientific discipline, computer vision is concerned with the theory behind artificial systems that extract information from images. The image data can take many forms, such as video sequences, views from multiple cameras, or multi-dimensional data from a medical scanner. Applications range from tasks such as industrial machine vision systems to research into artificial intelligence and computers or robots that can comprehend the world around them.

5. Computer Security and Cryptography

Computer security is a branch of computer technology, whose objective includes protection of information from unauthorized access, disruption, or modification while maintaining the accessibility and usability of the system for its intended users. Cryptography is the practice and study of hiding (**encryption**) and therefore deciphering (**decryption**) information. Modern cryptography is largely related to computer science, for many encryption and decryption algorithms are based on their computational complexity.

6. Concurrent, Parallel and Distributed Systems

Concurrency is a property of systems in which several computations are executing simultaneously, and potentially interacting with each other. A number of mathematical models have been developed for general concurrent computation including Petri nets, processing **calculi** and the Parallel Random Access Machine model. A distributed system extends the idea of concurrency onto multiple computers connected through a network. Computers within the same **distributed system** have their own private memory, and information is often exchanged amongst them to achieve a common goal.

7. Databases and Information Retrieval

A database is intended to organize, store, and retrieve large amounts of data easily. Digital databases are managed using database management systems to store, create, maintain, and search data, through **database models** and **query languages**.

8. Software Engineering

Software engineering is the study of designing, implementing, and modifying software in order to ensure it is of high quality, affordable, maintainable, and fast to build. It is a systematic approach to **software design**, involving the application of engineering practices to software. Software engineering deals with the organizing and analyzing of software— it doesn't just deal with the creation or manufacture of new software, but its internal maintenance and arrangement. Both computer applications software engineers and computer systems software engineers are projected to be among the fastest growing occupations from 2008 and 2018.

9. Computer Networks

A **computer networks** is a collection of computers and network hardware interconnected by **communication channels** that allow sharing of resources and information. This branch of computer science aims to manage networks between computers worldwide, it covers a broad range of topics, such as protocols, network architecture, routing, congestion control, wireless and mobility, performance evaluation. Computer networking can be considered a branch of electrical engineering, telecommunications, computer science, information technology or computer engineering, since it relies upon the theoretical and practical application of the related disciplines.

1.3 Is Computer Science Science?①

1.3.1 Common Understandings of Science

Our field was called computer science from its beginnings in the 1950s. Over the next four

① Author Peter J. Denning is the director of the Cebrowski Institute for Information Innovation and Superiority at the Naval Postgraduate School in Monterey, CA, and is a past president of ACM.

decades, we accumulated a set of principles that extended beyond its original mathematical foundations to include computational science, systems, engineering, and design. The 1989 report, Computing as a Discipline, defined the field as: "The discipline of computing is the systematic study of algorithmic processes that describe and transform information: their theory, analysis, design, efficiency, implementation, and application. The fundamental question underlying all of computing is, 'What can be (efficiently) automated?'"

Science, engineering, and mathematics combine into a unique and potent blend in our field. Some of our activities are primarily science—for example, experimental algorithms, experimental computer science, and computational science. Some are primarily engineering—for example, design, development, software engineering, and computer engineering. Some are primarily mathematics—for example, computational complexity, mathematical software, and numerical analysis. But most are combinations. All three sets of activities draw on the same fundamental principles. In 1989, we used the term "computing" instead of "computer science, mathematics and engineering." Today, computing science, engineering, mathematics, art, and all their combinations are grouped under the heading "computer science."

The scientific paradigm, which dates back to Francis Bacon, is the process of forming hypotheses and testing them through experiments; successful **hypotheses** become models that explain and predict phenomena in the world. Computing science follows this paradigm in studying information processes. The European synonym for computer science—**informatics**—more clearly suggests the field is about information processes, not computers.

The lexicographers offer two additional distinctions. One is between pure and applied science; pure science focuses on knowledge for its own sake and applied science focuses on knowledge of demonstrable utility. The other is between inexact (qualitative) and exact (quantitative) science; exact science deals with prediction and verification by observation, measurement and experiment.

(1-4) Computing research is rife with examples of the **scientific paradigm**. Cognition researchers, for example, hypothesize that much intelligent behavior is the result of information processes in brains and nervous systems; they build systems that implement hypothesized information processes and compare them with the real thing. The computers in these studies are tools to test the hypothesis; successful systems can be deployed immediately. Software engineering researchers hypothesize models for how programming is done and how defects arise; through testing they seek to understand which models work well and how to use them to create better programs with fewer defects. Experimental **algorithmics** studies the performance of real algorithms on real data sets and formulates models to predict their time and storage requirements; they may one day produce a more accurate theory than Big-O-Calculus and include a theory of locality. The nascent Human-Computer Interaction (HCI) field is examining the ways in which human information processes interact with automated processes.

By these definitions, computing qualifies as an exact science. It studies information processes, which occur naturally in the physical world; computer scientists work with an accepted, systematized body of knowledge; much computer science is applied; and computer science is used for prediction and verification.

The objection that computing is not a science because it studies man-made objects (technologies) is a red herring. Computer science studies information processes both artificial and natural. It helps other fields study theirs too. Physicists explain particle behavior with quantum information processes—some of which, like entanglement, are quite strange—and verify their theories with computer simulation experiments. Bioinformaticians explain DNA as encoded biological information and study how transcription enzymes read and act on it; computer models of these processes help customize therapies to individual patients. Pharmaceutical and materials labs create man-made molecules through computer simulations of the information processes underlying chemical compositions.

To help define the boundaries of science, lexicographers also contrast science with art. Art refers to the useful practices of a field, not to drawings or sculptures. Tab.1-1 lists some terms that are often associated with science and with art. Programming, design, software and hardware engineering, building and validating models, and building user interfaces are all "computing arts." If aesthetics is added, the computing arts extend to graphics, layout, drawings, photography, animation, music, games, and entertainment. All this computing art complements and enriches the science.

Tab.1-1 Science vs. Art

Science	Art
principles	practice
fundamental recurrences	skilled performance
explanation	action
discovery	invention
analysis	synthesis
dissection	construction

In his remarkable book about the workings of science, **Science in Action**, the philosopher Bruno Latour brings a note of caution to the distinction between science and art. Everything discussed in this column (a systematized body of knowledge, ability to make predictions, validation of models), is part of what he calls ready-made-science, science that is ready to be used and applied, science that is ready to support art. Much science-in-the-making appears as art until it becomes settled science.

Latour defines science-in-the making as the processes by which scientific facts are proposed, argued, and accepted. A new **proposition** is argued and studied in publications, conferences, letters, email correspondence, discussions, debates, practice, and repeated experiments. It becomes a "fact" only after it wins many allies among scientists and others using it. To win allies, a proposition must be independently verified by multiple observations and there must be no counterexamples. Latour sees science-in-themaking as a messy, political, human process, fraught with emotion and occasional polemics. The scientific literature bears him out.

Everything Latour says is consistent with the time-honored definition of the science paradigm. After sufficient time and validation, a model becomes part of the scientific body of knowledge.

1.3.2 Internal Disagreement

Computer scientists do not all agree whether computer science is science. Their judgment on this question seems to depend upon in which tradition they grew up. Hal Abelson and Gerry Sussman, who identify with the mathematical and engineering traditions of computing, said, "Computer science is not a science, and its ultimate significance has little to do with computers" They believe that the ultimate significance is with notations for expressing computations. Edsger Dijkstra, a mathematician who built exquisite software, frequently argued the same point, although he also believed computing is a mathematical science. Walter Tichy, an experimentalist and accomplished software builder, argues that computer science is science. David Parnas, an engineer, argues that the software part of computer science is really engineering. I myself have practiced in all three traditions of our field and do not see sharp boundaries.

Even the Computer Science and Technology Board of the National Research Council is not consistent. In 1994, a panel argued that experimental computer science is an essential aspect of the field. In 2004, another panel discussed the accomplishments of computer science research; aside from comments about abstraction in models, they say hardly a word about the experimental tradition. Paul Graham, a prominent member of the generation who grew up with computers, invented the Yahoo! store and early techniques for spam filters; he identifies with computing art. He says: "I never liked the term 'computer science'. … Computer science is a **grab bag** of tenuously related areas thrown together by an accident of history, like Yugoslavia. … Perhaps one day 'computer science' will, like Yugoslavia, get broken up into its component parts. That might be a good thing. Especially if it means independence for my native land, hacking". He is not arguing against computer science, but for an appellation like computer art that is more attractive to hackers (his term for elite programmers). Dana Gardner, of the Yankee Group, does not like this notion. He compares the current state of software development to the preindustrial Renaissance, when wealthy benefactors commissioned groups of highly trained artisans for single great works of art. He says, "Business people are working much closer to the realm of Henry Ford, where they are looking for reuse, interchangeable parts, automated processes, highly industrialized assembly lines."

1.3.3 Computer Science Thrives on Relationships

Horgan argued in 1996 that new scientific discoveries require mastering ever-greater amounts of complexity. In 2004 he repeated his main conclusion: "Science will never again yield revelations as monumental as the theory of evolution, general relativity, quantum mechanics, the big bang theory, DNA-based genetics. ··· Some farfetched goals of applied science—such as immortality, superluminal spaceships, and superintelligent machines—may forever elude us."

Has computer science already made all the big discoveries it's going to? Is incremental progress all that remains? Has computer science bubbled up at the end of the historical era of science?

I think not. Horgan argues that the number of scientific fields is limited and each one is slowly being exhausted. But computer science is going a different way. It is constantly forming relationships with other fields; each one opens up a new field. Paul Rosenbloom has put this eloquently in his recent analysis of computer science and engineering.

Rosenbloom charts the history of computer science by its relationships with the physical, life, and social sciences. With each one computer science has opened new fields by implementing, interacting, and embedding with those fields. Examples include autonomic systems, bioinformatics, biometrics, biosensors, cognitive prostheses, cognitive science, cyborgs, DNA computing, immersive computing, neural computing, and quantum computing. Rosenbloom believes that the constant birth and richness of new relationships guarantees a bright future for the field.

1.3.4 Validating Computer Science Claims

In a sample of 400 computer science papers published before 1995, Walter Tichy found that approximately 50% of those proposing models or hypotheses did not test them. In other fields of science the fraction of papers with untested hypotheses was about 10%. Tichy concluded that our failure to test more allowed many unsound ideas to be tried in practice and lowered the credibility of our field as a science. The relative youth of our field—barely 60 years old—does not explain the low rate of testing. Three generations seem sufficient time for computer scientists to establish that their principles are solid.

The perception of our field seems to be a generational issue. The older members tend to identify with one of the three roots of the field—science, engineering, or mathematics. The science paradigm is largely invisible within the other two groups. The younger generation, much less awed than the older one once was with new computing technologies, is more open to critical thinking. Computer science has always been part of their world; they do not question its validity. In their research, they are increasingly following the science paradigm. Tichy told me that the recent research literature shows a marked increase in testing.

The science paradigm has not been part of the mainstream perception of computer science. But soon it will be.

1.4 The Future of Computer Science[①]

1.4.1 Introduction

Modern computer science is undergoing a fundamental change. In the early years of the field, computer scientists were primarily concerned with the size, efficiency and reliability of computers. They attempted to increase the computational speed as well as reduce the physical size of computers, to make them more practical and useful. The research mainly dealt with hardware, programming languages, **compilers**, operating systems and data bases. Meanwhile, theoretical computer science developed an underlying mathematical foundation to support this research which in turn, led to the creation of automata theory, formal languages, computability and algorithm analysis. Through the efforts of these researchers, computers have shrunk from the size of a room to that of a dime, nearly every modern household has access to the internet and communications across the globe are virtually instantaneous. Computers can be found everywhere, from satellites hundreds of miles above us to pacemakers inside beating human hearts. The prevalence of computers, together with communication devices and data storage devices, has made vast quantities of data accessible. This data incorporates important information that reveals a closer approximation of the real world and is fundamentally different from what can be extracted from individual entities. Rather than analyzing and interpreting individual messages, we are more interested in understanding the complete set of information from a collective perspective. However, these large-scale data sets are usually far greater than can be processed by traditional means. Thus, future computer science research and applications will be less concerned with how to make computers work and more focused on the processing and analysis of such large amounts of data. Consider the following example of internet search. At the beginning of the internet era, users were required to know the **IP address** of the site to which they wished to connect. No form of search was available. As **websites** proliferated, online search services became necessary in order to make the internet navigable. The first internet search tool was developed in 1993, and dozens more were created over the next several years. Ask Jeeves, founded in 1996, relied partially on human editors who manually selected the best websites to return for various search queries. Given the huge number of websites available today, such a strategy is clearly no longer feasible. Google, founded in 1998, is a leader among today's search engines. It relies on a search algorithm that uses the structure of the internet to determine the most popular and thus, perhaps, the most reputable websites.

However, while Google's **search engine** was a major advance in search technology, there

---

① Hopcroft J E, Soundarajan S, Wang L R. The Future of Computer Science. Int J Software Informatics, 2011, 5(4):549-565.

will be more significant advances in the future. Consider a user who asks the question, "When was Einstein born?" Instead of returning hundreds of webpages to such a search, one might expect the answer: "Einstein was born at Ulm, in Wurttemberg Germany, on March 14, 1879", along with pointers to the source from which the answer was extracted. Other similar searches might be:

(1) Construct an annotated bibliography on graph theory.

(2) Which are the key papers in theoretical computer science?

(3) Which car should I buy?

(4) Where should I go to college?

(5) How did the field of computer science develop?

Search engine companies have saved billions of search records along with a whole archive of information. When we search for the answer to the question "Which car should I buy?", they can examine pages that other individuals who did similar searches have looked at, extract a list of factors that might be important (e.g. fuel economy, price, crash safety), and prompt the user to rank them. Given these priorities, the search engine will provide a list of automobiles ranked according to the preferences, as well as key specifications and hyperlinks to related articles about the recommended models. Another interesting question is, "Which are the key papers in theoretical computer science?" One would expect a list such as:

(1) Juris Hartmanis and Richard Stearns, "On the computational complexity of algorithms".

(2) Manuel Blum, "A machine-independent theory of the complexity of recursive Functions".

(3) Stephen Cook,"The complexity of theorem proving procedures".

(4) Richard Karp, "Reducibility among combinatorial problems".

With thousands of scientific papers published every year, information on which research areas are growing or declining would be of great help to rank the popularity and predict the evolutionary trend of various research topics. For example, Shaparenko *et al.* used sophisticated artificial intelligence techniques to **cluster** papers from the Neural Information Processing Systems (NIPS) conference held between 1987 and 2000 into several groups, as shown in Fig.1-3 since all papers presented at the NIPS conference are in digital format, one can use this information to plot the sizes of the clusters over time. Clusters 10 and 11 clearly show the two growing research areas in NIPS, namely "Bayesian methods" and "Kernel methods". The graph correctly indicates that the "Bayesian methods" cluster emerged before the "Kernel methods" cluster, with both topics starting to dominate the NIPS conference by 2000. In addition, Cluster 1 on neural networks, Cluster 4 on **supervised neural network** training, and Cluster 8 on biologically-inspired neural memories were popular in the early years of NIPS, but almost disappeared from the conference by 2000. With the help of advanced techniques, we should be able to accurately predict how important a paper will be when it is first published, as well as how a research area will evolve and who will be the key players.

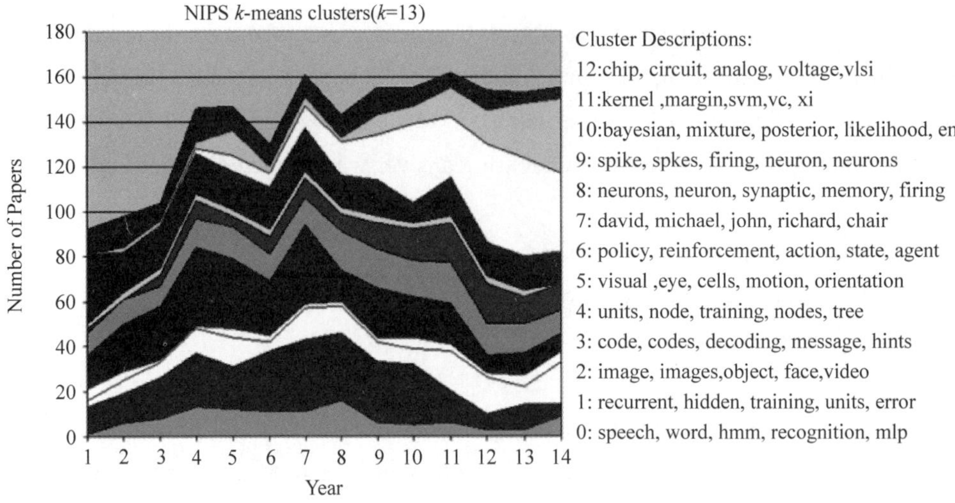

Fig.1-3 The distribution of $k = 13$ clusters of NIPS papers

In the beginning years of computer science, researchers established a mathematical foundation consisting of areas such as automata theory and **algorithm analysis** to support the applications of that time. As applications develop over time, so must the underlying theories develop accordingly. Surprisingly, the intuition and mathematics behind the theory of large or **high-dimensional** data are completely different from that of small or low dimensional data. **Heuristics** and methods that were effective merely a decade ago may already be outdated.

1.4.2 Innovative Research Projects

Traditional research in theoretical computer science has focused primarily on problems with small input sets. For instance, in the past, stores tracked the items purchased by each individual customer and gave that customer discounts for future purchases of those items. However, with the help of modern algorithms, service providers such as Netflix are now able to, not only make predictions based on a customer's past preferences, but amalgamate preferences from millions of customers to make accurate and intelligent suggestions and effectively increase sales revenue. The following subsections describe four ongoing research projects involving the analysis and interpretation of large data sets. Each represents a promising direction for rediscovering fundamental properties of large-scale networks that will reshape our understanding of the world.

1. Tracking Communities in Social Networks

A **social network** is usually modeled as a **graph** in which **vertices** represent entities and **edges** represent interactions between pairs of entities. In previous studies, a community was often defined to be a subset of vertices that are densely connected internally but sparsely connected to the rest of the network. Accordingly, the best community of the graph was typically a peripheral set of vertices barely connected to the rest of the network by a small number of edges. However, it is our view that for large-scale real-world societies, communities, though better connected internally than expected solely by chance, may also be well connected to the rest of the network. It is hard to imagine a small close-knit community with only a few edges connecting it to the outside world. Rather, members of a community, such as a computer science department, are likely to have many connections outside the community, such as family, religious groups, other academic departments and so on. Empirically, a community displays a higher than average edge to vertex squared ratio which reflects the probability of an edge between two randomly-picked vertices, but can also be connected to the rest of the network by a significant number of edges, which may even be larger than the number of its internal edges.

With this intuitive notion of community, two types of structures are defined: the "whiskers" and the "core". Whiskers are peripheral subsets of vertices that are barely connected to the rest of the network, while the core is the central piece that exclusively contains the type of community we are interested in. Then, the algorithm for finding a community can be reduced to two steps: ① identifying the core in which no whiskers exist, and ② identifying communities in the core. Further, extracting the exact core from both weighted and unweighted graphs has been proved to be NP-complete. Alternative heuristic algorithms have been developed, all of which are capable of finding an approximate core, and their performance can be justified by the experimental results based on various large-scale social graphs. In this way, one can obtain communities that are not only more densely connected than expected by chance alone, but also well connected to the rest of the network.

2. Tracking Flow of Ideas in Scientific Literature

Remarkable development in data storage has facilitated the creation of gigantic digital document collections available for searching and downloading. When navigating and seeking information in a digital document collection, the ability to identify topics with their time of appearance and predict their evolution over time, would be of significant help. Before starting research in a specific area, a researcher might quickly survey the area, determine how topics in the area have evolved, locate important ideas, and the papers that introduced those ideas. Knowing a specific topic, a researcher might find out whether it has been discussed in previous papers, or is a fairly new concept. As another example, a funding agency that administers a digital document collection might be interested in visualizing the landscape of topics in the collection to show the emergence and evolution of topics, the bursts of topics, and the interactions between different topics that change over time. Such information-seeking activities often require the ability to identify topics with their time of appearance and to follow their evolution. Recently, in their unpublished work, Jo *et al.* have developed a unique approach to achieving this goal in a **time-stamped document** collection with an underlying document network which represents a wide range of digital texts available over the internet. Examples are scientific paper collections, text collections associated with social networks such as blogs and Twitter, and more generally, web documents with **hyperlinks**. A document collection without an explicit network can be converted into this format by connecting textually similar documents to generate a document network.

3. Reconstructing Networks

The study of large networks has brought about many interesting questions, such as how to determine which members of a population to vaccinate in order to slow the spread of an infection, or where to place a limited number of sensors to detect the flow of a toxin through a water network. Most algorithms for solving such questions make the assumption that the structure of the underlying network is known. For example, detectives may want to use such algorithms to identify the leaders of a criminal network, and to decide which members to turn into informants. Unfortunately, the exact structure of the criminal network cannot be easily determined. However, it is possible that the police department has some information about the spread of a certain property through the network; for instance, some new drug may have first appeared in one neighborhood, and then in two other neighborhoods, and so on. The work by Soundarajan *et al.* attempts to create algorithms to recover the structure of a network given information about how some property, such as disease or crime, has spread through the network. This work begins by defining a model of contagion describing how some property has spread through a network. The model of contagion for information spread may be: "a vertex learns a piece of information in the time interval after one of its neighbors learns that information." A more complex model of contagion corresponding to the spread of belief may be: "a vertex adopts a new belief in the time interval after a proportion p of its neighbors adopts that belief." For example, a person will probably not join a political party as soon as one of his friends joins that party, but he may join it after two-thirds of his friends have joined it.

Next, the network recovery algorithm assumes that vertices are partitioned into **discrete** time intervals, corresponding to the time when they adopt the property. For a given model of contagion, the algorithm attempts to find a network over the set of vertices such that when the property in question (e.g. information, belief) is introduced to some vertices in the first time interval, and then spreads to other vertices in accordance with the model of contagion, every vertex adopts the property at an appropriate time. Initial work has created such algorithms for two models of contagion: the model corresponding to the spread of information, where a vertex adopts a property in the time interval after one of its neighbors has adopted that property, and the model corresponding to the spread of belief, where a vertex adopts a property in the time interval after at least half of its neighbors have adopted that property.

Future work will focus on finding algorithms for other models of contagion, especially the models in which a vertex adopts the property after a proportion p of its neighbors has adopted that property, for arbitrary values of p. Other directions include finding algorithms for networks in which there are two or more properties spreading through the network. This work also opens up questions about the types of graphs produced by these algorithms. For instance, do all possible graphs have some edges in common? Are there any edges that do not appear in any of the solution graphs? Which edges are the most or least likely?

4. Tracking Bird Migration in North America

Hidden Markov models (HMMs) assume a generative process for sequential data whereby a sequence of states (i.e. a sample path) is drawn from a **Markov chain** in a hidden experiment. Each state generates an output symbol from a given alphabet, and these output symbols constitute the sequential data (i.e. observations). The classic single path problem, solved by the Viterbi algorithm, is to find the most probable sample path given certain observations for a given Markov model.

Two generalizations of the single path problem for performing collective inference on Markov models are introduced, motivated by an effort to model bird migration patterns using a large database of static observations. The eBird database maintained by the Cornell Lab of Ornithology contains millions of bird observations from throughout North America reported by the general public using the eBird Web application. Recorded observations include location, date, species and number of birds observed. The eBird data set is very rich, and the human eye can easily discern migration patterns from animations showing the observations as they unfold overtime on a map of North America. However, the eBird data entries are static and movement is not explicitly recorded, only the distributions at different points in time. Conclusions about migration patterns are made by the human observer, and the goal is to build a mathematical framework to infer dynamic migration models from the static eBird data. Quantitative migration models are of great scientific and practical importance. For example, this problem comes from an interdisciplinary project at Cornell University to model the possible spread of avian influenza in North America through wild bird migration.

The migratory behavior of a species of birds can be modeled by a single **generative process** that independently governs how individual birds fly between locations. This gives rise to the following inference problem: a hidden experiment draws many independent sample paths simultaneously from a Markov chain, and the observations reveal collective information about the set of sample paths at each time step, from which the observer attempts to reconstruct the paths.

1.4.3 Theoretical Foundation

As demonstrated in the previous section, (1-5) the focus of modern computer science research is shifting to problems concerning large data sets. Thus, a theoretical foundation and science base is required for rigorously conducting studies in many related areas. The theory of large data sets is quite different from that of smaller data sets; when dealing with smaller data sets, discrete mathematics is widely used, but for large data sets, **asymptotic analysis** and **probabilistic methods** must be applied. Additionally, this change in the theoretical foundation requires a completely different kind of mathematical intuition.

Large graphs have become an increasingly important tool for representing real world data in modern computer science research. Many empirical experiments have been performed on large-scale graphs to reveal interesting findings. A computer network may have consisted of only a few hundred nodes in previous years, but now we must be able to deal with large-scale networks containing millions or even billions of nodes. Many important features of such large graphs remain constant when small changes are made to the network. Since the exact structure of large graphs is often unknown, one way to study these networks is to consider generative graph models instead, where a graph is constructed by adding vertices and edges in each time interval. Although such graphs typically differ from real-world networks in many important ways, researchers can use the similarities between the two types of networks to gain insight into real-world data sets.

A simple but commonly used model for creating random graphs is the Erdös-Renyi model, in which an edge exists between each pair of vertices with equal probability, independent of the other edges. A more realistic model is known as the "preferential attachment" model, in which the probability that an edge is adjacent to a particular vertex is proportional to the number of edges already adjacent to that vertex. In other words, a high degree vertex is likely to gain more edges than a low degree vertex. The preferential attachment model gives rise to the power-law degree distribution observed in many real-world graphs.

Consider the Erdös-Renyi random graph model in which each edge is added independently with equal probability. Suppose that we start with 1,000 vertices and zero edges. Then, there are clearly 1,000 components of size one. If we add one edge, we will have 998 components of size one and one component of size two. However, a giant component begins to emerge as more edges are added. This occurs because a component is more likely to attract additional vertices as its size increases.

Since random graph models mimic some vital features of real-world networks, it is often helpful to study the processes that generate these features in random graph models. An understanding of these processes can provide valuable insights for analyzing real-world data sets.

1.4.4 An Interview[①]

This is An Interview with Ken Calvert and Jim Griffioen on "The Future of Computer Science".

Ken Calvert, Ph.D. and chair of the Department of Computer Science, University of Kentucky. **Jim Griffioen**, Ph.D., professor of computer science and director of the Laboratory for Advanced Networking.

Q: What are shaping up to be the greatest areas of opportunity in the computer science field over the next few years?

K.C. I think this is an exciting time in computer science. Hardware has become so cheap that both compute cycles and storage bytes have essentially become commoditized. We're seeing this right now with the cloud computing model. A company can now pay someone a relatively low monthly fee to run their Web server instead of shelling out thousands of dollars for hardware, software and maintenance. It's basically the same transition that happened with electric power 100 years ago. Nicholas Carr's book, ***The Big Switch***, describes how, back then, factories had to

① http://www.engr.uky.edu/news/2012/11/the-future-of-computer-science-an-interview-with-ken-calvert-and-jim-griffioen/.

be located next to big streams because that's where they got the power to run their machines. When electric power grids came along, generation of power became centralized. The same exact centralization is happening with the advent of cloud computing. It makes a lot more sense to have one big centralized data center run by people who know what they're doing than for every little company to run its own.

J.G. Historically, computer scientists have created technology without fully knowing how it's going to play out. The Internet was built so machines could communicate back and forth and share information. Well, then users came along and said, "I need this to be easy to use. I need a Web interface. I need a browser." None of those uses were part of the original design. Now we have virtualization through cloud computing as well as ubiquitous networking—you can be on the network at all times. In addition, we also have a very mobile society. Devices which can maximize the benefits of the cloud will need to be developed. I think we're on the edge of some of these things just exploding and once it explodes, we'll have a whole new set of issues to address—how to secure such a world, etc.

K.C. What virtualization also means is that software is going to be king. Evcrything is going to be about software because hardware is so cheap. I think the opportunities in software are tremendous. However, as Jim mentioned, we now have to consider questions such as: how do I keep control of my information? How do I know what information people are collecting about me? Businesses already know a lot about us and they are going to try to monetize that any way they can. Why do Facebook and Twitter have such astronomical valuations? I believe it's because they know who is talking to whom and what they're saying. Privacy is a huge issue going forward and it's not just "old people" who are concerned about it. We need to understand how to maximize the benefits of virtualization without the Big Brother risks.

Q: What does the future look like on the security front?

J.G. When everyday users weigh the prospective gain of a new application against the possible security risks, they almost always accept the tradeoff. It is difficult to keep up with potential threats and understand the risks because the landscape changes so quickly. On the positive side, though, industry has finally recognized that security is not an afterthought. In the past, companies created products and tacked security onto the back end of the development process. Often, that made it hard to add the security because it wasn't present from the start. Now, computer scientists are asking, "How do I design the architecture so that if it doesn't have security now, it is amenable to it later?" There are discussions going on right now about the next generation of the Internet. Naturally, security is a central topic.

K.C. As long as we have the Internet architecture, we're not going to solve many of the current problems. The architecture doesn't have the things we need to solve them, and there's just too much inertia to counteract. So it's hard to say what the future is going to look like there. But again, almost as important as security is privacy. When it comes to the leaders in software and social media, people aren't given a choice to use the product and still maintain their privacy. Those companies say, "Here are our policies, take them or leave them." And people agree, even though the policies are not in their favor, because they want to use the product. I printed out the iTunes license agreement once. It was 29 pages of 9 point font. No one is going to read that! That's why I think we really need more collaboration between experts in computer science and experts in psychology. As systems get more and more complex and everyday people have to make decisions about privacy settings on their computer or home router, we need to design systems and educate users so the consequences of each decision they have to make is much clearer. That is certainly not the case right now. Unfortunately, until software providers accept accountability for their products—until they have incentive to change—the situation will remain challenging.

Q. What areas in the field besides security and privacy need attention?

K.C. We need to focus on parallelism. You often hear that Moore's Law is running out of gas. On the contrary, Moore's Law is still going strong; but the dividends of Moore's Law are now being paid in parallelism, not in faster sequential computation. Rather than doing each step of the computation faster, you can do multiple steps at once in the same amount of time.

J.G. As far as teaching parallelism in the classroom, we have to change our approach. We've been teaching the students a step-by-step process; basically, that's how computer scientists have always conceived writing programs. Well, now we have multiple processors running on chips and we have to start thinking, "How do I write a program that does three things at once? Eight things at once?" What happens when the chips allow us to do hundreds of things at once? We need to start changing the mindset of everyone in our program and challenge them to think, "I'm going to do lots of things at once."

K.C. If you're only doing one thing at a time, you cannot take advantage of the additional power that Moore's Law is giving you. So, like Jim said, we have to be able to figure out how to do multiple things at once, like putting meat on the stove to brown and, while that's happening, mixing other ingredients. That's the way we need to think about things all the time. It's not trivial. We want to turn out graduates who can master doing things in parallel because this is the way it's going to be from now on. Right now, though, the tools we have for taking advantage of Moore's Law and parallelism aren't very good, so it's definitely an area that needs attention.

Q. How much of a challenge is it to stay on the leading edge of an industry where technology changes so rapidly, let alone translate those changes into your curricula?

K.C. It's almost impossible. We could spend all of our time just trying to keep up. It's a catch-22: we have to show our students technology and let them get their hands dirty, but the reality is whatever we show them as freshmen will have changed and might even be obsolete by the time they are seniors. Five years ago, everybody was using Perl and CGI scripts on the Web. Now those tools have been replaced by a new generation of languages and platforms. So, our task is to teach fundamental principles and I think we do a good job of that. Fortunately, students quickly adapt to the rate of change. They're fearless and not afraid to pick up new technology and play with it. I consider that a good thing and we need to try to leverage it in the classroom.

J.G. At the same time, we faculty have to make the purpose of learning fundamental concepts and principles clear to them. They have to know that chances are whatever programming language we teach them their freshman year will probably be out of date by the time they graduate. The turnaround times really are that short.

K.C. That actually seems to make it easier to motivate our students to learn the fundamentals, though, because incoming students have seen the short life cycles of various technologies several times already. It's pretty obvious to them now that if they don't focus on the stuff that doesn't change, they're not going to be able to adapt when they're forced to.

J.G. Even though I'm a longstanding faculty member, I often learn from the students. There is so much software out there, so many programs, so many computing languages, that I can't play with them all. Students will come to me and tell me about a program and I'll say, "Explain it to me. How does it work? What does it do?" I learn a lot from interacting with them.

K.C. The only way to stay on the leading edge is to invent everything. We have a weekly "Keeping Current" seminar, where students share what they've learned or some new technology they've discovered. They're always coming in and telling us about stuff we've never heard of. It's a volunteer thing, very informal, but a lot of fun. There are so many tools around, it's just unbelievable.

Q. How does the future of computer science look from the perspective of college students choosing it as a career?

K.C. It couldn't be better. In the early 2000s, people were afraid all the computer science jobs were going to be outsourced overseas. That hasn't happened. In fact, the Bureau of Labor projects software engineering jobs will grow by 38% over the next ten years—one of the top professions as far as growth. Our students are in demand and will continue to be in demand for a long time. I am constantly being contacted by people wanting to hire our graduates. It's clear there are more jobs than people to do them, and I don't see that changing.

J.G. I was contacted by a mid-sized company the other day that decided they were going to get into the mobile world, but didn't have a clue as to how to go about it and wanted to know if any of our students or graduates could help them figure it out. Companies need people who know how to take advantage of the technology, not just throw around terms. One aspect that will change in light of the switch to cloud computing, however, will be the kinds of jobs available. There won't be as much need for systems administrator jobs if everything is run through a centralized data center. So what a graduate might do once they're in the marketplace might change, but the demand is still very high.

K.C. Our goal is to equip students to be able to adapt to change. We teach them how to think and how to learn because that's the only way they're going to survive. If they think they're going to learn C++, graduate and be a C++ programmer all their lives, it's just not going to happen.

Q. What are some myths and misconceptions about the computer science industry?

J.G. One myth I often hear is that all the exciting stuff is happening in industry. "Companies are where the exciting things are happening," someone will say, downplaying the need for education in the field. While it's now true that bright high school kids can get programming jobs with big companies right away, we still believe in the importance of developing a skill set based on the fundamentals that will last a long time.

K.C. I think another myth is that computer science is all about programming. Computing professionals need to have an understanding of programming, but it's even more important to have a broad understanding of the business you're in: social networking, data mining, business concepts, etc. The future is about applications and applying computing to problems in biology, medicine, engineering, the environment, business, entertainment and other industries—it's a great time to be a software entrepreneur! Another myth is that computer science is something only guys would want to do. The stereotypical image of scruffy-haired guys with beards staring at computer screens needs to be replaced by one which illustrates the openness of the field to anyone who wants to get in on the opportunities available.

1.5 Key Terms and Review Questions

1. Technical Terms

accumulator	累加器	1.1
analytical engine	分析机	1.1
application	应用（程序）	1.1
arithmetic operation	算术运算	1.1
arithmometer	四则运算器	1.1
binary system	二进制系统	1.1
capacitor	电容	1.1
conditional control	条件控制	1.1
crystal diodes	二极管	1.1
decimal place	小数位	1.1
decimal system	十进制系统	1.1
difference engine	差分机	1.1
finite state machine	有限状态机	1.1
graphical user interface	图形用户界面	1.1
hand-soldered joint	手工焊点	1.1
hypertext markup language	超文本标记语言	1.1
instruction program	指令程序	1.1
integral	积分的	1.1
interface	接口，界面	1.1
Jacquard loom	提花织布机	1.1
logarithm	对数	1.1
machine language	机器语言	1.1

odometer	里程表	1.1
operating system	操作系统	1.1
Pascaline	加法机	1.1
perforated strip	打孔带	1.1
prototype	原型	1.1
punched card	穿孔卡片	1.1
relay	继电器	1.1
resistors	电阻	1.1
steam-powered	蒸气驱动的	1.1
stored-program technique	存储程序技术	1.1
subroutines	子程序	1.1
tabulator	制表机	1.1
technophobia	技术恐惧	1.1
theorize	建立学说或理论	1.1
trigonometric	三角	1.1
Turing machine	图灵机	1.1
vacuum tube	真空管	1.1
applied mathematics	应用数学	1.2
computational geometry	计算几何	1.2
calculi	微积分学	1.2
communication channel	通信信道	1.2
computability theory	可计算性理论	1.2
computational complexity	计算复杂性	1.2
computer graphics	计算机图形学	1.2
computer network	计算机网络	1.2
computer vision	计算机视觉	1.2
concurrency	并发性	1.2
congestion control	拥塞控制	1.2
cross-disciplinary	跨学科	1.2
cryptography	密码学	1.2
cybernetics	控制论	1.2
data compression	数据压缩	1.2
data structure	数据结构	1.2
database	数据库	1.2
database model	数据模型	1.2
decryption	解密	1.2
deterministic	确定性的	1.2
distributed system	分布式系统	1.2

electrical engineering	电子工程	1.2
encryption	加密	1.2
image processing	图像处理	1.2
network architecture	网络体系结构	1.2
neurophysiology	神经生物学	1.2
performance evaluation	性能评价	1.2
protocols	协议	1.2
quantification of information	信息量化	1.2
query language	查询语言	1.2
routing	路由	1.2
semiotics	符号学	1.2
signal processing	信号处理	1.2
software design	软件设计	1.2
software engineering	软件工程	1.2
storing and retrieving	存储和检索	1.2
symbolic logic	符号逻辑	1.2
Turing Test	图灵测试	1.2
well-defined instructions	定义明确的指令	1.2
proposition	命题	1.3
algorithmics	算法学	1.3
Big-O-Calculus	大 O 运算符	1.3
hypotheses	假定	1.3
informatics	信息学	1.3
scientific paradigm	科学范式	1.3
search engine	搜索引擎	1.4
asymptotic analysis	渐近分析	1.4
cluster	聚类	1.4
compiler	编译器	1.4
discrete	离散的	1.4
generative process	生成式过程	1.4
graph	图	1.4
heuristics	启发式	1.4
Hidden Markov models	隐马尔可夫模型	1.4
high-dimensional	高维	1.4
hyperlink	超链接	1.4
IP address	网络地址	1.4
Markov chain	马尔可夫链	1.4
probabilistic method	概率方法	1.4

social network	社交网络	1.4
supervised neural network	有监督的神经网络	1.4
time-stamped document	带时间标记的文件	1.4
vertices	顶点	1.4
website	站点	1.4

2. Translation Exercises

(1-1) An application is any program that a computer runs that enables you to get things done. This includes things like word processors for creating text, graphics packages for drawing pictures, and communication packages for moving data around the globe.

(1-2) Starting from an initial state and initial input (perhaps empty), the instructions describe a computation that, when executed, proceeds through a finite number of well-defined successive states, eventually producing "output" and terminating at a final ending state. The transition from one state to the next is not necessarily deterministic; some algorithms, known as randomized algorithms, incorporate random input.

(1-3) The starting-point in the late 1940s was Alan Turing's question "Can computers think?", and the question remains effectively unanswered although the "Turing Test" is still used to assess computer output on the scale of human intelligence. But the automation of evaluative and predictive tasks has been increasingly successful as a substitute for human monitoring and intervention in domains of computer application involving complex real-world data.

(1-4) Computing research is rife with examples of the scientific paradigm. Cognition researchers, for example, hypothesize that much intelligent behavior is the result of information processes in brains and nervous systems; they build systems that implement hypothesized information processes and compare them with the real thing.

(1-5) the focus of modern computer science research is shifting to problems concerning large data sets. Thus, a theoretical foundation and science base is required for rigorously conducting studies in many related areas. The theory of large data sets is quite different from that of smaller data sets; when dealing with smaller data sets, discrete mathematics is widely used, but for large data sets, asymptotic analysis and probabilistic methods must be applied. Additionally, this change in the theoretical foundation requires a completely different kind of mathematical intuition.

References

[1] The history of computer science[EB/OL]. (2006-07-30). http://lecture.eingang.org /forefathers.html.

[2] Chase G C. History of mechanical computing machinery[J]. IEEE Annals of the History of Computing, 1980, 2(3): 204-205.

[3] Martin C K. Computer: A history of the information machine[M]. 2nd ed. Boulder: Westview Press, 2004.

[4] Parker S P. McGraw-Hill Dictionary of scientific & technical terms[M]. The McGraw-Hill Companies, 2003.

[5] Computer science[EB/OL]. (2017-07-10). http://en.wikipedia.org/wiki/Computer_science.

[6] Knuth D. The Art of Computer Programming[M].3rd ed. Addison-Wesley, 1997.

[7] Shapiro L G. Stockman G C. Computer vision[M]. Prentice Hall, 2003.

[8] Denning P J. Is computer science science?[J].Communications of the ACM, 2005, 49(4).

[9] Hopcroft J E, Soundarajans, Wang L R. The future of computer science[J]. Int J Software Informatics, 2011, 5 (4): 549-565.

[10] The future of computer science an interview[EB/OL]. http://www.engr.uky.edu/news/ 2012/ 11./the-future -of- computer-science-an-interview-with-ken-calvert-and-jim-griffioen/.

Chapter 2 Computer Architecture and Networks

2.1 Introduction

2.1.1 Computer Architecture

Computer architecture means the structure and organization of a computer's **hardware** or system **software**. The term "architecture" in computer literature can be traced back to the work of Lyle R. Johnson, Mohammad Usman Khan and Frederick P. Brooks, Jr., members in 1959 of the Machine Organization department in IBM's main research center. Johnson had the opportunity to write a proprietary research communication about the Stretch, an IBM-developed supercomputer for Los Alamos Scientific Laboratory. To describe the level of detail for discussing the luxuriously embellished computer, he noted that his description of formats, instruction types, hardware parameters, and speed enhancements were at the level of "system architecture"—a term that seemed more useful than "machine organization".

The art of computer architecture has three main subcategories:①

(1) Instruction set architecture, or ISA. The ISA defines the codes that a **central processor** reads and acts upon. It is the machine language (or assembly language), including the instruction set, word size, memory address modes, processor registers, and address and data formats.

(2) **Microarchitecture**, also known as computer organization describes the data paths, data processing elements and data storage elements, and describes how they should implement the ISA. The size of a computer's CPU cache for instance, is an organizational issue that generally has nothing to do with the ISA.

(3) System Design includes all of the other hardware components within a computing system. These include: Data paths, such as **computer buses** and switches, memory controllers and hierarchies, data processing other than the CPU, such as direct memory access (DMA) and other miscellaneous issues such as **virtualization, multiprocessing** and software features.

---

① Hennessy J L, Patterson D A. **Computer Architecture: A Quantitative Approach** (Third Edition). Morgan Kaufmann Publishers.

2.1.2 Design Goals

Computer architectures usually trade off standards, cost, **memory capacity, latency** (latency is the amount of time that it takes for information from one node to travel to the source) and **throughput**. Sometimes other considerations, such as features, size, weight, reliability, expandability and power consumption are factors.

(2-1) The most common scheme carefully chooses the bottleneck that most reduces the computer's speed. Ideally, the cost is allocated proportionally to assure that the data rate is nearly the same for all parts of the computer, with the most costly part being the slowest. This is how skillful commercial integrators optimize personal computers such as smart cellphones.

Modern computer performance is often described in MIPS per MHz (millions of instructions per millions of cycles of clock speed). This measures the efficiency of the architecture at any clock speed. Since a faster clock can make a faster computer, this is a useful, widely applicable measurement. Historic computers had MIPs/MHz as low as 0.1. Simple modern processors easily reach near 1. **Superscalar processors** may reach three to five by executing several instructions per clock cycle. Multicore and vector processing CPUs can multiply this further by acting on a lot of data per instruction, since they have several CPU cores executing in parallel.

Historically, many people measured a computer's speed by the clock rate (usually in MHz or GHz). This refers to the cycles per second of the main clock of the CPU. However, this metric is somewhat misleading, as a machine with a higher clock rate may not necessarily have higher performance. As a result manufacturers have moved away from clock speed as a measure of performance.

Other factors influence speed, such as the mix of functional units, bus speeds, available **memory**, and the type and order of instructions in the programs being run.

In a typical home computer, the simplest, most reliable way to speed performance is usually to add random access memory (RAM). More RAM increases the likelihood that needed data or a program is in RAM—so the system is less likely to need to move memory data from the disk. The disk is often ten thousand times slower than RAM because it has mechanical parts that must move to access its data.

There are two main types of speed: latency and throughput. Latency is the time between the start of a process and its completion. Throughput is the amount of work done per unit time. **Interrupt latency** is the guaranteed maximum response time of the system to an electronic event.

Performance is affected by a very wide range of design choices — for example, **pipelining** a processor usually makes latency worse (slower) but makes throughput better. Computers that control machinery usually need low interrupt latencies. These computers operate in a real-time environment and fail if an operation is not completed in a specified amount of time. For example, computer-controlled anti-lock brakes must begin braking within a predictable, short time after the brake pedal is sensed.

The performance of a computer can be measured using other metrics, depending upon its application domain. A system may be CPU bound (as in numerical calculation), I/O bound (as in a webserving application) or memory bound (as in video editing). Power consumption has become important in servers and portable devices like laptops.

Benchmarking tries to take all these factors into account by measuring the time a computer takes to run through a series of test programs. Although benchmarking shows strengths, it may not help one to choose a computer. Often the measured machines split on different measures. For example, one system might handle scientific applications quickly, while another might play popular video games more smoothly. Furthermore, designers may add special features to their products, in hardware or software, that permit a specific benchmark to execute quickly but don't offer similar advantages to general tasks.

Power consumption is another measurement that is important in modern computers. Power efficiency can often be traded for speed or lower cost. The typical measurement in this case is MIPS/W (millions of instructions per second per watt).

Modern circuits have more power per transistor as the number of transistors per chip grows. Therefore, power efficiency has increased in importance. Recent processor designs such as the Intel Core 2 put more emphasis on increasing power efficiency. Also, in the world of embedded computing, power efficiency has long been and remains an important goal next to throughput and latency.

2.2 Computer System

(2-2) A computer system is one that is able to take a set of inputs, process them and create a set of outputs. This is done by a combination of hardware and software. Computer hardware consists of the physical components that make a computer go. Software, on the other hand, is the programming that tells all those components what to do. **Windows** and **Photoshop** and **Web browsers** are software. Knowing how to operate software is a bit like knowing how to drive a car: It's what you use the computer for on a daily basis. But understanding hardware is like knowing how the car works.

2.2.1 Hardware

We all know hardware describes the physical pieces that make a computer hum with life. But what are those individual pieces? Well, you've probably heard of the processor, or central processing unit (CPU). That's the heart of the computer. It's a **chip** that takes instructions from programs (software), makes calculations and spits out the results. It may be the most important part, but it's certainly not the only one — and like understanding the parts of a car, understanding computer hardware could help you repair one when things go wrong.

1. The Building Blocks of a Computer

By the 1980s, computers were small enough to fit into our homes, but still too expensive and specialized for the average person to put together. That really changed in the 1990s and 2000s, and now computers are shockingly easy to assemble with the right parts, a little patience and a screwdriver.

There are some basic pieces that go into every computer. A **case**, or **tower**, holds all the **components**, with a large open area that fits a **motherboard**. Think of the motherboard as the computer's nervous system: It's a big slab of fiberglass etched with **circuitry** that connects each component of a computer together. Every piece of computer hardware will connect to the motherboard.

Cases also include fans for keeping a computer cool, and room for a big **power supply unit**, or PSU, that handles power conversion for all the parts of a computer. Random access memory (RAM) is an integrated circuit that stores data in such a way that it's quickly accessible to the processor. **Hard drives** and solid state drives store **gigabytes** or **terabytes** of data using different technologies. A graphics card is its own little ecosystem, with a processor dedicated to different tasks than the CPU and high performance RAM. And that's just about all it takes to make a computer go. When the processor is plugged into the motherboard, a heat sink rests on top to keep it cool.

Today's parts are better labeled, cases are more accessible, and computer hardware is cheaper than ever, but the actual makeup of a PC really hasn't changed much.

2. The Evolution of Computer Hardware

The basic components of a personal computer are more or less the same today as they were in the 1990s. Well, perhaps "less" rather than "more". Parts still perform the same overall functions as they once did. The motherboard still serves as the computer's central **hub**, with everything connecting to it; the processor still follows instructions; RAM still stores data for quick access, and hard drives still store data long-term. The way those pieces are connected and how quickly they operate has changed tremendously, however.

Many people who talk about improvements in computers reference **Moore's Law**, which essentially states the number of **integrated circuits** in **microprocessors** will double within every two years. The more integrated circuits, or **transistors**, a chip has, the faster it's going to be. But that's only one thing that makes computers faster and better. For example, the magnetic storage of **hard drive** disks has increased tremendously since the 1990s. We measure drives in terabytes when we used to measure them in megabytes. New interfaces for transmitting data also make a big difference. The Parallel ATA systems topped out at a speed of 133 MB per second, while Serial ATA, or SATA, currently supports up to 6 gigabits per second (768 MB).

Recently, computers have begun to use **solid state** or **flash memory** technology to store data instead of hard drives, enabling computers to access data even faster. Since the rise of the smartphone, computer hardware has gotten smaller than ever. But even in the smartphone space, a lot of the same components are doing the same jobs they do in full-size computers.

3. Different Types of Computers

Laptops, desktops, smartphones, tablets: Their use cases couldn't be much different, could they? We use computers in more places and ways than ever before. But the internal components that make that possible are very similar. In most cases, they're just smaller.

Intel manufactures ULV, or ultra low voltage, processors for thin-and-light notebooks which run on less wattage than its regular laptop chips. Laptops also use smaller RAM and hard drives than desktops. Some laptop makers, like Apple, even solder solid state memory right onto the motherboard instead of including a hard drive, which saves even more space.

Phones and tablets have to be incredibly compact. Instead of a motherboard, the heart of a mobile device is a system-on-a-chip, or SoC. The SoC integrates everything, processor, graphics processor, RAM, interfaces like USB, interfaces for audio, and more, onto a single board. Of course, touch devices include some hardware that desktop computers don't, like touch controllers for sensing our fingers. Instead of internal power supplies, laptops and mobile devices contain batteries.

But for the most part, they're all computers — the hardware simply comes in different shapes and sizes.

2.2.2 Software

Software refers to one or more computer programs and data held in the storage of the computer. In other words, software is a set of programs, procedures, algorithms and its documentation concerned with the operation of a data processing system. Program software performs the function of the program it implements, either by directly providing instructions to the digital electronics or by serving as input to another piece of software. The term was coined to contrast to the term hardware (meaning physical devices). In contrast to hardware, software "cannot be touched".

Practical computer systems divide software systems into three major classes: **system software**, **programming software** and **application software**, although the distinction is arbitrary, and often blurred.

1. System Software

System software is computer software designed to operate the computer hardware, to provide basic functionality, and to provide a platform for running application software. System software includes device drivers, operating systems, servers, utilities, and window systems.

System software is responsible for managing a variety of independent hardware components, so that they can work together harmoniously. Its purpose is to unburden the application software programmer from the often complex details of the particular computer being used, including such accessories as communications devices, printers, device readers, displays and keyboards, and also to partition the computer's resources such as memory and processor time in a safe and stable manner.

2. Programming Software

Programming software includes tools in the form of programs or applications that software developers use to create, debug, maintain, or otherwise support other programs and applications. The term usually refers to relatively simple programs such as compilers, **debuggers**, **interpreters**, **linkers**, and text editors, that can be combined together to accomplish a task, much as one might use multiple hand tools to fix a physical object. Programming tools are intended to assist a programmer in writing computer programs, and they may be combined in an integrated development environment (IDE) to more easily manage all of these functions.

3. Application Software

Application software is all the computer software that causes a computer to perform useful tasks (compare with computer viruses) beyond the running of the computer itself. A specific instance of such software is called a software application, application or app.

The term is used to contrast such software with system software, which manages and integrates a computer's capabilities but does not directly perform tasks that benefit the user. The system software serves the application, which in turn serves the user.

Application software falls into two general categories; horizontal applications and vertical applications. Horizontal applications are the most popular and widespread in departments or companies. Vertical applications are niche products, designed for a particular type of organization. For example, a particular type of business (manufacturing versus banking) or a department within a company (accounting versus customer service).

There are many types of application software:

(1) An **application suite** consists of multiple applications bundled together. They usually have related functions, features and user interfaces, and may be able to interact with each other, e.g. open each other's files. Business applications often come in suites, e.g. Microsoft Office, LibreOffice and iWork, which bundle together a word processor, a spreadsheet, etc.; but suites exist for other purposes, e.g. graphics or music.

(2) **Enterprise software** addresses the needs of an entire organization's processes and data flow, across most all departments, often in a large distributed environment. (Examples include financial systems, customer relationship management (CRM) systems and supply chain management (SCM) software). Departmental Software is a sub-type of enterprise software with a focus on smaller organizations and/or groups within a large organization. (Examples include travel expense management and IT Helpdesk.)

(3) Enterprise infrastructure software provides common capabilities needed to support enterprise software systems. (Examples include databases, email servers, and systems for managing networks and security.)

(4) Information worker software lets users create and manage information, often for individual projects within a department, in contrast to enterprise management. Examples include time management, resource management, documentation tools, analytical, and collaborative. Word processors, spreadsheets, email and blog clients, personal information system, and

individual media editors may aid in multiple tasks of information worker.

(5) Content access software is used primarily to access content without editing, but may include software that allows for content editing. Such software addresses the needs of individuals and groups to consume digital entertainment and published digital content. (Examples include media players, Web browsers, and help browsers.)

(6) Educational software is related to content access software, but has the content and/or features adapted for use in by educators or students. For example, it may deliver evaluations (tests), track progress through material, or include collaborative capabilities.

(7) **Simulation software** simulates physical or abstract systems for research, training or entertainment purposes.

(8) Media development software generates print and electronic media for others to consume, most often in a commercial or educational setting. This includes graphic-art software, desktop publishing software, multimedia development software, HTML editors, digital-animation editors, digital audio and video composition, and many others.

(9) Product engineering software is used in developing hardware and software products. This includes computer-aided design (CAD), computer-aided engineering (CAE), computer language editing and compiling tools, integrated development environments, and application programmer interfaces.

Applications can also be classified by computing platform such as a particular operating system, delivery network such as in cloud computing and web 2.0 applications or delivery devices such as mobile apps for mobile devices.

(2-3) Computer software has to be "loaded" into the computer's storage (such as the hard drive or memory). Once the software has loaded, the computer is able to execute the software. This involves passing instructions from the application software, through the system software, to the hardware which ultimately receives the instruction as machine code. Each instruction causes the computer to carry out an operation—moving data, carrying out a computation, or altering the control flow of instructions.

2.3 Computer Networking

A computer network is a **telecommunications** network that allows computers to exchange data. The physical connection between networked computing devices is established using either **cable media** or **wireless** media. The best-known computer network is the Internet.

Network devices that originate, route and terminate the data are called network nodes. Nodes can include hosts such as **servers** and personal computers, as well as networking hardware. Two devices are said to be networked when a process in one device is able to exchange information with a process in another device.

Computer networks support applications such as access to the World Wide Web, shared use of application and storage servers, printers, and fax machines, and use of email and **instant**

messaging applications. The remainder of this article discusses local area network technologies and classifies them according to the following characteristics: the physical media used to transmit signals, the communications protocols used to organize network traffic, along with the network's size, its **topology** and its organizational intent.

2.3.1 Network Hardware

Apart from the physical communications media described above, networks comprise additional basic hardware building blocks, such as **network interface controller cards**, **repeaters, hubs, bridges, switches, routers,** and **firewalls**.

1. Network Interfaces

A network interface controller (NIC) is a hardware accessory that provides a computer with both a physical interface for accepting a network cable connector and the ability to process low-level network information.

In Ethernet networks, each network interface controller has a unique media access control (MAC) address which is usually stored in the card's permanent memory. **MAC address** uniqueness is maintained and administered by the Institute of Electrical and Electronics Engineers (IEEE) in order to avoid address conflicts between devices on a network. The size of an **Ethernet** MAC address is six octets. The 3 most significant **octets** are reserved to identify card manufacturers. The card manufacturers, using only their assigned prefixes, uniquely assign the 3 least-significant octets of every Ethernet card they produce.

2. Repeaters and Hubs

A repeater is an electronic device that receives a network signal, cleans it of unnecessary noise, and regenerates it. The signal is **retransmitted** at a higher power level, or to the other side of an obstruction, so that the signal can cover longer distances without **degradation**. In most **twisted pair** Ethernet **configurations**, repeaters are required for cable that runs longer than 100 meters. A repeater with multiple ports is known as a hub. Repeaters work on the **physical layer** of the **OSI model**. Repeaters require a small amount of time to regenerate the signal. This can cause a **propagation delay** which can affect network performance. As a result, many network architectures limit the number of repeaters that can be used in a row, e.g., the Ethernet 5-4-3 rule.

Repeaters and hubs have been mostly obsoleted by modern switches.

3. Bridges

A network bridge connects multiple network segments at the **data link layer** (layer 2) of the OSI model to form a single network. Bridges **broadcast** to all ports except the **port** on which the broadcast was received. However, bridges do not **promiscuously** copy traffic to all ports, as hubs do. Instead, bridges learn which MAC addresses are reachable through specific ports. Once the bridge associates a port with an address, it will send traffic for that address to that port only.

Bridges learn the association of ports and addresses by examining the **source address** of **frames** that it sees on various ports. Once a frame arrives through a port, the bridge assumes that the MAC address is associated with that port and stores its source address. The first time a bridge sees a previously unknown **destination address**, the bridge will **forward** the frame to all ports other than the one on which the frame arrived. Bridges come in three basic types:

Local bridges: Directly connect LANs.

Remote bridges: Can be used to create a wide area network (WAN) link between LANs. Remote bridges, where the connecting link is slower than the end networks, largely have been replaced with routers.

Wireless bridges: Can be used to join LANs or connect remote devices to LANs.

4. Switches

A network switch is a device that forwards and **filters** OSI layer 2 **datagrams** between ports based on the MAC addresses in the **packets**. A switch is distinct from a hub in that it only forwards the frames to the ports involved in the communication rather than all ports connected. A switch breaks the collision domain but represents itself as a broadcast domain. Switches make decisions about where to forward frames based on MAC addresses. A switch normally has numerous ports, facilitating a **star topology** for devices, and cascading additional switches. Multi-layer switches are capable of routing based on layer 3 addressing or additional logical levels. The term switch is often used loosely to include devices such as routers and bridges, as well as devices that may distribute **traffic** based on load or based on application content (e.g., a Web URL identifier).

5. Routers

A router is an internetworking device that forwards packets between networks by processing the routing information included in the packet or datagram (Internet protocol information from layer 3). The routing information is often processed in conjunction with the routing table (or forwarding table). A router uses its routing table to determine where to forward packets. (A destination in a routing table can include a "null" interface, also known as the "black hole" interface because data can go into it, however, no further processing is done for said data.)

6. Firewalls

A firewall is a network device for controlling network security and access rules. Firewalls are typically configured to reject access requests from unrecognized sources while allowing actions from recognized ones. The vital role firewalls play in network security grows in parallel with the constant increase in **cyber attacks**.

2.3.2 Network Protocols

In the context of data communication, a network protocol is a formal set of rules, conventions and data structure that governs how computers and other network devices exchange information over a network. In other words, protocol is a standard procedure and format that two data communication devices must understand, accept and use to be able to talk to each other.

(2-4)In modern protocol design, protocols are "layered" according to the OSI 7 layer model or a similar layered model. Layering is a design principle which divides the protocol design into a number of smaller parts, each part accomplishing a particular sub-task and interacting with the other parts of the protocol only in a small number of well-defined ways. Layering allows the parts of a protocol to be designed and tested without a combinatorial explosion of cases, keeping each design relatively simple. Layering also permits familiar protocols to be adapted to unusual circumstances.

The header and/or trailer at each layer reflect the structure of the protocol. Detailed rules and procedures of a protocol or protocol group are often defined by a lengthy document. For example, IETF uses RFCs (Request for Comments) to define protocols and updates to the protocols.

A wide variety of communication protocols exists. These protocols were defined by many different standard organizations throughout the world and by technology vendors over years of technology evolution and development. One of the most popular protocol suites is TCP/IP, which is the heart of internetworking communications. The IP, the Internet Protocol, is responsible for exchanging information between routers so that the routers can select the proper path for network traffic, while TCP is responsible for ensuring the data packets are transmitted across the network reliably and error free. LAN and WAN protocols are also critical protocols in network communications. The LAN protocols suite is for the physical and data link layers of communications over various LAN media such as Ethernet wires and wireless radio waves. The WAN protocol suite is for the lowest three layers and defines communication over various wide-area media, such as fiber optic and copper cables.

Network communication has slowly evolved. Today's new technologies are based on the accumulation over years of technologies, which may be either still existing or obsolete. Because of this, the protocols which define the network communication are highly inter-related. Many protocols rely on others for operation. For example, many routing protocols use other network protocols to exchange information between routers.

In addition to standards for individual protocols in transmission, there are now also interface standards for different layers to talk to the ones above or below (usually operating system specific). For example: Winsock and Berkeley **sockets** between layers 4 and 5; NDIS and ODI between layers 2 and 3.

The protocols for data communication cover all areas as defined in the OSI model. However, the OSI model is only loosely defined. A protocol may perform the functions of one or more of the OSI layers, which introduces complexity to understanding protocols relevant to the OSI 7 layer model. In real-world protocols, there is some argument as to where the distinctions between layers are drawn; there is no one black and white answer.

To develop a complete technology that is useful for the industry, very often a group of protocols is required in the same layer or across many different layers. Different protocols often describe different aspects of a single communication; taken together, these form a protocol suite. For example, Voice over IP (VoIP), a group of protocols developed by many vendors and standard organizations, has many protocols across the 4 top layers in the OSI model.

Protocols can be implemented either in hardware or software or a mixture of both. Typically, the lower layers are implemented in hardware, with the higher layers being implemented in software.

Protocols could be grouped into suites (or families, or stacks) by their technical functions, or origin of the protocol introduction, or both. A protocol may belong to one or multiple protocol suites, depending on how you categorize it. For example, the Gigabit Ethernet protocol IEEE 802.3z is a LAN (Local Area Network) protocol and it can also be used in MAN (Metropolitan Area Network) communications.

Most recent protocols are designed by the IETF for Internetworking communications and by the IEEE for local area networking (LAN) and metropolitan area networking (MAN). The ITU-T contributes mostly to wide area networking (WAN) and telecommunications protocols. ISO has its own suite of protocols for internetworking communications, which is mainly deployed in European countries.

2.3.3 Internet and TCP/IP

The Internet is a global system of interconnected computer networks that use the standard Internet protocol suite (TCP/IP) to serve several billion users worldwide. According to Internet World Stats, as of December 31, 2011, there was an estimated 2,267,233,742 Internet users worldwide. This represents 32.7% of the world's population.

(2-5) No one actually owns the Internet, and no single person or organization controls the Internet in its entirety. It is a network of networks that consists of millions of private, public, academic, business, and government networks, of local to global scope, that are linked by a broad array of electronic, wireless and optical networking technologies. The Internet carries an extensive range of information resources and services, such as the inter-linked hypertext documents of the World Wide Web (WWW), the **infrastructure** to support email, and peer-to-peer networks.

The Internet is a globally distributed network comprising many voluntarily interconnected autonomous networks. It operates without a central governing body. However, to maintain interoperability, the principal name spaces of the Internet are administered by the Internet Corporation for Assigned Names and Numbers (ICANN), headquartered in Marina del Rey, California. ICANN is the authority that coordinates the assignment of unique identifiers for use on the Internet, including **domain names**, **Internet Protocol** (IP) addresses, application port numbers in the **transport protocols**, and many other parameters. Globally unified name spaces, in which names and numbers are uniquely assigned, are essential for maintaining the global reach of the Internet. ICANN is governed by an international board of directors drawn from across the Internet technical, business, academic, and other non-commercial communities.

ICANN's role in coordinating the assignment of unique identifiers distinguishes it as perhaps the only central coordinating body for the global Internet. The government of the United States continues to have a primary role in approving changes to the DNS root zone that lies at the heart of the domain name system. On 16 November 2005, the United Nations-sponsored World Summit on the Information Society, held in Tunis, established the Internet Governance Forum

(IGF) to discuss Internet-related issues.

The technical underpinning and standardization of the Internet's core protocols (IPv4 and IPv6) is an activity of the Internet Engineering Task Force (IETF), a non-profit organization of loosely affiliated international participants that anyone may associate with by contributing technical expertise.

The Internet protocol suite is the networking model and a set of communications protocols used for the Internet and similar networks. It is commonly known as TCP/IP, because its most important protocols, the Transmission Control Protocol (TCP) and the Internet Protocol (IP) were the first networking protocols defined in this standard. It is occasionally known as the DoD model due to the foundational influence of the ARPANET in the 1970s (operated by DARPA, an agency of the United States Department of Defense).

The Internet protocol suite uses **encapsulation** to provide abstraction of protocols and services. Encapsulation is usually aligned with the division of the protocol suite into layers of general functionality. In general, an application (the highest level of the model) uses a set of protocols to send its data down the layers, being further encapsulated at each level.

The layers of the protocol suite near the top are logically closer to the user application, while those near the bottom are logically closer to the physical transmission of the data. Viewing layers as providing or consuming a service is a method of abstraction to isolate upper layer protocols from the details of transmitting bits over, for example, Ethernet and **collision detection**, while the lower layers avoid having to know the details of each and every application and its protocol.

Even when the layers are examined, the assorted architectural documents—there is no single architectural model such as ISO 7498, the Open Systems Interconnection (OSI) model—have fewer and less rigidly defined layers than the OSI model, and thus provide an easier fit for real-world protocols. One frequently referenced document, RFC 1958, does not contain a stack of layers. The lack of emphasis on layering is a major difference between the IETF and OSI approaches. It only refers to the existence of the internetworking layer and generally to upper layers; this document was intended as a 1996 snapshot of the architecture: "The Internet and its architecture have grown in evolutionary fashion from modest beginnings, rather than from a Grand Plan. While this process of evolution is one of the main reasons for the technology's success, it nevertheless seems useful to record a snapshot of the current principles of the Internet architecture."

RFC 1122, entitled Host Requirements, is structured in paragraphs referring to layers, but the document refers to many other architectural principles not emphasizing layering. It loosely defines a four-layer model, with the layers having names, not numbers, as follows:

(1) Application layer (process-to-process): This is the scope within which applications create user data and communicate this data to other processes or applications on another or the same host. The communications partners are often called peers. This is where the higher level protocols such as SMTP, FTP, SSH, HTTP, etc. operate.

(2) Transport layer (host-to-host): The transport layer constitutes the networking regime between two network hosts, either on the local network or on remote networks separated by routers. The transport layer provides a uniform networking interface that hides the actual topology (layout) of the underlying network connections. This is where flow-control, error-correction, and connection protocols exist, such as TCP. This layer deals with opening and maintaining connections between Internet hosts.

(3) Internet layer: The internet layer has the task of exchanging datagrams across network boundaries. It is therefore also referred to as the layer that establishes internetworking, indeed, it defines and establishes the Internet. This layer defines the addressing and routing structures used for the TCP/IP protocol suite. The primary protocol in this scope is the Internet Protocol, which defines IP addresses. Its function in routing is to transport datagrams to the next IP router that has the connectivity to a network closer to the final data destination.

(4) Link layer: This layer defines the networking methods within the scope of the local network link on which hosts communicate without intervening routers. This layer describes the protocols used to describe the local network topology and the interfaces needed to effect transmission of internet layer datagrams to next-neighbor hosts.

The Internet protocol suite and the layered protocol stack design were in use before the OSI model was established. Since then, the TCP/IP model has been compared with the OSI model in books and classrooms, which often results in confusion because the two models use different assumptions and goals, including the relative importance of strict layering.

This abstraction also allows upper layers to provide services that the lower layers do not provide. While the original OSI model was extended to include connectionless services (OSIRM CL), IP is not designed to be reliable and is a best effort delivery protocol. This means that all transport layer implementations must choose whether or how to provide reliability. UDP provides data integrity via a checksum but does not guarantee **delivery**; TCP provides both **data integrity** and delivery guarantee by retransmitting until the receiver acknowledges the **reception** of the packet.

This model lacks the formalism of the OSI model and associated documents, but the IETF does not use a formal model and does not consider this a limitation, as illustrated in the comment by David D. Clark, "We reject: kings, presidents and voting. We believe in: rough consensus and running code." Criticisms of this model, which have been made with respect to the OSI model, often do not consider ISO's later extensions to that model.

For **multiaccess** links with their own addressing systems (e.g. Ethernet), an address mapping protocol is needed. Such protocols can be considered to be below IP but above the existing link system. While the IETF does not use the **terminology**, this is a subnetwork dependent **convergence** facility according to an extension to the OSI model, the internal organization of the network layer (IONL).

ICMP & IGMP operate on top of IP but do not transport data like UDP or TCP. Again, this functionality exists as layer management extensions to the OSI model in its Management

Framework (OSIRM MF).

The library operates above the transport layer (uses TCP) but below application protocols. Again, the SSL/TLS re was no intention, on the part of the designers of these protocols, to comply with OSI architecture.

The link is treated like a black box. The IETF explicitly does not intend to discuss transmission systems, which is a less academic but practical alternative to the OSI model.

2.4 Wireless Network

Wireless networks utilize **radio waves** and/or **microwaves** to maintain **communication channels** between computers. Wireless networking is a more modern alternative to wired networking that relies on **copper** and/or **fiber optic cabling** between network devices.

A wireless network offers advantages and disadvantages compared to a wired network. Advantages of wireless include mobility and elimination of unsightly cables. Disadvantages of wireless include the potential for **radio interference** due to weather, other wireless devices, or obstructions like walls.

2.4.1 Wireless LAN Networking Basics

Wireless is rapidly gaining in popularity for both home and business networking. Wireless technology continues to improve, and the cost of wireless products continues to decrease. Popular wireless local area networking (WLAN) products conform to the 802.11 ("Wi-Fi") standards. The gear a person needs to build wireless networks includes network adapters (NICs), access points (APs), and routers.

Most wireless networks are based on the IEEE 802.11 standards. A basic wireless network consists of multiple stations communicating with radios that broadcast in either the 2.4GHz or 5GHz **band**, though this varies according to the locale and is also changing to enable communication in the 2.3GHz and 4.9GHz ranges.

(2-6) 802.11 networks are organized in two ways. In **infrastructure mode**, one station acts as a master with all the other stations associating to it. The master station is termed an access point (AP) and all communication passes through the AP; even when one station wants to communicate with another wireless station, messages must go through the AP. In the second form of network, there is no master and stations communicate directly. This form of network is commonly known as an **ad-hoc network**.

802.11 networks were first deployed in the 2.4GHz band using protocols defined by the IEEE 802.11 and 802.11b standard. These specifications include the operating frequencies and the MAC layer characteristics, including framing and **transmission rates**, as communication can occur at various rates. Later, the 802.11a standard defined operation in the 5GHz band, including different signaling mechanisms and higher transmission rates. Still later, the 802.11g standard defined the use of 802.11a signaling and transmission mechanisms in the 2.4GHz band in such a way as to be backwards compatible with 802.11b networks.

Separate from the underlying transmission techniques, 802.11 networks have a variety of security mechanisms. The original 802.11 specifications defined a simple security protocol called WEP. This protocol uses a fixed **pre-shared key** and the RC4 cryptographic cipher to encode data transmitted on a network. Stations must all agree on the fixed key in order to communicate. This scheme was shown to be easily broken and is now rarely used except to discourage transient users from joining networks. Current security practice is given by the IEEE® 802.11i specification that defines new cryptographic ciphers and an additional protocol to authenticate stations to an access point and exchange keys for data communication. Cryptographic keys are periodically refreshed and there are mechanisms for detecting and countering **intrusion** attempts. Another security protocol specification commonly used in wireless networks is termed WPA, which was a precursor to 802.11i. WPA specifies a subset of the requirements found in 802.11i and is designed for implementation on legacy hardware. Specifically, WPA requires only the TKIP cipher that is derived from the original WEP cipher. 802.11i permits use of TKIP but also requires support for a stronger cipher, AES-CCM, for encrypting data. The AES cipher was not required in WPA because it was deemed too computationally costly to be implemented on legacy hardware.

The other standard to be aware of is 802.11e. It defines protocols for deploying multimedia applications, such as streaming video and voice over IP (VoIP), in an 802.11 network. Like 802.11i, 802.11e also has a precursor specification termed WME (later renamed WMM) that has been defined by an industry group as a subset of 802.11e that can be deployed now to enable multimedia applications while waiting for the final ratification of 802.11e. The most important thing to know about 802.11e and WME/WMM is that it enables prioritized traffic over a wireless network through Quality of Service (QoS) protocols and enhanced media access protocols. Proper implementation of these protocols enables high speed bursting of data and prioritized **traffic flow**.

FreeBSD supports networks that operate using 802.11a, 802.11b, and 802.11g. The WPA and 802.11i security protocols are likewise supported (in conjunction with any of 11a, 11b, and 11g) and QoS and traffic prioritization required by the WME/WMM protocols are supported for a limited set of wireless devices.

2.4.2 Mobile Network

(2-7) A **cellular network** or **mobile network** is a radio network distributed over land areas called cells, each served by at least one fixed-location **transceiver**, known as a **cell site** or **base station**. In a cellular radio system, a land area to be supplied with radio service is divided into regular shaped cells, which can be hexagonal, square, circular or some other regular shapes, although hexagonal cells are conventional. Each of these cells is assigned multiple frequencies which have corresponding radio base stations. The group of frequencies can be reused in other cells, provided that the same frequencies are not reused in adjacent neighboring cells as that would cause **co-channel interference**.

The increased capacity in a cellular network, compared with a network with a single transmitter, comes from the fact that the same radio frequency can be reused in a different area for a completely different transmission. If there is a single plain transmitter, only one transmission can be used on any given frequency. Unfortunately, there is inevitably some level of interference from the signal from the other cells which use the same frequency. This means that, in a standard FDMA(Frequency Division Multiple Access) system, there must be at least a one cell gap between cells which reuse the same frequency.

In the simple case of the taxi company, each radio had a manually operated channel selector knob to tune to different frequencies. As the drivers moved around, they would change from channel to channel. The drivers knew which frequency covered approximately what area. When they did not receive a signal from the transmitter, they would try other channels until they found one that worked. The taxi drivers would only speak one at a time, when invited by the base station operator (this is, in a sense, time division multiple access (TDMA)).

A cellular network is used by the mobile phone operator to achieve both coverage and capacity for their subscribers. Large geographic areas are split into smaller cells to avoid line-of-sight signal loss and to support a large number of active phones in that area. All of the cell sites are connected to **telephone exchanges** (or switches), which in turn connect to the public telephone network. In cities, each cell site may have a range of up to approximately ½ mile, while in rural areas, the range could be as much as 5 miles. It is possible that in clear open areas, a user may receive signals from a cell site 25 miles away.

Since almost all mobile phones use cellular technology, including GSM, CDMA, and AMPS (analog), the term "**cell phone**" is in some regions, notably the US, used interchangeably with "**mobile phone**". However, **satellite phones** are mobile phones that do not communicate directly with a ground-based cellular tower, but may do so indirectly by way of a satellite.

As the phone user moves from one cell area to another cell while a call is in progress, the mobile station will search for a new channel to attach, in order not to drop the call. Once a new channel is found, the network will command the mobile unit to switch to the new channel and at the same time switch the call onto the new channel.

With CDMA, multiple CDMA **handsets** share a specific radio channel. The signals are separated by using a pseudonoise code (PN code) specific to each phone. As the user moves from one cell to another, the handset sets up radio links with multiple cell sites (or sectors of the same site) simultaneously. This is known as "soft handoff" because, unlike with traditional cellular technology, there is no one defined point where the phone switches to the new cell.

If there is no ongoing communication or the communication can be interrupted, it is possible for the mobile unit to spontaneously move from one cell to another and then notify the base station with the strongest signal.

2.4.3 Wireless Sensor Network

A wireless sensor network (WSN) consists of spatially distributed autonomous sensors to monitor physical or environmental conditions, such as temperature, sound, pressure, etc. and to cooperatively pass their data through the network to a main location. The more modern networks are **bi-directional**, also enabling control of sensor activity. The development of wireless sensor networks was motivated by military applications such as battlefield surveillance; today such networks are used in many industrial and consumer applications, such as industrial process monitoring and control, machine health monitoring, air pollution monitoring, forest fire detection, natural disaster prevention, accurate agriculture , smart home monitoring, and so on.

The WSN is built of "nodes"—from a few to several hundreds or even thousands, where each node is connected to one (or sometimes several) sensors. Each such sensor network node has typically several parts: a radio transceiver with an internal antenna or connection to an external antenna, a microcontroller, an electronic circuit for interfacing with the sensors and an energy source, usually a battery or an embedded form of **energy harvesting**. A sensor node might vary in size from that of a shoebox down to the size of a grain of dust. The cost of sensor nodes is similarly variable, ranging from a few to hundreds of dollars, depending on the complexity of the individual sensor nodes. Size and cost constraints on sensor nodes result in corresponding constraints on resources such as energy, memory, computational speed and communications bandwidth. The topology of the WSNs can vary from a simple star network to an advanced multi-hop wireless mesh network. The propagation technique between the hops of the network can be routing or flooding.

The main characteristics of a WSN include:

(1) Power consumption constrains for nodes using batteries or energy harvesting.

(2) Ability to cope with node failures.

(3) Mobility of nodes.

(4) Communication failures.

(5) **Heterogeneity** of nodes.

(6) **Scalability** to large scale of deployment.

(7) Ability to withstand harsh environmental conditions.

(8) Ease of use.

Sensor nodes can be imagined as small computers, extremely basic in terms of their interfaces and their components. They usually consist of a processing unit with limited computational power and limited memory, sensors or MEMS (Micro Electro-Mechanical Systems), a communication device (usually radio transceivers or alternatively optical), and a power source usually in the form of a battery. Other possible inclusions are energy harvesting modules, secondary ASIC (Application Specific Integrated Circuit), and possibly secondary communication interface (e.g. RS-232 or USB).

The base stations are one or more components of the WSN with much more computational, energy and communication resources. They act as a **gateway** between sensor nodes and the end user as they typically forward data from the WSN on to a server. Other special components in routing based networks are routers, designed to compute, calculate and distribute the routing tables.

Despite the innumerable applications of WSNs, these networks have several restrictions, e.g., limited energy supply, limited computing power, and limited bandwidth of the wireless links connecting sensor nodes. One of the main design goals of WSNs is to carry out data communication while trying to prolong the lifetime of the network and prevent connectivity degradation by employing aggressive energy management techniques. The design of routing protocols in WSNs is influenced by many challenging factors. These factors must be overcome before efficient communication can be achieved in WSNs. In the following, we summarize some of the routing challenges and design issues that affect routing process in WSNs.

(1) Node deployment: Node deployment in WSNs is application dependent and affects the performance of the routing protocol. The deployment can be either deterministic or randomized. In deterministic deployment, the sensors are manually placed and data is routed through pre-determined paths. However, in random node deployment, the sensor nodes are scattered randomly creating an infrastructure in an ad hoc manner. If the resultant distribution of nodes is not uniform, optimal clustering becomes necessary to allow connectivity and enable energy efficient network operation. Inter-sensor communication is normally within short transmission ranges due to energy and bandwidth limitations. Therefore, it is most likely that a route will consist of multiple wireless hops.

(2) Energy consumption without losing accuracy: sensor nodes can use up their limited supply of energy performing computations and transmitting information in a wireless environment. As such, energy-conserving forms of communication and computation are essential. Sensor node lifetime shows a strong dependence on the battery lifetime. In a multi-hop WSN, each node plays a dual role as data sender and data router. The malfunctioning of some sensor nodes due to power failure can cause significant topological changes and might require rerouting of packets and reorganization of the network.

(3) Node/Link heterogeneity: In many studies, all sensor nodes were assumed to be homogeneous, i.e., having equal capacity in terms of computation, communication, and power. However, depending on the application, a sensor node can have different role or capability. The existence of heterogeneous set of sensors raises many technical issues related to data routing. For example, some applications might require a diverse mixture of sensors for monitoring temperature, pressure and humidity of the surrounding environment, detecting motion via acoustic signatures, and capturing the image or video tracking of moving objects. These special sensors can be either deployed independently or the different functionalities can be included in the same sensor nodes. Even data reading and reporting can be generated from these sensors at different rates, subject to diverse quality of service constraints, and can follow multiple data reporting models. For example, hierarchical protocols designate a **cluster-head** node different from the normal sensors. These cluster-heads can be chosen from the deployed sensors or can be more powerful than other sensor nodes in terms of energy, bandwidth, and memory. Hence, the burden of transmission to the base station is handled by the set of cluster-heads.

(4) **Fault tolerance**: Some sensor nodes may fail or be blocked due to lack of power,

physical damage, or environmental interference. The failure of sensor nodes should not affect the overall task of the sensor network. If many nodes fail, MAC and routing protocols must accommodate formation of new links and routes to the data collection base stations. This may require actively adjusting transmit powers and signaling rates on the existing links to reduce energy consumption, or rerouting packets through regions of the network where more energy is available. Therefore, multiple levels of **redundancy** may be needed in a fault-tolerant sensor network.

(5) Scalability: The number of sensor nodes deployed in the sensing area may be in the order of hundreds or thousands, or more. Any routing scheme must be able to work with this huge number of sensor nodes. In addition, sensor network routing protocols should be scalable enough to respond to events in the environment. Until an event occurs, most of the sensors can remain in the sleep state, with data from the few remaining sensors providing a coarse quality.

(6) Network dynamics: Most of the network architectures assume that sensor nodes are stationary. However, mobility of either base stations or sensor nodes is sometimes necessary in many applications. Routing messages from or to moving nodes is more challenging since route stability becomes an important issue, in addition to energy, bandwidth etc. Moreover, the sensed phenomenon can be either dynamic or static depending on the application, e.g., it is dynamic in a target detection/tracking application, while it is static in forest monitoring for early fire prevention. Monitoring static events allows the network to work in a reactive mode, simply generating traffic when reporting. Dynamic events in most applications require periodic reporting and consequently generate significant traffic to be routed to the BS.

(7) Transmission Media: In a multi-hop sensor network, communicating nodes are linked by a wireless medium. The traditional problems associated with a wireless channel (e.g., fading, high error rate) may also affect the operation of the sensor network. In general, the required bandwidth of sensor data will be low, on the order of 1~100kb/s. Related to the transmission media is the design of medium access control (MAC). One approach of MAC design for sensor networks is to use TDMA based protocols that conserve more energy compared to contention based protocols like CSMA (e.g., IEEE 802.11). **Bluetooth** technology can also be used.

(8) Connectivity: High node density in sensor networks precludes them from being completely isolated from each other. Therefore, sensor nodes are expected to be highly connected. This, however, may not prevent the network topology from being variable and the network size from being shrinking due to sensor node failures. In addition, connectivity depends on the possibly random, distribution of nodes.

(9) Coverage: In WSNs, each sensor node obtains a certain view of the environment. A given sensor's view of the environment is limited both in range and in accuracy; it can only cover a limited physical area of the environment. Hence, area coverage is also an important design parameter in WSNs.

(10) **Data Aggregation**: Since sensor nodes may generate significant redundant data, similar packets from multiple nodes can be aggregated so that the number of transmissions is

reduced. Data aggregation is the combination of data from different sources according to a certain aggregation function, e.g., duplicate suppression, minima, maxima and average. This technique has been used to achieve energy efficiency and data transfer optimization in a number of routing protocols. Signal processing methods can also be used for data aggregation. In this case, it is referred to as data fusion where a node is capable of producing a more accurate output signal by using some techniques such as beamforming to combine the incoming signals and reducing the noise in these signals.

(11) Quality of Service: In some applications, data should be delivered within a certain period of time from the moment it is sensed; otherwise the data will be useless. Therefore bounded latency for data delivery is another condition for time-constrained applications. However, in many applications, conservation of energy, which is directly related to network lifetime, is considered relatively more important than the quality of data sent. As the energy gets depleted, the network may be required to reduce the quality of the results in order to reduce the energy dissipation in the nodes and hence lengthen the total network lifetime. Hence, energy-aware routing protocols are required to capture this requirement.

2.5 Key Terms and Review Questions

1. Technical Terms

central processor	中央处理器	2.1
computer architecture	计算机体系结构	2.1
computer bus	计算机总线	2.1
hardware	硬件	2.1
interrupt latency	中断等待时间	2.1
latency	延迟	2.1
memory	内存	2.1
memory capacity	内存容量	2.1
microarchitecture	微体系结构	2.1
multiprocessing	多重处理	2.1
pipeline	流水线	2.1
software	软件	2.1
superscalar processor	超级标量处理器	2.1
throughput	吞吐量	2.1
virtualization	虚拟化	2.1
power supply unit	供电单元	2.2
application	应用（软件）	2.2
application software	应用软件	2.2
application suite	应用套件	2.2
battery	电池	2.2

case	机箱	2.2
chip	芯片	2.2
circuitry	电路	2.2
component	元件、组件	2.2
debuggers	调试器（程序）	2.2
desktop	台式计算机	2.2
digital-animation	数字动画	2.2
enterprise software	企业软件	2.2
flash memory	闪存	2.2
gigabytes	GB，10 亿字节	2.2
hard drives	硬盘驱动器	2.2
heat sink	散热器	2.2
hub	集线器	2.2
integrated circuits	集成电路	2.2
interpreters	解释器（解释程序）	2.2
laptops	膝上电脑、笔记本	2.2
linkers	连接程序	2.2
microprocessors	微处理器	2.2
Moore’s Law	摩尔定律	2.2
motherboard	计算机主板	2.2
Photoshop	图像处理软件名	2.2
programming software	编程软件	2.2
simulation software	仿真软件	2.2
smartphones	智能手机	2.2
solid state memory	固态存储器	2.2
system software	系统软件	2.2
tablets	平板电脑	2.2
terabytes	TB，1 万亿字节	2.2
touch devices	触摸设备	2.2
tower	塔式机箱	2.2
transistors	晶体管	2.2
Web browser	网页浏览器	2.2
Windows	一种操作系统的名字	2.2
sockets	套接字	2.3
star topology	星状拓扑	2.3
bridge	网桥，桥接	2.3
broadcast	广播	2.3
cable media	电缆介质	2.3
collision detection	碰撞检测	2.3

configurations	配置	2.3
convergence	聚合、会聚	2.3
cyber attacks	网络攻击	2.3
data integrity	数据完整性	2.3
data link layer	数据链路层	2.3
datagram	数据报文	2.3
degradation	退化	2.3
delivery	交付、送达	2.3
destination address	目的地址	2.3
domain names	域名	2.3
encapsulation	封装	2.3
Ethernet	以太网	2.3
filter	过滤	2.3
firewall	防火墙	2.3
forward	转发	2.3
frames	帧	2.3
infrastructure	基础设施	2.3
instant messaging applications	即时消息应用	2.3
Internet Protocol	互联网协议	2.3
MAC address	介质访问控制地址	2.3
multiaccess	多路访问	2.3
network interface controller card	网络接口控制卡，简称网卡	2.3
octet	8 位字节	2.3
OSI model	开放系统互连模型	2.3
packet	数据包	2.3
physical layer	物理层	2.3
port	端口	2.3
propagation delay	传输延迟	2.3
reception	接收	2.3
repeater	中继器	2.3
retransmit	转发	2.3
router	路由器	2.3
routing table	路由表	2.3
server	服务器	2.3
source address	源地址	2.3
switch	交换机	2.3
telecommunication	远程通信	2.3
terminology	术语	2.3
topology	拓扑	2.3

traffic	流量，通信量	2.3
Transmission Control Protocol	传输控制协议	2.3
transport protocols	传输协议	2.3
twisted pair	双绞线	2.3
wireless	无线	2.3
microwaves	微波	2.4
communication channel	通信信道	2.4
copper calble	铜缆	2.4
fiber optic cable	光缆	2.4
radio interference	无线电干扰	2.4
radio waves	无线电波	2.4
wireless networks	无线网络	2.4
ad-hoc network	自组织网络	2.4
band	波段	2.4
infrastructure mode	基础架构模式	2.4
intrusion	入侵	2.4
pre-shared key	预共享密钥	2.4
transmission rates	传输速率	2.4
cellular network	蜂窝网络	2.4
mobile network	移动网络	2.4
transceiver	收发器	2.4
cell site	蜂窝基站	2.4
base station	基站	2.4
co-channel interference	同波道干扰	2.4
telephone exchange	电话交换台	2.4
wireless sensor network	无线传感器网络	2.4
bi-directional	双向的	2.4
multi-hop wireless mesh network	多跳无线网状网络	2.4
flooding	洪泛	2.4
energy harvesting	能量采集	2.4
heterogeneity	异质性	2.4
scalability	可扩展性	2.4
gateway	网关	2.4
cluster-head	簇头	2.4
fault tolerance	容错	2.4
redundancy	冗余	2.4
bluetooth	蓝牙	2.4
data aggregation	数据聚合	2.4

2. Translation Exercises

(2-1) The most common scheme carefully chooses the bottleneck that most reduces the computer's speed. Ideally, the cost is allocated proportionally to assure that the data rate is nearly the same for all parts of the computer, with the most costly part being the slowest. This is how skillful commercial integrators optimize personal computers such as smart cellphones.

(2-2) A computer system is one that is able to take a set of inputs, process them and create a set of outputs. This is done by a combination of hardware and software. Computer hardware consists of the physical components that make a computer go. Software, on the other hand, is the programming that tells all those components what to do. **Windows** and **Photoshop** and **Web browsers** are software. Knowing how to operate software is a bit like knowing how to drive a car: It's what you use the computer for on a daily basis. But understanding hardware is like knowing how the car works.

(2-3) Computer software has to be "loaded" into the computer's storage (such as the hard drive or memory). Once the software has loaded, the computer is able to execute the software. This involves passing instructions from the application software, through the system software, to the hardware which ultimately receives the instruction as machine code. Each instruction causes the computer to carry out an operation– moving data, carrying out a computation, or altering the control flow of instructions.

(2-4) In modern protocol design, protocols are "layered" according to the OSI 7 layer model or a similar layered model. Layering is a design principle which divides the protocol design into a number of smaller parts, each part accomplishing a particular sub-task and interacting with the other parts of the protocol only in a small number of well-defined ways. Layering allows the parts of a protocol to be designed and tested without a combinatorial explosion of cases, keeping each design relatively simple. Layering also permits familiar protocols to be adapted to unusual circumstances.

(2-5) No one actually owns the Internet, and no single person or organization controls the Internet in its entirety. It is a network of networks that consists of millions of private, public, academic, business, and government networks, of local to global scope, that are linked by a broad array of electronic, wireless and optical networking technologies. The Internet carries an extensive range of information resources and services, such as the inter-linked hypertext documents of the World Wide Web (WWW), the infrastructure to support email, and peer-to-peer networks.

(2-6) 802.11 networks are organized in two ways. In infrastructure mode, one station acts as a master with all the other stations associating to it. The master station is termed an access point (AP) and all communication passes through the AP; even when one station wants to communicate with another wireless station, messages must go through the AP. In the second form of network, there is no master and stations communicate directly. This form of network is commonly known as an ad-hoc network.

(2-7) A cellular network or mobile network is a radio network distributed over land areas called cells, each served by at least one fixed-location transceiver, known as a cell site or base station. In a cellular radio system, a land area to be supplied with radio service is divided into regular shaped cells, which can be hexagonal, square, circular or some other regular shapes, although hexagonal cells are conventional. Each of these cells is assigned multiple frequencies which have corresponding radio base stations. The group of frequencies can be reused in other cells, provided that the same frequencies are not reused in adjacent neighboring cells as that would cause co-channel interference.

References

[1] Hennessy J L, Patterson D A. **Computer Architecture: A Quantitative Approach** (Third Edition)[M]. Morgan Kaufmann Publishers,2011.

[2] What is computer hardware [EB/OL]. (2017-05-23). http://computer.howstuffworks.com/what-is- computer -hardware.htm.

[3] What is software? - Definition from Whatis.com [EB/OL].(2012-05-13). http://Searchsoa.techtarget.com.

[4] Sutton C. Internet Began 35 Years Ago at UCLA with First Message Ever Sent Between Two Computers[A]. UCLA, 2008.

[5] Kurose J F, Ross K W. Computer Networking: A Top-Down Approach[M]. Pearson, 2008.

[6] Metcalfe R M, Boggs D R. Ethernet: Distributed Packet Switching for Local Computer Networks[J]. Communications of the ACM,1976,19(5): 395-404.

[7] Curtin M. Introduction to Network Security[EB/OL].[2017-05-24]. http://www.interhack.net/ pubs /network-security.

[8] Security of the Internet[G/OL].[2017-05-24]. http://www.cert.org/encyc article/tocencyc.html.

[9] Dargie W, Poellabauer C. Fundamentals of wireless sensor networks: theory and practice[M], John Wiley and Sons, 2010.

[10] Al-Karaki J N, Kamal A E. Routing Techniques in Wireless Sensor Networks: A Survey[J]. IEEE Wireless Communications, 2004, 11(6): 6-28.

Chapter 3 Operating System

3.1 Definition and Function

When you turn on your computer, it's nice to think that you're in control. There's the trusty computer mouse, which you can move anywhere on the screen, summoning up your music library or Internet browser at the slightest whim. Although it's easy to feel like a director in front of your desktop or laptop, there's a lot going on inside, and the real man behind the curtain handling the necessary tasks is the **operating system**.

3.1.1 What is Operating System?

The operating system (OS) is the most important program that runs on a computer. Every general-purpose computer must have an operating system to run other programs and applications. Computer operating systems perform basic tasks (see Fig.3-1), such as recognizing input from the keyboard, sending output to the display screen, keeping track of files and directories on the disk, and controlling **peripheral devices** such as printers. In short, an operating system enables user interaction with computer systems by acting as an interface between users or application programs and the computer hardware.

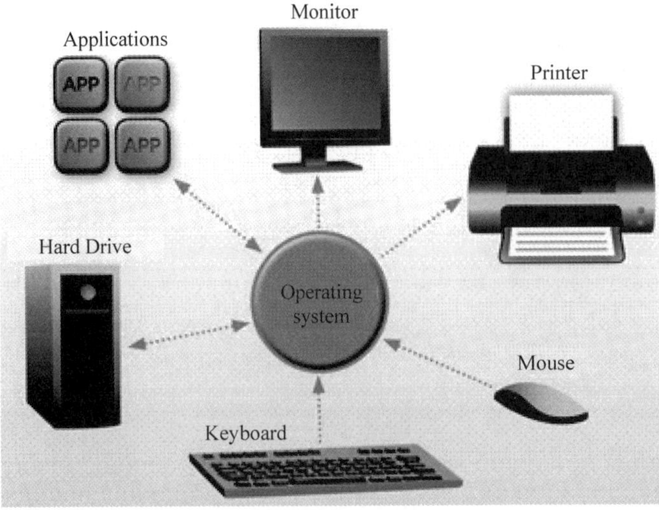

Fig.3-1 Basic tasks of operating system

Most desktop or laptop PCs come pre-loaded with Microsoft Windows. Macintosh computers come pre-loaded with Mac OS X. Many corporate servers use the Linux or UNIX operating systems. The operating system (OS) is the first thing loaded onto the computer — without the operating system, a computer is useless.

More recently, operating systems have started to pop up in smaller computers as well. If you like to tinker with electronic devices, you're probably pleased that operating systems can now be found on many of the devices we use every day, from cell phones to wireless access points. The computers used in these little devices have gotten so powerful that they can now actually run an operating system and applications. The computer in a typical modern cell phone is now more powerful than a desktop computer from 20 years ago, so this progression makes sense and is a natural development.

The purpose of an operating system is to organize and control hardware and software so that the device it lives in behaves in a flexible but predictable way. In this article, we'll tell you what a piece of software must do to be called an operating system, show you how the operating system in your desktop computer works and give you some examples of how to take control of the other operating systems around you.

It is vital to understand that an operating system (OS) is just a program—a very large, very complex program, but still just a program. The OS provides support for the loading and execution of other programs (which we will refer to below as "application programs"), and the OS will set things up so that it has some special privileges which user programs don't have, but in the end, the OS is simply a program.

3.1.2 Functions of Operating System

At the simplest level, an operating system does two things:

(1) It manages the hardware and software resources of the system. In a desktop computer, these resources include such things as the processor, memory, disk space and more (on a cell phone, they include the keypad, the screen, the address book, the phone dialer, the battery and the network connection).

(2) It provides a stable, consistent way for applications to deal with the hardware without having to know all the details of the hardware.

(3-1) The first task, managing the hardware and software resources, is very important, as various programs and input methods compete for the attention of the central processing unit (CPU) and demand memory, storage and input/output (I/O) bandwidth for their own purposes. In this capacity, the operating system plays the role of the good parent, making sure that each application gets the necessary resources while playing nicely with all the other applications, as well as husbanding the limited capacity of the system to the greatest good of all the users and applications.

The second task, providing a consistent application interface, is especially important if there is to be more than one of a particular type of computer using the operating system, or if the hardware making up the computer is ever open to change. A consistent application program interface (API) allows a software developer to write an application on one computer and have a high level of confidence that it will run on another computer of the same type, even if the amount of memory or the quantity of storage is different on the two machines.

Even if a particular computer is unique, an operating system can ensure that applications continue to run when hardware upgrades and updates occur. This is because the operating system — not the application — is charged with managing the hardware and the distribution of its resources. One of the challenges facing developers is keeping their operating systems flexible enough to run hardware from the thousands of vendors manufacturing computer equipment. Today's systems can accommodate thousands of different printers, disk drives and special peripherals in any possible combination.

What does it mean for a program to be "running" anyway? Recall that the CPU is constantly performing its fetch/execute/fetch/execute/··· cycle. For each fetch, it fetches whatever instruction the Program Counter (PC) is pointing to. If the PC is currently pointing to an instruction in your program, then your program is running! Each time an instruction of your program executes, the circuitry in the CPU will update the PC, having it point to either the next instruction (the usual case) or an instruction located elsewhere in your program (in the case of jumps).

The point is that the only way your program can stop running is the PC changed to point to another program, say the OS. How might this happen? Other than cases involving **bugs** in your program, there are only two ways this can occur:

Your program can voluntarily relinquish the CPU to the OS. It does this via a system call, which is a call to some function in the operating system that provides some useful service.

For example, suppose the C source file from which a.out was compiled had a call to scanf(). The scanf() function is a C library function, which was linked into a.out during compilation of a.out. But scanf() itself calls read(), a function within the OS. So, when a.out reaches the scanf() call, that will result in a call to the OS, but after the OS does the read from the keyboard, the OS will return to a.out.

The other possibility is that a hardware interrupt occurs. This is a signal—a **physical pulse** of current along an **interrupt-request** line in the bus—from some input/output (I/O) device such as the keyboard to the CPU. The circuitry in the CPU is designed to then jump to a place in memory which we designated upon **bootup** of the machine. This will be a place in the OS, so the OS will now run. The OS will attend to the I/O device, e.g. record the keystroke in the case of the keyboard, and then return to the interrupted program.

Note in our keystroke example that the keystroke may not have been made by you. While your program is running, some other user of the machine may hit a key. The interrupt will cause your program to be suspended; the OS will run the device driver for whichever device caused the interrupt — the keyboard, if the person was sitting at the console of the machine, or the network interface card, if the person was logged in remotely, say via telnet — which will record the keystroke in the buffer belonging to that other user, and the OS will do IRET, causing your program to resume.

So, when your program is running, it is king. The OS has no power to stop it. The only ways your program can stop running is if it voluntarily does so or is forced to stop by action occurring at an I/O device.

3.1.3 Types of Operating Systems

Here is an overview of the different types of operating systems.

1. Real-time Operating System

It is a **multitasking** operating system that aims at executing real-time applications. **Real-time** operating systems often use specialized scheduling algorithms so that they can achieve a deterministic nature of behavior. The main object of real-time operating systems is their quick and predictable response to events. They either have an **event-driven** design or a **time-sharing** one. An event-driven system switches between tasks based of their priorities while time-sharing operating systems switch tasks based on clock interrupts. Windows CE, OS-9, Symbian and LynxOS are some of the commonly known real-time operating systems.

2. Multi-user and Single-user Operating Systems

Computer operating systems of this type allow multiple users to access a computer system simultaneously. Time-sharing systems can be classified as multi-user systems as they enable a multiple user access to a computer through time sharing. Single-user operating systems, as opposed to a multi-user operating system, are usable by only one user at a time. Being able to have multiple accounts on a Windows operating system does not make it a multi-user system. Rather, only the network administrator is the real user. But for a UNIX-like operating system, it is possible for two users to log in at a time and this capability of the OS makes it a multi-user operating system. Windows 95, Windows 2000, Mac OS and Palm OS are examples of single-user operating systems. UNIX and OpenVMS are examples of multi-user operating systems.

3. Multi-tasking and Single-tasking Operating Systems

When a single program is allowed to run at a time, the system is grouped under the single-tasking system category, while in case the operating system allows for execution of multiple tasks at a time, it is classified as a multi-tasking operating system. Multi-tasking can be of two types namely, **pre-emptive** or **co-operative**. In pre-emptive multitasking, the operating system slices the CPU time and dedicates one slot to each of the programs. UNIX-like operating systems such as Solaris and Linux support pre-emptive multitasking. If you are aware of the **multi-threading** terminology, you can consider this type of multi-tasking as similar to interleaved multi-threading. Cooperative multitasking is achieved by relying on each process to give time to the other processes in a defined manner. This kind of multi-tasking is similar to the idea of block multi-threading in which one **thread** runs till it is blocked by some other event. MS Windows prior to Windows 95 used to support cooperative multitasking. Palm OS for Palm handheld is a single-task operating system. Windows 9x support multi-tasking. DOS Plus is a relatively less-known multi-tasking operating system. It can support the multi-tasking of a maximum of four CP/M-86 programs.

4. Distributed Operating System

An operating system that manages a group of independent computers and makes them appear to be a single computer is known as a **distributed operating system**. The development of networked computers that could be linked and made to communicate with each other, gave rise to distributed computing. Distributed computations are carried out on more than one machine. When computers in a group work in cooperation, they make a distributed system. Amoeba, Plan9 and LOCUS (developed during the 1980s) are some examples of distributed operating systems.

5. Embedded System

The operating systems designed for being used in embedded computer systems are known as **embedded operating systems**. They are designed to operate on small machines like PDAs with less autonomy. They are able to operate with a limited number of resources. They are very compact and extremely efficient by design. Windows CE, FreeBSD and Minix 3 are some examples of embedded operating systems. The use of Linux in embedded computer systems is referred to as Embedded Linux.

6. Mobile Operating System

Though not a functionally distinct kind of operating system, mobile OS is definitely an important mention in the list of operating system types. A mobile OS controls a mobile device and its design supports wireless communication and mobile applications. It has built-in support for mobile multimedia formats. **Tablet PCs** and smartphones run on mobile operating systems. Blackberry OS, Google's Android and Apple's iOS are some of the most known names of mobile operating systems.

7. Batch Processing and Interactive Systems

Batch processing refers to execution of computer programs in "batches" without manual intervention. In batch processing systems, programs are collected, grouped and processed on a later date. There is no prompting the user for inputs as input data are collected in advance for future processing. Input data are collected and processed in batches, hence the name batch processing. IBM's z/OS has batch processing capabilities. As against this, **interactive operating** requires user intervention. The process cannot be executed in the user's absence.

8. Online and Offline Processing

In online processing of data, the user remains in contact with the computer and processes are executed under control of the computer's central processing unit. When processes are not executed under direct control of the CPU, the processing is referred to as offline. Let's take the example of batch processing. Here, the batching or grouping of data can be done without user and CPU intervention; it can be done offline. But the actual process execution may happen under direct control of the processor, that is online.

Operating systems contribute to simplifying human interaction with the computer hardware. They are responsible for linking application programs with the hardware, thus achieving easy user access to computers. Ever imagined a computer without an OS? It wouldn't be so user-friendly then!

3.2 Tasks of an Operating System

When you turn on the power to a computer, the first program that runs is usually a set of instructions kept in the computer's read-only memory (ROM). This code examines the system hardware to make sure everything is functioning properly. This power-on self test (POST) checks the CPU, memory, and basic input-output systems (BIOS) for errors and stores the result in a special memory location. Once the POST has successfully completed, the software loaded in ROM (sometimes called the BIOS or firmware) will begin to activate the computer's disk drives. In most modern computers, when the computer activates the hard disk drive, it finds the first piece of the operating system: the **bootstrap loader**.

The bootstrap loader is a small program that has a single function: It loads the operating system into memory and allows it to begin operation. In the most basic form, the bootstrap loader sets up the small driver programs that interface with and control the various hardware subsystems of the computer. It sets up the divisions of memory that hold the operating system, user information and applications. It establishes the data structures that will hold the myriad signals, flags and **semaphores** that are used to communicate within and between the subsystems and applications of the computer. Then it turns control of the computer over to the operating system.

The operating system's tasks, in the most general sense, fall into six categories:

(1) Processor management;

(2) Memory management;

(3) Device management;

(4) Storage management;

(5) Application interface;

(6) User interface.

While there are some who argue that an operating system should do more than these six tasks, and some operating-system vendors do build many more utility programs and auxiliary functions into their operating systems, these six tasks define the core of nearly all operating systems. Next, let's look at the tools the operating system uses to perform each of these functions.

3.2.1 Processor Management

The heart of managing the processor comes down to two related issues:

(1) Ensuring that each process and application receives enough of the processor's time to function properly.

(2) Using as many processor cycles as possible for real work.

The basic unit of software that the operating system deals with in scheduling the work done by the processor is either a process or a thread, depending on the operating system.

It's tempting to think of a process as an application, but that gives an incomplete picture of how processes relate to the operating system and hardware. The application you see (word processor, spreadsheet or game) is, indeed, a process, but that application may cause several other processes to begin, for tasks like communications with other devices or other computers. There are also numerous processes that run without giving you direct evidence that they ever exist. For example, Windows XP and UNIX can have dozens of background processes running to handle the network, memory management, disk management, virus checks and so on.

A **process**, then, is software that performs some action and can be controlled — by a user, by other applications or by the operating system.

It is processes, rather than applications, that the operating system controls and schedules for execution by the CPU. In a single-tasking system, the schedule is straightforward. The operating system allows the application to begin running, suspending the execution only long enough to deal with interrupts and user input.

(3-2) Interrupts are special signals sent by hardware or software to the CPU. It's as if some part of the computer suddenly raised its hand to ask for the CPU's attention in a lively meeting. Sometimes the operating system will schedule the priority of processes so that interrupts are masked — that is, the operating system will ignore the interrupts from some sources so that a particular job can be finished as quickly as possible. There are some interrupts (such as those from error conditions or problems with memory) that are so important that they can't be ignored. These non-maskable interrupts (NMIs) must be dealt with immediately, regardless of the other tasks at hand.

While interrupts add some complication to the execution of processes in a single-tasking system, the job of the operating system becomes much more complicated in a multi-tasking system. Now, the operating system must arrange the execution of applications so that you believe that there are several things happening at once. This is complicated because the CPU can only do one thing at a time. Today's multi-core processors and multi-processor machines can handle more work, but each processor core is still capable of managing one task at a time.

In order to give the appearance of lots of things happening at the same time, the operating system has to switch between different processes thousands of times a second. Here's how it happens:

(1) A process occupies a certain amount of RAM. It also makes use of registers, stacks and queues within the CPU and operating-system memory space.

(2) When two processes are multi-tasking, the operating system allots a certain number of CPU execution cycles to one program.

(3) After that number of cycles, the operating system makes copies of all the registers, stacks and queues used by the processes, and notes the point at which the process paused in its execution.

(4) It then loads all the **registers**, **stacks** and **queues** used by the second process and allows it a certain number of CPU cycles.

(5) When those are complete, it makes copies of all the registers, stacks and queues used by the second program, and loads the first program.

3.2.2 Process Management

All of the information needed to keep track of a process when switching is kept in a data package called a **process control block**. The process control block typically contains:

(1) An ID number that identifies the process.

(2) Pointers to the locations in the program and its data where processing last occurred.

(3) Register contents.

(4) States of various flags and switches.

(5) Pointers to the upper and lower bounds of the memory required for the process.

(6) A list of files opened by the process.

(7) The priority of the process.

(8) The status of all I/O devices needed by the process.

Each process has a status associated with it. Many processes consume no CPU time until they get some sort of input. For example, a process might be waiting for a keystroke from the user. While it is waiting for the keystroke, it uses no CPU time. While it's waiting, it is "**suspended**". When the keystroke arrives, the OS changes its status. When the status of the process changes, from pending to active, for example, or from suspended to running, the information in the process control block must be used like the data in any other program to direct execution of the task-switching portion of the operating system.

This process swapping happens without direct user interference, and each process gets enough CPU cycles to accomplish its task in a reasonable amount of time.

(3-3)Trouble can begin if the user tries to have too many processes functioning at the same time. The operating system itself requires some CPU cycles to perform the saving and swapping of all the registers, queues and stacks of the application processes. If enough processes are started, and if the operating system hasn't been carefully designed, the system can begin to use the vast majority of its available CPU cycles to swap between processes rather than run processes. When this happens, it's called **thrashing**, and it usually requires some sort of direct user intervention to stop processes and bring order back to the system.

One way that operating-system designers reduce the chance of thrashing is by reducing the need for new processes to perform various tasks. Some operating systems allow for a "process-lite", called a thread, which can deal with all the CPU-intensive work of a normal process, but generally does not deal with the various types of I/O and does not establish structures requiring the extensive process control block of a regular process. A process may start many threads or other processes, but a thread cannot start a process.

So far, all the scheduling we've discussed has concerned a single CPU. In a system with two or more CPUs, the operating system must divide the workload among the CPUs, trying to balance the demands of the required processes with the available cycles on the different CPUs. Asymmetric operating systems use one CPU for their own needs and divide application processes among the remaining CPUs. Symmetric operating systems divide themselves among the various CPUs, balancing demand versus CPU availability even when the operating system itself is all that's running.

If the operating system is the only software with execution needs, the CPU is not the only resource to be scheduled. Memory management is the next crucial step in making sure that all processes run smoothly.

3.2.3 Memory and Storage Management

When an operating system manages the computer's memory, there are two broad tasks to be accomplished:

(1) Each process must have enough memory in which to execute, and it can neither run into the memory space of another process nor be run into by another process.

(2) The different types of memory in the system must be used properly so that each process can run most effectively.

The first task requires the operating system to set up memory boundaries for types of software and for individual applications.

As an example, let's look at an imaginary small system with 1 megabyte (1024 kilobyte) of RAM. During the boot process, the operating system of our imaginary computer is designed to go to the top of available memory and then "back up" far enough to meet the needs of the operating system itself. Let's say that the operating system needs 300 kilobytes to run. Now, the operating system goes to the bottom of the pool of RAM and starts building up with the various driver software required to control the hardware subsystems of the computer. In our imaginary computer, the drivers take up 200 kilobytes. So after getting the operating system completely loaded, there are 500 kilobytes remaining for application processes.

When applications begin to be loaded into memory, they are loaded in block sizes determined by the operating system. If the block size is 2 kilobytes, then every process that's loaded will be given a chunk of memory that's a multiple of 2 kilobytes in size. Applications will be loaded in these fixed block sizes, with the blocks starting and ending on boundaries established by words of 4 or 8 bytes. These blocks and boundaries help to ensure that applications won't be loaded on top of one another's space by a poorly calculated bit or two. With that ensured, the larger question is what to do when the 500-kilobyte application space is filled.

In most computers, it's possible to add memory beyond the original capacity. For example, you might expand RAM from 1 to 2 gigabytes. This works fine, but can be relatively expensive. It also ignores a fundamental fact of computing — most of the information that an application stores in memory is not being used at any given moment. A processor can only access memory one location at a time, so the vast majority of RAM is unused at any moment. Since disk space is cheap compared to RAM, then moving information in RAM to hard disk can greatly expand RAM space at no cost. This technique is called **virtual memory management**.

Disk storage is only one of the memory types that must be managed by the operating system, and it's also the slowest. Ranked in order of speed, the types of memory in a computer system are:

(1) **High-speed cache** — This is fast, relatively small amounts of memory that are available to the CPU through the fastest connections. Cache controllers predict which pieces of data the CPU will need next and pull it from main memory into high-speed cache to speed up system performance.

(2) Main memory — This is the RAM that you see measured in megabytes when you buy a computer.

(3) **Secondary memory** — This is most often some sort of rotating magnetic storage that keeps applications and data available to be used, and serves as virtual RAM under the control of the operating system.

The operating system must balance the needs of the various processes with the availability of the different types of memory, moving data in blocks (called pages) between available memory as the schedule of processes dictates, see Fig.3-2.

Memory Management

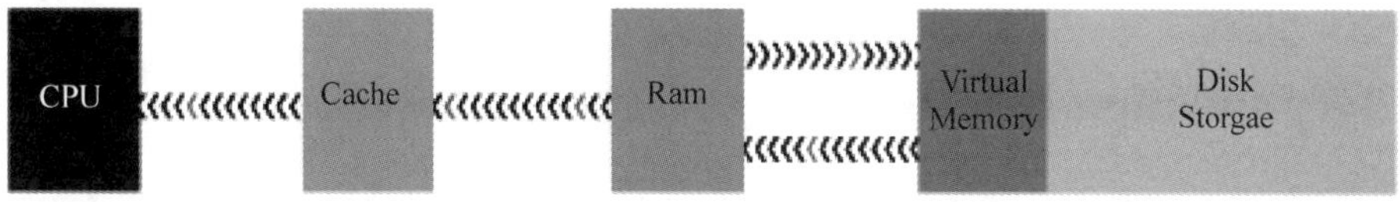

Fig.3-2 Relations of different types of memory in a computer system

3.2.4 Device Management

The path between the operating system and virtually all hardware not on the computer's motherboard goes through a special program called a **driver**. Much of a driver's function is to be the translator between the electrical signals of the hardware subsystems and the high-level programming languages of the operating system and application programs. Drivers take data that the operating system has defined as a file and translate them into streams of bits placed in specific locations on storage devices, or a series of laser pulses in a printer.

Because there are such wide differences in the hardware, there are differences in the way that the driver programs function. Most run when the device is required, and function much the same as any other process. The operating system will frequently assign high-priority blocks to drivers so that the hardware resource can be released and readied for further use as quickly as possible.

(3-4) One reason that drivers are separate from the operating system is so that new functions can be added to the driver — and thus to the hardware subsystems — without requiring the operating system itself to be modified, recompiled and redistributed. Through the development of new hardware device drivers, development often performed or paid for by the manufacturer of the subsystems rather than the publisher of the operating system, input/output capabilities of the overall system can be greatly enhanced.

Managing input and output is largely a matter of managing queues and **buffers**, special storage facilities that take a stream of bits from a device, perhaps a keyboard or a serial port, hold those bits, and release them to the CPU at a rate with which the CPU can cope. This function is especially important when a number of processes are running and taking up processor time. The operating system will instruct a buffer to continue taking input from the device, but to stop sending data to the CPU while the process using the input is suspended. Then, when the process requiring input is made active once again, the operating system will command the buffer to send data. This process allows a keyboard or a **modem** to deal with external users or computers at a high speed even though there are times when the CPU can't use input from those sources.

Managing all the resources of the computer system is a large part of the operating system's function and, in the case of real-time operating systems, may be virtually all the functionality required. For other operating systems, though, providing a relatively simple, consistent way for applications and humans to use the power of the hardware is a crucial part of their reason for existing.

3.2.5 Application Interface

Just as drivers provide a way for applications to make use of hardware subsystems without having to know every detail of the hardware's operation, application program interfaces (APIs) let application programmers use functions of the computer and operating system without having to directly keep track of all the details in the CPU's operation. Let's look at the example of creating a hard disk file for holding data to see why this can be important.

A programmer writing an application to record data from a scientific instrument might want to allow the scientist to specify the name of the file created. The operating system might provide an API function named MakeFile for creating files. When writing the program, the programmer would insert a line that looks like this:

MakeFile [1, %Name, 2]

In this example, the instruction tells the operating system to create a file that will allow random access to its data (signified by the 1 — the other option might be 0 for a serial file), will have a name typed in by the user (%Name) and will be a size that varies depending on how much data is stored in the file (signified by the 2 — other options might be zero for a fixed size, and 1 for a file that grows as data is added but does not shrink when data is removed). Now, let's look at what the operating system does to turn the instruction into action.

The operating system sends a query to the disk drive to get the location of the first available free storage location.

With that information, the operating system creates an entry in the file system showing the beginning and ending locations of the file, the name of the file, the file type, whether the file has been archived, which users have permission to look at or modify the file, and the date and time of the file's creation.

The operating system writes information at the beginning of the file that identifies the file, sets up the type of access possible and includes other information that ties the file to the application. In all of this information, the queries to the disk drive and addresses of the beginning and ending point of the file are in formats heavily dependent on the manufacturer and model of the disk drive.

Because the programmer has written the program to use the API for disk storage, the programmer doesn't have to keep up with the instruction codes, data types and response codes for every possible hard disk and tape drive. The operating system, connected to drivers for the various hardware subsystems, deals with the changing details of the hardware. The programmer must simply write code for the API and trust the operating system to do the rest.

APIs have become one of the most hotly contested areas of the computer industry in recent years. Companies realize that programmers using their API will ultimately translate this into the ability to control and profit from a particular part of the industry. This is one of the reasons that so many companies have been willing to provide applications like readers or viewers to the public at no charge. They know consumers will request that programs take advantage of the free readers, and application companies will be ready to pay royalties to allow their software to provide the functions requested by the consumers.

3.2.6 User Interface

Every computer that is to be operated by an individual requires a user interface. The user interface is usually referred to as a shell and is essential if human interaction is to be supported. The user interface views the directory structure and requests services from the operating system that will acquire data from input hardware devices, such as a keyboard, mouse or credit card reader, and requests operating system services to display prompts, status messages and such on output hardware devices, such as a video monitor or printer. The two most common forms of a user interface have historically been the **command-line interface**, where computer commands are typed out line-by-line, and the **graphical user interface**, where a visual environment (most commonly a WIMP) is present.

Most of the modern computer systems support graphical user interfaces (GUI), and often include them. In some computer systems, such as the original implementation of the classic Mac OS, the GUI is integrated into the kernel.

While technically a graphical user interface is not an operating system service, incorporating support for one into the operating system kernel can allow the GUI to be more responsive by reducing the number of context switches required for the GUI to perform its output functions. Other operating systems are modular, separating the graphics subsystem from the kernel and the Operating System. In the 1980s UNIX, VMS and many others had operating systems that were built this way. Linux and Mac OS are also built this way. Modern releases of Microsoft Windows such as Windows Vista implement a graphics subsystem that is mostly in user-space; however the graphics drawing routines of versions between Windows NT 4.0 and Windows Server 2003 exist mostly in kernel space. Windows 9x had very little distinction between the interface and the kernel.

Many computer operating systems allow the user to install or create any user interface they desire. The X Window System in conjunction with GNOME or KDE Plasma 5 is a commonly found setup on most UNIX and UNIX-like (BSD, Linux, Solaris) systems. A number of Windows shell replacements have been released for Microsoft Windows, which offer alternatives to the included Windows shell, but the shell itself cannot be separated from Windows.

Numerous UNIX-based GUIs have existed over time, most derived from X11. Competition among the various vendors of UNIX (HP, IBM, Sun) led to much fragmentation, though an effort to standardize in the 1990s to COSE and CDE failed for various reasons, and were eventually eclipsed by the widespread adoption of GNOME and K Desktop Environment. Prior to free software-based toolkits and desktop environments, Motif was the prevalent toolkit/desktop combination (and was the basis upon which CDE was developed).

Graphical user interfaces evolve over time. For example, Windows has modified its user interface almost every time a new major version of Windows is released, and the Mac OS GUI changed dramatically with the introduction of Mac OS X in 1999.

In human–computer interaction, WIMP is the mostly used system, which stands for "windows, icons, menus, pointer", denoting a style of interaction using these elements of the user interface. It was coined by Merzouga Wilberts in 1980. Other expansions are sometimes used, substituting "mouse" and "mice" or "pull-down menu" and "pointing", for menus and pointer, respectively.

Though the term has fallen into disuse, some use it as an approximate synonym for graphical user interface (GUI). Any interface that uses graphics can be called a GUI, and WIMP systems derive from such systems. However, while all WIMP systems use graphics as a key element (the icon and pointer elements), and therefore are GUIs, the reverse is not true. Some GUIs are not based in windows, icons, menus, and pointers. For example, most mobile phones represent actions as icons, and some may have menus, but very few include a pointer or run programs in a window.

WIMP interaction was developed at Xerox PARC (see Xerox Alto, developed in 1973) and popularized with Apple's introduction of the Macintosh in 1984, which added the concepts of the "menu bar" and extended window management.[8]

In a WIMP system:

(1) A window runs a self-contained program, isolated from other programs that (if in a multi-program operating system) run at the same time in other windows.

(2) An icon acts as a shortcut to an action the computer performs (e.g., execute a program or task).

(3) A menu is a text or icon-based selection system that selects and executes programs or tasks.

(4) The pointer is an onscreen symbol that represents movement of a physical device that the user controls to select icons, data elements, etc.

This style of system improves human–computer interaction (HCI) by emulating real-world interactions and providing better ease of use for non-technical people. Users can carry skill at a standardized interface from one application to another.

3.3 Examples of Popular Modern Operating Systems

3.3.1 UNIX and UNIX-like Operating Systems

Most operating systems can be grouped into two different families. Aside from Microsoft's Windows NT-based operating systems, nearly everything else traces its heritage back to UNIX.

Linux, Mac OS X, Android, iOS, Chrome OS, Orbis OS used on the PlayStation 4, whatever firmware is running on your router — all of these operating systems are often called "UNIX-like" operating systems.

UNIX was developed in AT&T's Bell Labs back in the mid-to-late 1960's. The initial release of UNIX had some important design attributes that live on today.

One is the "UNIX philosophy" of creating small, modular utilities that do one thing and do them well. If you're familiar with using a Linux terminal, this should be familiar to you — the system offers a number of utilities that can be combined in different ways through pipes and other features to perform more complex tasks. Even graphical programs are likely calling simpler utilities in the background to do the heavy lifting. This also makes it easy to create shell scripts, stringing together simple tools to do complicated things.

UNIX also had a single file system that programs use to communicate with each other. This is why "everything is a file"on Linux – including hardware devices and special files that provide system information or other data. It's also why only Windows has drive letters, which it inherited from DOS — on other operating systems, every file on the system is part of a single directory hierarchy.

Like any history going back over 40 years, the history of UNIX and its descendants is messy, as shown in Fig.3-3. To simplify things, we can roughly group UNIX's descendants into two groups.

One group of UNIX descendants were developed in academia. The first was BSD (Berkeley Software Distribution), an open-source, UNIX-like operating system. BSD lives on today through FreeBSD, NetBSD, and OpenBSD. NeXTStep was also based on the original BSD, Apple's Mac OS X was based on NeXTStep, and iOS was based on Mac OS X. Many other operating systems, including the Orbis OS used on the PlayStation 4, are derived from types of BSD operating systems.

Richard Stallman's GNU project was also started as a reaction to AT&T's increasingly restrictive UNIX software licensing terms. MINIX was a UNIX-like operating system created for educational purposes, and Linux was inspired by MINIX. The Linux we know today is really GNU/Linux, as it's made up of the Linux kernel and a lot of GNU utilities. GNU/Linux isn't directly descended from BSD, but it is descended from UNIX's design and has its roots in academia. Many operating systems today, including Android, Chrome OS, Steam OS, and a huge amount of embedded operating systems for devices, are based on Linux.

On the other hand, there were the commercial UNIX operating systems. AT&T UNIX, SCO UnixWare, Sun Microsystems Solaris, HP-UX, IBM AIX, SGI IRIX — many big corporations wanted to create and license their own versions of UNIX. These aren't quite as common today, but some of them are still out there.

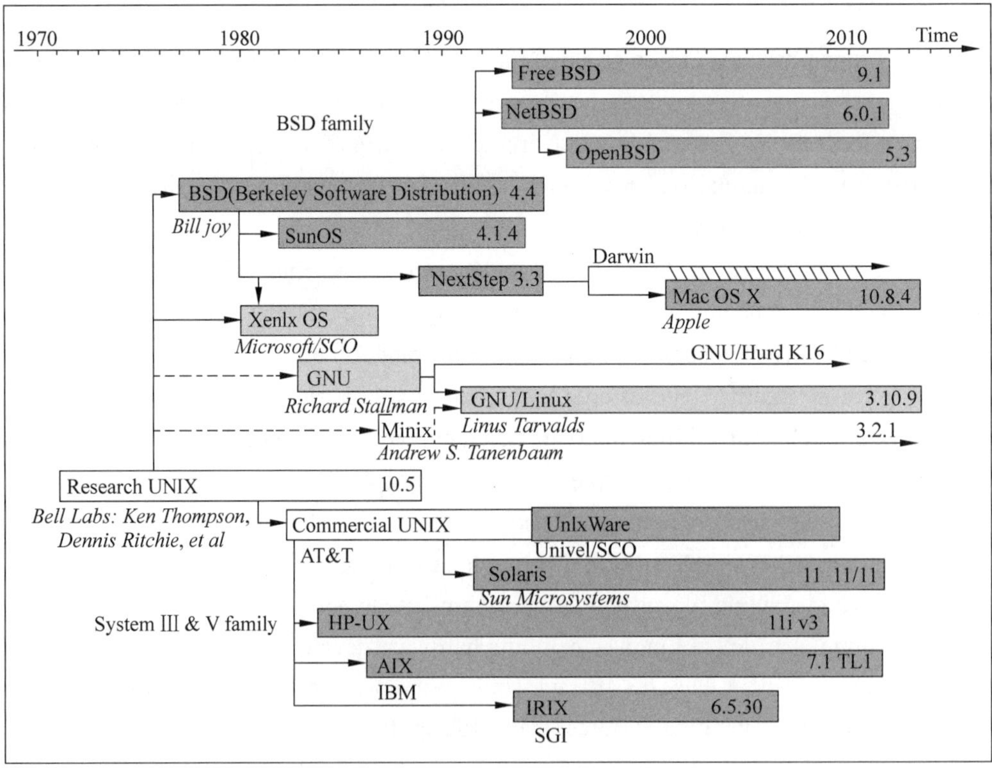

Fig.3-3 History of UNIX

The architecture of UNIX can be divided into three levels of functionality, as shown in Fig.3-4.

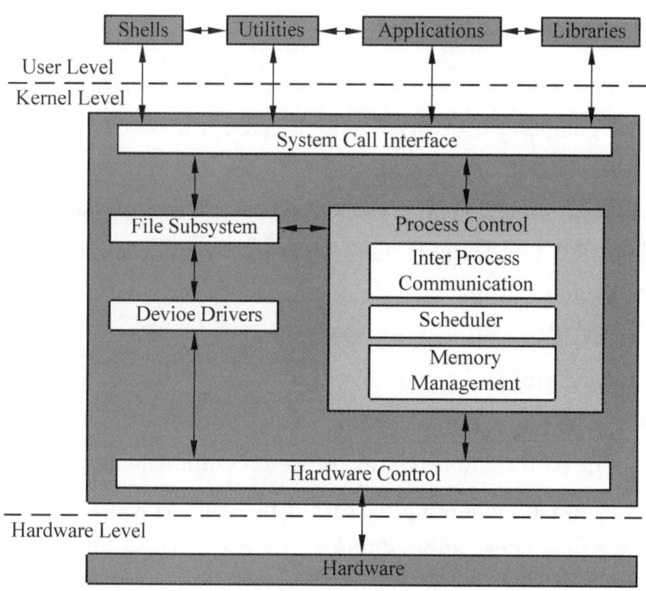

Fig.3-4 UNIX architecture

(1) The lowest level is the kernel, which schedules tasks, manages resources, and controls security.

(2) The next level is the shell, which acts as the UI, interpreting user commands and starting applications.

(3) The tools are at the highest level. These provide utility functions such as ls, vi, and cat.

The UNIX operating system supports the following features and capabilities:

(1) Multitasking and multiuser.

(2) Kernel written in high-level language.

(3) Programming interface.

(4) Use of files as abstractions of devices and other objects.

(5) Character-based default UI.

(6) Built-in networking. (TCP/IP is standard.)

(7) Persistent system service processes called "daemons" and managed by init or inetd.

3.3.2 Microsoft Windows

Many people expected UNIX to become the industry standard operating system, but DOS and "IBM PC compatible" computers eventually exploded in popularity. Microsoft's DOS became the most successful DOS of them all. DOS was never based on UNIX at all, which is why Windows uses a backslash for file paths while everything else uses a forward slash. This decision was made back in the early days of DOS, and later versions of Windows inherited it, just as BSD, Linux, Mac OS X, and other UNIX-like operating systems inherited many aspects of UNIX's design.

Windows 3.1, Windows 95, Windows 98, and Windows ME were all based on DOS underneath. Microsoft was developing a more modern and stable operating system at the time, which they named Windows NT — for "Windows New Technology". Windows NT eventually made its way to regular computer users as Windows XP, but it was available for corporations as Windows 2000 and Windows NT before that.

All of Microsoft's operating systems are based on the Windows NT kernel today. Windows 7, Windows 8, Windows RT, Windows Phone 8, Windows Server, and the Xbox One's operating system all use the Windows NT kernel. Unlike most other operating systems, Windows NT wasn't developed as a UNIX-like operating system.

Microsoft didn't start with a completely clean slate, of course. To maintain compatibility with DOS and old Windows software, Windows NT inherited many DOS conventions like drive letters, backslashes for file paths, and forward slashes for command-line switches.

The user mode includes application processes, which are typically Windows programs and a set of protected subsystems. These subsystems are referred to as "protected" because each of these subsystems is a separate process with its own protected virtual address space. Of these, the most important subsystem is the Microsoft Win32® subsystem, which supplies much of the Windows functionality. The Windows application programming interface (API) is useful for developing both 32-bit and 64-bit Windows-based applications.

Another important subsystem, particularly with respect to migration of UNIX applications, is the Portable Operating System Interface (POSIX) for computing environments. This is a set of international standards for implementing UNIX-like interfaces. The POSIX subsystem implements these standards-based interfaces and allows application developers to easily port their applications to Windows from another operating system. The POSIX subsystem is not implemented on Windows Server 2003 but comes as a part of the Interix (Windows Services for UNIX 3.5) installation on Windows.

(3-5) The kernel mode is a highly privileged mode of operation in which the program code has direct access to the virtual memory. This includes the address spaces of all user mode processes and applications and their hardware. The kernel mode is also known as the supervisor mode, protected mode, or Ring 0. The kernel mode of Windows Server 2003 contains the Windows NT executive as well as the system kernel. The Windows NT executive exports generic services that protected subsystems call to obtain basic operating system services, such as file operations, input/output (I/O), and synchronization services. Partitioning of the protected subsystems and the system kernel simplifies the base operating system design and makes it possible to extend the features of an individual protected subsystem without affecting the kernel. The kernel controls how the operating system uses the processors. Its operations include scheduling, multiprocessor synchronization, and providing objects that the executive can use or export to applications.

The Windows operating system supports the following features and capabilities:

(1) Multitasking.

(2) Flexibility to choose a programming interface (user and kernel APIs).

(3) A graphical user interface (GUI) and a command-line interface for users and administrators. (The default UI is graphical.)

(4) Built-in networking. (Transmission Control Protocol/Internet Protocol [TCP/IP] is standard.)

(5) Persistent system service processes called “Windows Services” and managed by the Windows Service Control Manager (SCM).

(6) Single compatible implementation irrespective of the vendor from whom it is purchased.

3.4 Comparison of Windows and UNIX Environments

This section compares the Windows and UNIX architectures, emphasizing the areas that directly affect software development in a migration project.

1. Kernels and APIs

As with most operating systems, Windows and UNIX both have kernels. The kernel provides the base functionality of the operating system. The major functionality of the kernel includes process management, memory management, thread management, scheduling, I/O management, and power management.

In UNIX, the API functions are called system calls. System calls are a programming interface common to all implementations of UNIX. The kernel is a set of functions that are used by processes through system calls.

Windows has an API for programming calls to the executive. In addition to this, each subsystem provides a higher-level API. This approach allows Windows operating systems to provide different APIs, some of which mimic the APIs provided by the kernels of other operating systems. The standard subsystem APIs include the Windows API (the Windows native API) and the POSIX API (the standards-based UNIX API).

2. Windows Subsystems

A subsystem is a portion of the Windows operating system that provides a service to application programs through a callable API.

Subsystems come in two varieties, depending upon the application program that finally handles the request:

(1) Environment subsystems. These subsystems run in a user mode and provide functions through a published API. The Windows API subsystem provides an API for operating system services, GUI capabilities, and functions to control all user input and output. The Win32 subsystem and POSIX subsystem are part of the environment subsystems and are described as follows.

① Win32 subsystem. The Win32 environment subsystem allows applications to benefit from the complete power of the Windows family of operating systems. The Win32 subsystem has a vast collection of functions, including those required for advanced operating systems, such as security, synchronization, virtual memory management, and threads. You can use Windows

APIs to write 32-bit and 64-bit applications that run on all versions of Windows.

② POSIX subsystem and Windows Services for UNIX. To provide more comprehensive support for UNIX programs, Windows uses the Interix subsystem. Interix is a multiuser UNIX environment for a Windows-based computer. Interix conforms to the POSIX.1 and POSIX.2 standards. It provides all features of a traditional UNIX operating system, including pipes, hard links, symbolic links, UNIX networking, and UNIX graphical support through the X Windows system. It also includes case-sensitive file names, job control tools, compilation tools, and more than 300 UNIX commands and tools, such as KornShell, C Shell, awk, and vi.

(2) Integral subsystems. These subsystems perform key operating system functions and run as a part of the executive or kernel. Examples are the user-mode subsystems, Local Security Authority subsystem (LSASS), and Remote Procedure Call subsystem (RPCSS).

3. Kernel Objects and Handles

Kernel objects are used to manage and manipulate resources—such as files, synchronization objects, and pipes—across the processes. These **kernel objects** are owned by the kernel and not by the process. Handles are the opaque numbers or the data type used to represent the objects and to uniquely identify the objects. To interact with an object, you must obtain a **handle** to the object.

In UNIX, the kernel object can be created using the system calls and it returns an unsigned integer. There is no exact equivalent of handles in UNIX.

In Windows, Windows APIs are used to create the kernel object and it returns a Windows-specific data type called HANDLE.

4. Hardware Drivers

Hardware drivers are system software used to interact the hardware devices with the operating system.

In UNIX, there are several different ways to manage drivers. Some UNIX implementations allow dynamic loading and unloading of drivers, whereas other implementations do not. The UNIX vendor usually provides drivers. The range of Intel hardware supported for UNIX is typically smaller than that for Windows.

In Windows, the Windows driver model provides a platform for developing drivers for industry-standard hardware devices attached to a Windows-based system. The key to developing a good driver package is to provide reliable setup and installation procedures and interactive GUI tools for configuring devices after the installation. In addition, the hardware must be compatible with Windows Plug and Play technology in order to provide a user-friendly hardware installation.

5. Process Management

A process is usually defined as the instance of the running program. Process management describes how the operating systems manage the multiple processes running at a particular instance of time. Multitasking operating systems such as Windows and UNIX must manage and control many processes simultaneously. Both Windows and UNIX operating systems support processes and threads.

The following sections provide more details on process management in both UNIX and Windows:

1) Multitasking

UNIX is a multiprocessing, multiuser system. At any given point, you can have many processes running on UNIX. Consequently, UNIX is very efficient at creating processes.

Windows has evolved greatly from its predecessors, such as Digital Equipment Corporation's VAX/VMS. It is now a preemptive multitasking operating system. As a result, Windows relies more heavily on threads than processes. (A thread is a construct that enables parallel processing within a single process.) Creating a new process is a relatively expensive operation while creating a new thread is not as expensive in terms of system resources like memory and time. Hence, multiprocess-oriented applications on UNIX typically translate to multithreaded applications on the Windows platform, thus saving such system resources as memory and time.

2) Multiple Users

One key difference between UNIX and Windows is in the creation of multiple user accounts on a single computer.

When you log on to a computer running UNIX, a shell process is started to service your commands. The UNIX operating system keeps track of the users and their processes and prevents processes from interfering with one another. The operating system does not come with any default interface for user interaction. However, the shell process on the computer running UNIX can connect to other computers to load third-party UIs.

When a user logs on interactively to a computer running Windows, the Win32 subsystem's Graphical Identification and Authentication dynamic-link library (GINA) creates the initial process for that user, which is known as the user desktop, where all user interaction or activity takes place. The desktop on the user's computer is loaded from the server. Only the user who is logged on has access to the desktop. Other users are not allowed to log on to that computer at the same time. However, if a user employs Terminal Services or Citrix, Windows can operate in a server-centric mode just as UNIX does.

The Windows operating system supports multiple users simultaneously through the command line and a GUI. The latter requires the use of Windows Terminal Services. The UNIX operating system supports multiple simultaneous users through the command line and through a GUI. The latter requires the use of X Windows. Windows comes with a default command shell (cmd.exe); UNIX typically includes several shells and encourages each user to choose a shell as the user's default shell. Both operating systems provide complete isolation between simultaneous users. There are some key differences between the two: Windows comes with a "single user" version that allows one user at a time (Windows XP) as well as a multiuser server version (Windows Server). It is rare for a Windows Server system to have multiple simultaneous command-line users.

3) Multithreading

Most new UNIX kernels are multithreaded to take advantage of symmetric multiprocessing (SMP) computers. Initially, UNIX did not expose threads to programmers. However, POSIX has user-programmable threads. There is a POSIX standard for threads (called Pthreads) that all current versions of UNIX support.

In Windows, creating a new thread is very efficient. Windows applications are capable of using threads to take advantage of SMP computers and to maintain interactive capabilities when some threads take a long time to execute.

4) Process Hierarchy

When a UNIX-based application creates a new process, the new process becomes a child of the process that created it. This process hierarchy is often important, and there are system calls for manipulating child processes.

Unlike UNIX, Windows processes do not share a hierarchical relationship. The creating process receives the process handle and ID of the process it created, so a hierarchical relationship can be maintained or simulated if required by the application. However, the operating system treats all processes like they belong to the same generation. Windows provides a feature called Job Objects, which allows disparate processes to be grouped together and adhere to one set of rules.

6. Signals, Exceptions, and Events

In both operating systems, events are signaled by software interrupts. A signal is a notification mechanism used to notify a process that some event has taken place or to handle the interrupt information from the operating system. An event is used to communicate between the processes. Exceptions occur when a program executes abnormally because of conditions outside the program's control.

In UNIX, signals are used for typical events, simple interprocess communication (IPC), and abnormal conditions such as floating point exceptions.

Windows has the following mechanisms.

(1) Windows supports some UNIX signals and others can be implemented using Windows API and messages.

(2) An event mechanism that handles expected events, such as communication between two processes.

(3) An exception mechanism that handles nonstandard events, such as the termination of a process by the user, **page faults**, and other execution violations.

7. Daemons and Services

A **daemon** is a process that detaches itself from the terminal and runs disconnected in the background, waiting for requests and responding to them. A service is a special type of application that is available on Windows and runs in the background with special privileges.

In UNIX, a daemon is a process that the system starts to provide a service to other applications. Typically, the daemon does not interact with users. UNIX daemons are started at boot time from init or rc scripts. To modify such a script, it needs to be opened in a text editor and the values of the variables in the script need to be physically changed. On UNIX, a daemon runs with an appropriate user name for the service that it provides or as a root user.

A Windows service is the equivalent of a UNIX daemon. It is a process that provides one or more facilities to client processes. Typically, a service is a long-running, Windows-based application that does not interact with users and, consequently, does not include a UI. Services may start when the system restarts and then continue running across logon sessions. Windows has a registry that stores the values of the variables used in the services. Control Panel provides a UI that allows users to set the variables with the valid values in the registry. The security context of that user determines the capabilities of the service. Most services run as either Local Service or Network Service. The latter is required if the service needs to access network resources and must run as a domain user with enough privileges to perform the required tasks.

8. Virtual Memory Management

Virtual memory is a method of extending the available physical memory or RAM on a computer. In a virtual memory system, the operating system creates a pagefile, or swapfile, and divides memory into units called pages. Virtual memory management implements virtual addresses and each application is capable of referencing a physical chunk of memory, at a specific virtual address, throughout the life of the application.

Both UNIX and Windows use virtual memory to extend the memory available to an application beyond the actual physical memory installed on the computer. For both operating systems, on 32-bit architecture, each process gets a private 2GB of virtual address space. This is called user address space or process address space. The operating system uses 2GB of virtual address space, called system address space or operating system memory. On 64-bit architecture, each process gets 8 terabytes of user address space.

9. File Systems and Networked File Systems

This section describes the file system characteristics of UNIX and Windows. Both UNIX and Windows support many different types of file system implementations. Some UNIX implementations support Windows file system types, and there are products that provide Windows support for some UNIX file system types.

File system characteristics and interoperability of file names between UNIX and Windows are discussed as follows.

(1) File names and path names.

Everything in the file system is either a file or a directory. UNIX and Windows file systems are both hierarchical, and both operating systems support long file names of up to 255 characters. Almost any character is valid in a file name, except the following:

① In UNIX: /.

② In Windows: ?, ", /, \, >, <, *, |, and :.

UNIX file names are case sensitive while Windows file names are not.

In UNIX, a single directory known as the root is at the top of the hierarchy. You locate all files by specifying a path from the root. UNIX makes no distinction between files on a local hard drive partition, CD-ROM, floppy disk, or network file system (NFS). All files appear in one tree under the same root.

The Windows file system can have many hierarchies, for example, one for each **partition** and one for each network drive. A UNIX system provides a single file system tree, with a single root, to the applications it hosts. Mounted volumes (whether local devices or network shares) are "spliced" into that tree at locations determined by the system administrator. The Windows operating system exposes a forest of file system trees, each with its own root, to the applications it hosts. Mounted volumes appear as separate trees ("drive letters") as determined by the administrator or user. Both UNIX and Windows provide a tree view of all network-accessible shares. UNIX provides an administrator-defined view of these shares through an automounter mechanism, while Windows provides a full view through the Universal Naming Convention (UNC) pathname syntax.

(2) Server message block (SMB) and Common Internet File System (CIFS).

One of the earliest implementations of network resource sharing for the Microsoft MS-DOS® platform was network basic input/output system (NetBIOS). The features of NetBIOS allow it to accept disk I/O requests and direct them to file shares on other computers. The protocol used for this was named server message block (SMB). Later, additions were made to SMB to apply it to the Internet, and the protocol is now known as Common Internet File System (CIFS).

UNIX supports this through a software option called Samba. Samba is an open-source, freeware, server-side implementation of a UNIX CIFS server.

In Windows, the server shares a directory, and the client then connects to the Universal Naming Convention (UNC) to connect to the shared directory. Each network drive usually appears with its own drive letter, such as X.

(3) Windows and UNIX NFS interoperability.

UNIX and Windows can interoperate using NFS on Windows or CIFS on UNIX. There are a number of commercial NFS products for Windows. For UNIX systems, Samba is an alternative to installing NFS client software on Windows-based computers for interoperability with UNIX-based computers. It also implements NetBIOS-style name resolution and browsing. Microsoft offers a freely downloadable NFS Server, Client, and Gateway as part of Windows Services for UNIX 3.5 installation. Windows Services for UNIX also provide a number of services for interoperability between Windows-based and UNIX-based computers.

10. Security

This section describes some of the security implementation details and differences between UNIX and Windows.

1) User Authentication

A user can log on to a computer running UNIX by entering a valid user name and password. A UNIX user can be local to the computer or known on a Network Information System (NIS) domain (a group of cooperating computers). In most cases, the NIS database contains little more than the user name, password, and group.

A user can log on to a computer running Windows by entering a valid user name and password. A Windows user can be local to the computer, can be known on a Windows domain, or be known in the Microsoft Active Directory® **directory service.** The Windows domain contains only a user name, the associated password, and some user groups. *Active Directory* contains the same information as the Windows domain and may contain the contact information of the user, organizational data, and certificates.

2) UNIX versus Windows Security

UNIX uses a simple security model. The operating system applies security by assigning permissions to files. This model works because UNIX uses files to represent devices, memory, and even processes. When a user logs on to the system with a user name and a password, UNIX starts a shell with the UID and GID of that user. From then on, the permissions assigned to the UID and GID, or the process, control all access to files and other resources. Most UNIX vendors can provide Kerberos support, which raises their sophistication to about that of Windows.

Windows uses a unified security model that protects all objects from unauthorized access. The system maintains security information for:

(1) Users. System users are people who log on to the system, either interactively by entering a set of credentials (typically user name and password) or remotely through the network. Each user's security context is represented by a logon session.

(2) Objects. These are the secured resources (for example, files, computers, synchronization objects, and applications) that a user can access.

(3) Active Directory. Windows Server 2003 uses the Active Directory directory service to store information about objects. These objects include users, computers, printers, and every domain on one or more wide area networks (WANs). Active Directory can scale from a single computer to many large computer networks. It provides a store for all the domain security policy and account information.

11. Networking

Networking basically provides the communication between two or more computers. It defines various sets of protocols, configures the domains, IP addresses, and ports, and communicates with the external devices like telephones or modems and data transfer methods. It also provides the standard set of API calls to allow applications to access the networking features.

The primary networking protocol for UNIX and Windows is TCP/IP. The standard programming API for TCP/IP is called sockets. The Windows implementation of sockets is known as Winsock (formally known as Windows Sockets). Winsock conforms well to the Berkeley implementation, even at the API level. The key difference between UNIX sockets and Winsock exists in asynchronous network event notification. There is also a difference in the remote procedure calls (RPC) implementation in UNIX and Windows.

12. User Interfaces

The user interface (UI) provides a flexible way of communicating between the users, applications, and the computer.

The UNIX UI was originally based on a character-oriented command line, whereas the Windows UI was based on GUI. UNIX originated when graphic terminals were not available. However, the current versions of UNIX support the graphical user interface using the X Windows system. Motif is the most common windowing system, library, and user-interface style built on X Windows. This allows the building of graphical user interface applications on UNIX.

The Windows user interface was designed to take advantage of advancements in the graphics capabilities of computers. It can be used by all applications—including both client side and server side—and can also be used for tasks such as service administration. Windows contains the Graphics Device Interface (GDI) engine to support the graphical user interface.

13. System Configuration

(3-6) UNIX users generally configure a system by editing the configuration files with any of the available text editors. The advantage of this mechanism is that the user does not need to learn how to use a large set of configuration tools, but must only be familiar with an editor and possibly a scripting language. The disadvantage is that the information in the files comes in various formats; hence the user must learn the various formats in order to change the settings. UNIX users often employ scripts to reduce the possibility of repetition and error. In addition, they can also use NIS to centralize the management of many standard configuration files. Although different versions of UNIX have GUI management tools, such tools are usually specific to each version of UNIX.

Windows has GUI tools for configuring the system. The advantage of these tools is that they can offer capabilities depending on what is being configured. In recent years, Microsoft Management Console (MMC) has provided a common tool and UI for creating configuration tools. Windows provides a scripting interface for most configuration needs through the Windows Script Host (WSH). Windows provides WMI (Windows Management Instrumentation), which can be used from scripts. Windows also includes extensive command-line tools for controlling system configuration. In Windows Server 2003, anything that can be done to manage a system through a GUI can be done through a command-line tool as well.

14. Intercrosses Communication

An operating system designed for multitasking or multiprocessing must provide mechanisms for communicating and sharing data between applications. These mechanisms are called interprocess communication (IPC).

(3-7) UNIX has several IPC mechanisms that have different characteristics and which are appropriate for different situations. Shared memory, **pipes**, and **message queues** are all suitable for processes running on a single computer. Shared memory and message queues are suitable for communicating among unrelated processes. Pipes are usually chosen for communicating with a child process through standard input and output. For communications across the network, sockets are usually the chosen technique.

Windows also has many IPC mechanisms. Like UNIX, Windows has shared memory, pipes, and events (equivalent to signals). These are appropriate for processes that are local to a computer. In addition to these mechanisms, Windows supports clipboard/Dynamic Data Exchange (DDE), and Component Object Model (COM). Winsock and Microsoft Message Queuing are good choices for cross-network tasks. Other cross-network IPC mechanisms for Windows include remote procedure calls (RPCs) and mail slots. RPC has several specifications and implementations, developed by third-party vendors, which support client server applications in distributed environments. The most prevalent RPC specifications are Open Network Computing (ONC) by Sun Microsystems and Distributed Computing Environment (DCE) by Open Software Foundation. UNIX systems support interoperability with Windows RPC. UNIX does not implement mailslots.

15. Synchronization

In a multithreaded environment, threads may need to share data between them and perform various actions. These operations require a mechanism to synchronize the activity of the threads. These synchronization techniques are used to avoid race conditions and to wait for signals when resources are available.

UNIX environments use several techniques in the Pthreads implementation to achieve synchronization. They are:

(1) **Mutexes**

(2) Condition variables

(3) Semaphores

(4) Interlocked exchange

Similarly, the synchronization techniques available in the Windows environment are:

(1) Spinlocks

(2) Events

(3) Critical sections

(4) Semaphores

(5) Mutexes

(6) Interlocked exchange

16. DLLs and Shared Libraries

Windows and UNIX both have a facility that allows the application developer to put common functionality in a separate code module. This feature is called a shared library in UNIX and a dynamic-link library (DLL) in Windows. Both allow application developers to link together object files from different compilations and to specify which symbols will be exported from the library for use by external programs. The result is the capability to reuse code across applications. The Windows operating system and most Windows programs use many DLLs.

Windows DLLs do not need to be compiled to position-independent code (PIC), while UNIX shared objects must be compiled to PIC. However, the exact UNIX behavior can be emulated in Windows by pre-mapping different DLLs at different fixed addresses.

17. Component-based Development

Component-based development is an extension to the conventional software development where the software is assembled by integrating several components. The components themselves may be written in different technologies and languages, but each has a unique identity, and each of them exposes common interfaces so that they can interact with other components.

UNIX supports CORBA as the main component-based development tool. However, it is not a standard component of the UNIX system; you have to obtain a CORBA implementation from another source.

On the other hand, the Windows environment offers a wide range of component-based development tools and technologies. This includes:

(1) COM

(2) COM+

(3) Distributed COM (DCOM)

(4).NET components

18. Middleware

This section compares the various **middleware** solutions available for UNIX-based and Windows-based applications. Middleware technologies are mostly used to connect the presentation layer with the back-end business layers or data sources. One of the most prominent middleware technologies used in applications is a message queuing system.

Message queuing is provided as a feature in AT&T System V UNIX and can be achieved through sockets in Berkeley UNIX versions. These types of memory queues are most often used for IPC and do not meet the requirements for persistent store and forward messaging.

To meet these requirements, versions of the IBM MQSeries and the BEA Systems MessageQ (formally the DEC MessageQ) are available for UNIX. Microsoft provides similar functionality with Message Queuing for Windows.

IBM and BEA Systems also provide versions of their queuing systems for Windows.

19. Transaction Processing Monitors

A transaction processing monitor (TP monitor) is a subsystem that groups sets of related database updates and submits them to a database as a transaction. These transactions, often monitored and implemented by the TP monitors, are known as online transaction processing (OLTP). OLTP is a group of programs that allow real-time updating, recording, and retrieval of data to and from a networked system. OLTP systems have been implemented in the UNIX environments for many years. These systems perform such functions as resource management, threading, and distributed transaction management.

Although OLTP was originally developed for UNIX, many OLTP systems have Windows versions. Windows also ships with its own transaction manager. In addition, gateways exist to integrate systems that use different transaction monitors.

20. Shells and Scripting

A shell is a command-line **interpreter** that accepts typed commands from a user and executes the resulting request. Every shell has its own set of programming features known as scripting languages. Programs written through the programming features of a shell are called shell scripts. As with shell scripts, scripting languages are interpreted.

Windows and UNIX support a number of shells and scripting languages, some of which are common to both operating systems. Examples are: Rexx, Python, and Tcl/Tk.

21. Command-Line Shells

A number of standard shells are provided in the UNIX environment as part of the standard installation. They are:

(1) Bourne shell (sh)

(2) C shell (csh)

(3) Korn shell (ksh)

(4) Bourne Again Shell (bash)

On the Windows platform, Cmd.exe is the command prompt or the shell.

Windows versions of the Korn shell and the C shell are delivered with the Windows Services for UNIX 3.5, MKS, and Cygwin products.

22. Scripting Languages

In UNIX, there are three main **scripting languages** that correspond to the three main shells: the Bourne shell, the C shell, and the Korn shell. Although all the scripting languages are developed from a common core, they have certain shell-specific features to make them slightly different from each other. These scripting languages mainly use a group of UNIX commands for execution without the need for prior compilation. Some of the external scripting languages that are also supported on the UNIX environment are Perl, Rexx, and Python.

On the Windows environment, WSH is a language-independent environment for running scripts and is compatible with the standard command shell. It is often used to automate administrative tasks and logon scripts. WSH provides objects and services for scripts, establishes security, and invokes the appropriate script engine, depending on the script language. Objects and services supplied allow the script to perform such tasks as displaying messages on the screen, creating objects, accessing network resources, and modifying environment variables and registry keys. WSH natively supports VBScript and JScript. Other languages that are available for this environment are Perl, Rexx, and Python. WSH is built-in to all current versions of Windows and can be downloaded or upgraded from Microsoft.

23. Development Environments

The generic UNIX development environment uses a set of command-line tools. In addition, there are many third-party integrated development environments (IDEs) available for UNIX. Most of the character-based and visual IDEs provide the necessary tools, libraries, and headers needed for application development.

The Windows development environment can be of two types: a standard Windows development environment or a UNIX-like development environment such as Windows Services for UNIX.

The standard Windows development environment uses the Microsoft Platform Software Development Kit (SDK) and Microsoft Visual Studio® .NET. The platform SDK delivers documentation for developing Windows-based applications, libraries, headers, and definitions needed by language compilers. It also includes a rich set of command-line and stand-alone tools for building, debugging, and testing applications. It is available at no cost from the MSDN Web site.

The Microsoft Visual Studio .NET environment delivers a complete set of tools for application development, including the development of multitier components, user interface design, and database programming and design. It also provides several language tools, editing tools, debugging tools, performance analysis tools, and application installation tools.

The development environment of Windows Services for UNIX contains documentation, tools, API libraries, and headers needed by language compilers. It also comes with a UNIX development environment, with tools such as GNU gcc, g++, g77 compilers, and a gdb debugger, which makes compilation of UNIX applications possible on the Windows environment.

24. Application Architectures

The following sections introduce and discuss different application architectures and how these applications are implemented on UNIX and Windows platforms.

1) Distributed Applications

Distributed applications are logically partitioned into multiple tiers: the view or the **presentation tier**, **the business tier**, and the data storage and access tier. A simple distributed application model consists of a client that communicates with the middle tier, which consists of the application server and an application containing the business logic. The application, in turn, communicates with a database that stores and supplies the data.

The most commonly used databases in UNIX applications are Oracle and DB2. You can either run the application's current database (for example, Oracle) on Windows (that is, migrate the existing database from UNIX to Windows) or migrate the database to Microsoft SQL Server™. In some cases, the best migration decision is a conversion to SQL Server.

Another option available is Oracle. It offers a range of features available to both Windows Server 2003 and UNIX. You may choose to use these features with the current Oracle database. By separating the platform migration from the database migration, you have greater flexibility in migrating database applications.

Next is the presentation tier, which provides either a thick or a thin client interface to the application. UNIX applications may use XMotif to provide a thick client interface. In Windows, a thick client can be developed using the Win32 API, GDI+. The .NET Framework provides Windows Forms that help in rapid development of thick clients. Either one of these two methods can be used while migrating the thick client from UNIX to Windows.

2) Workstation Applications

Workstation-based applications run at the UNIX workstation (desktop) computer and access data that resides on network file shares or database servers. Workstation-based applications have the following architectural characteristics.

(1) They can be single-process applications or multiple-process applications.

(2) They use character-based or GUI-based (for example, X Windows or OpenGL) UIs.

(3) They access a file server (through NFS) or a database server for data resources.

(4) They access a computer server for computer-intensive services (for example, finite element models for structural analysis).

The Windows environment supports a similar workstation application using client/server technology.

3) Web Applications

Web applications from a UNIX Web server are normally one of the following types.

(1) Common Gateway Interface (CGI). CGI is a standard for connecting external applications with information servers, such as Hypertext Transfer Protocol (HTTP) or Web servers. CGI has long been out of favor because of performance problems.

(2) Java Server Page (JSP). Java Server Page (JSP) technology also allows for the development of dynamic Web pages. JSP technology uses Extensible Markup Language (XML)—like tags and scriptlets written in the Java programming language to encapsulate the logic that generates the content for the page. The application logic can reside in server-based resources (for example, the JavaBean component architecture) that the page accesses by using these tags and scriptlets. All HTML or XML formatting tags are passed directly back to the response page.

(3) HTML. Hypertext Markup Language is the authoring language used to create documents on the World Wide Web.

(4) PHP. PHP is a widely used, general-purpose scripting language that is especially suited for Web development and can be embedded into HTML.

(5) JavaScript. JavaScript is an extension of HTML. JavaScript is a script language (a system of programming codes) created by Netscape that can be embedded into the HTML of a Web page to add functionality.

Web applications on UNIX can be used on Windows with minor configuration changes on the Web server. You can also migrate Java-based Web applications on UNIX to Microsoft Web technologies on Windows.

4) Graphics-Intensive Applications

Graphics-intensive applications may support additional UI standards, such as OpenGL. OpenGL has become a widely used and supported two-dimensional and three-dimensional graphics API. OpenGL fosters innovation and provides for faster application development by incorporating a broad set of rendering, texture mapping, special effects, and other powerful visualization functions. Developers can take advantage of the capabilities of OpenGL across all popular desktop and workstation platforms, ensuring wide application deployment.

OpenGL runs on every major operating system, including Macintosh OS, OS/2, UNIX, Windows Server 2003, Windows XP, Linux, OPENStep, and BeOS. OpenGL also works with every major windowing system, including Win32, Macintosh OS, Presentation Manager, and X-Windows. OpenGL includes a full complement of graphics functions. The OpenGL standard has language bindings for C, C++, FORTRAN, Ada, and Java; therefore, applications that use OpenGL functions are typically portable across a wide array of platforms.

5) System-Level Programs or Device Drivers

The existence of an appropriate device driver on the Windows platform is often a gating factor for migrating applications that make use of nonstandard device drivers. Typically, customers do not have access to device-driver source code and are therefore unable to migrate a UNIX device driver to Windows. The topic of migrating device drivers is highly specialized and is not covered in this guide. The Windows Driver Development Kit (DDK) can be used to develop drivers on the Windows environment.

3.5 Key Terms and Review Questions

1. Technical Terms

operating system	操作系统	3.1
peripheral device	外围设备	3.1
bug	漏洞、错误	3.1
physical pulse	物理脉冲	3.1
interrupt-request	中断请求	3.1
bootup	启动	3.1
multitasking	多重任务处理	3.1
real-time	实时	3.1
event-driven	事件驱动	3.1
time-sharing	分时	3.1
pre-emptive	抢占式的	3.1
co-operative	合作的	3.1
multi-threading	多线程	3.1
distributed operating system	分布式操作系统	3.1
embedded operating system	嵌入式操作系统	3.1
tablet PC	平板电脑	3.1
batch processing	批处理	3.1
interactive operating	交互操作	3.1
bootstrap loader	引导装入程序	3.2
semaphore	信号量	3.2
register	寄存器	3.2
stack	堆栈	3.2

queue	队列	3.2
process control block	进程控制块	3.2
suspended	暂停、挂起	3.2
thrashing	超负荷	3.2
process	进程	3.2
thread	线程	3.2
virtual memory management	虚拟内存管理	3.2
high-speed cache	高速缓存	3.2
secondary memory	辅助存储	3.2
driver	驱动程序	3.2
buffer	缓冲区	3.4
modem	调制解调器	3.4
command-line interface	命令行界面	3.4
graphical user interface	图形用户界面	3.4
daemon	守护进程	3.4
kernel object	内核对象	3.4
handle	句柄（表示对象）	3.4
page faults	分页错误	3.4
partition	分区	3.4
directory service	目录服务	3.4
active directory	动态目录	3.4
pipe	管道	3.4
message queues	消息队列	3.4
synchronization	同步	3.4
mutexe	互斥	3.4
middleware	中间件	3.4
interpreter	解释器（解释程序）	3.4
scripting languages	脚本语言	3.4
presentation tier	表示层	3.4
business tier	业务层	3.4
data storage and access tier	数据存储和访问层	3.4

2. Translation Exercises

(3-1) The first task, managing the hardware and software resources, is very important, as various programs and input methods compete for the attention of the central processing unit (CPU) and demand memory, storage and input/output (I/O) bandwidth for their own purposes. In this capacity, the operating system plays the role of the good parent, making sure that each application gets the necessary resources while playing nicely with all the other applications, as well as husbanding the limited capacity of the system to the greatest good of all the users and applications.

(3-2) Interrupts are special signals sent by hardware or software to the CPU. It's as if some part of the computer suddenly raised its hand to ask for the CPU's attention in a lively meeting. Sometimes the operating system will schedule the priority of processes so that interrupts are masked — that is, the operating system will ignore the interrupts from some sources so that a particular job can be finished as quickly as possible.

(3-3) Trouble can begin if the user tries to have too many processes functioning at the same time. The operating system itself requires some CPU cycles to perform the saving and swapping of all the registers, queues and stacks of the application processes. If enough processes are started, and if the operating system hasn't been carefully designed, the system can begin to use the vast majority of its available CPU cycles to swap between processes rather than run processes. When this happens, it's called **thrashing**, and it usually requires some sort of direct user intervention to stop processes and bring order back to the system.

(3-4) One reason that drivers are separate from the operating system is so that new functions can be added to the driver — and thus to the hardware subsystems — without requiring the operating system itself to be modified, recompiled and redistributed. Through the development of new hardware device drivers, development often performed or paid for by the manufacturer of the subsystems rather than the publisher of the operating system, input/output capabilities of the overall system can be greatly enhanced.

(3-5) The kernel mode is a highly privileged mode of operation in which the program code has direct access to the virtual memory. This includes the address spaces of all user mode processes and applications and their hardware.

(3-6) UNIX users generally configure a system by editing the configuration files with any of the available text editors. The advantage of this mechanism is that the user does not need to learn how to use a large set of configuration tools, but must only be familiar with an editor and possibly a scripting language. The disadvantage is that the information in the files comes in various formats; hence the user must learn the various formats in order to change the settings. UNIX users often employ scripts to reduce the possibility of repetition and error.

(3-7) UNIX has several IPC mechanisms that have different characteristics and which are appropriate for different situations. Shared memory, **pipes**, and **message queues** are all suitable for processes running on a single computer. Shared memory and message queues are suitable for communicating among unrelated processes. Pipes are usually chosen for communicating with a child process through standard input and output. For communications across the network, sockets are usually the chosen technique.

References

[1] Alina B. Operating systems.[M/OL]. [2016-08-22]. http://www.iterating.com/productclasses/Operating-Systems.

[2] Gustavo D. How computers boot up.[EB/OL]. (2008-06-05) [2016-08-22]. http://duartes.org/

gustavo/blog/post/how-computers-boot-up.

[3] David K. Basic concepts of real-time operating systems. [EB/OL]. (2003-11-18) [2016-08-22]. http://LinuxDevices.com.

[4] Madison N. What is an operating system? [EB/OL]. [2016-08-22]. http://www.wisegeek.com/what-is-an-operating-system.htm.

[5] Mehler R W. ECE425 Microprocessor Systems: Interrupts and Resets.[R/OL].[2016-08-23]. http://www.csun.edu/~rmehler/mehler_files/ece425/425lecture13-15.pdf.

[6] Russinovich M E, David A S. Processes, Threads and Jobs. Microsoft Windows Internals, Fourth Edition: Microsoft Windows Server 2003, Windows XP, and Windows 2000[M/OL]. [2016-08-21]. http://download.microsoft.com/download/5/b/3/5b38800c-ba6e-4023-9078-6e9ce2383e65/C06X1116607.pdf.

Chapter 4 Algorithms, Data Structures and Software Engineering

4.1 Algorithm①

4.1.1 Introduction

Consider the following three examples. What do they all have in common?

1. Chocolate Cream Pie

Heat milk, marshmallows and chocolate in 3-quart saucepan over low heat, stir constantly, until chocolate and marshmallows are melted and blended. Refrigerate about 20 minutes, stirring occasionally until mixture mounds slightly when dropped from a spoon.

Beat whipping cream in chilled small bowl with electric mixer on high speed until soft peaks form. Fold chocolate mixture into whipped cream. Pour into pie shell. Refrigerate uncovered about 8 hours or until set. Garnish with milk chocolate curls and whipped cream.

2. Directions to John's House

From the Quik Mart, you should follow Saddle road for four miles until you reach a stoplight. Then make a left-hand turn at the stop light. Now you will be on Hollow Street. Continue driving on Hollow Street for one mile. You should drive past four blocks until you reach the post office. Once you are at the post office, turn right onto Jackson road. Then stay on Jackson for about 10 miles. Eventually you will pass the Happy Meadow farm on your right. Just after Happy Meadow, you should turn left onto Brickland drive. My house is the first house on your left.

3. How to Change Your Motor Oil

(1) Place the oil pan underneath the oil plug of your car.

(2) Unscrew the oil plug.

(3) Drain oil.

(4) Replace the oil plug.

(5) Remove the oil cap from the engine.

---

① http://courses.cs.vt.edu/csonline/Algorithms/.

(6) Pour in 4 quarts of oil.

(7) Replace the oil cap.

Each of these examples is an algorithm, a set of instructions for solving a problem. Once we have created an algorithm, we no longer need to think about the principles on which the algorithm is based. For example, once you have the directions to John's house, you do not need to look at a map to decide where to make the next turn. The intelligence needed to find the correct route is contained in the algorithm. All you have to do is following the directions. This means that algorithms are a way of capturing intelligence and sharing it with others. Once you have encoded the necessary intelligence to solve a problem in an algorithm, many people can use your algorithm without needing to become experts in a particular field.

(4-1) Algorithms are especially important to computers because computers are really general purpose machines for solving problems. But in order for a computer to be useful, we must give it a problem to solve and a technique for solving the problem. Through the use of algorithms, we can make computers "intelligent" by programming them with various algorithms to solve problems. Because of their speed and accuracy, computers are well-suited for solving tedious problems such as searching for a name in a large telephone directory or adding a long column of numbers. However, the usefulness of computers as problem solving machines is limited because the solutions to some problems cannot be stated in an algorithm.

Much of the study of computer science is dedicated to discovering efficient algorithms and representing them so that they can be understood by computers. During our study of algorithms, we will discuss what defines an algorithm, how to represent algorithms, and what makes algorithms efficient.

4.1.2 Definition of Algorithms

In the introduction, we gave an informal definition of an algorithm as "a set of instructions for solving a problem" and we illustrated this definition with a recipe, directions to a friend's house, and instructions for changing the oil in a car engine. You also created your own algorithm for putting letters and numbers in order. While these simple algorithms are fine for us, they are much too ambiguous for a computer. In order for an algorithm to be applicable to a computer, it must have certain characteristics. We will specify these characteristics in our formal definition of an algorithm.

An algorithm is a well-ordered collection of unambiguous and effectively computable operations that when executed produces a result and halts in a finite amount of time[①].

With this definition, we can identify five important characteristics of algorithms.

(1) Algorithms are well-ordered.

(2) Algorithms have unambiguous operations.

(3) Algorithms have effectively computable operations.

① Schneider M, Gersting J. An Invitation to Computer Science. West Publishing Company, New York, NY, 1995.

(4) Algorithms produce a result.

(5) Algorithms halt in a finite amount of time.

These characteristics need a little more explanation, so we will look at each one in detail.

(1) Algorithms are well-ordered.

Since an algorithm is a collection of operations or instructions, we must know the correct order in which to execute the instructions. If the order is unclear, we may perform the wrong instruction or we may be uncertain which instruction should be performed next. This characteristic is especially important for computers. A computer can only execute an algorithm if it knows the exact order of steps to perform.

(2) Algorithms have unambiguous operations.

Each operation in an algorithm must be sufficiently clear so that it does not need to be simplified. Given a list of numbers, you can easily order them from largest to smallest with the simple instruction "Sort these numbers." A computer, however, needs more detail to sort numbers. It must be told to search for the smallest number, how to find the smallest number, how to compare numbers together, etc. The operation "Sort these numbers" is ambiguous to a computer because the computer has no basic operations for sorting. Basic operations used for writing algorithms are known as **primitive operations** or primitives. When an algorithm is written in computer primitives, then the algorithm is unambiguous and the computer can execute it.

(3) Algorithms have effectively computable operations.

Each operation in an algorithm must be doable, that is, the operation must be something that is possible to do. Suppose you were given an algorithm for planting a garden where the first step instructed you to remove all large stones from the soil, this instruction may not be doable if there is a four ton rock buried just below ground level. For computers, many mathematical operations such as division by zero or finding the square root of a negative number are also impossible. These operations are not effectively computable so they cannot be used in writing algorithms.

(4) Algorithms produce a result.

In our simple definition of an algorithm, we stated that an algorithm is a set of instructions for solving a problem. Unless an algorithm produces some results, we can never be certain whether our solution is correct. Have you ever given a command to a computer and discovered that nothing changed? What was your response? You probably thought that the computer was **malfunctioning** because your command did not produce any type of result. Without some visible change, you have no way of determining the effect of your command. The same is true with algorithms. Only algorithms which produce results can be verified as either right or wrong.

(5) Algorithms halt in a finite amount of time.

Algorithms should be composed of a finite number of operations and they should complete their execution in a finite amount of time. Suppose we wanted to write an algorithm to print all the integers greater than 1. Our steps might look something like this:

Print the number 2.

Print the number 3.

Print the number 4.

...

While our algorithm seems to be pretty clear, we have two problems. First, the algorithm must have an infinite number of steps because there are an infinite number of integers greater than one. Second, the algorithm will run forever trying to count to **infinity**. These problems violate our definition that an algorithm must halt in a finite amount of time. Every algorithm must reach some operation that tells it to stop.

4.1.3 Specifying Algorithms

When writing algorithms, we have several choices of how we will specify the operations in our algorithm. One option is to write the algorithm using plain English. An example of this approach is the directions to John's house given in the introduction lesson. Although plain English may seem like a good way to write an algorithm, it has some problems that make it a poor choice. First, plain English is too wordy. When we write in plain English, we must include many words that contribute to correct grammar or style but do nothing to help communicate the algorithm. Second, plain English is too ambiguous. Often an English sentence can be interpreted in many different ways. Remember that our definition of an algorithm requires that each operation be unambiguous.

Another option for writing algorithms is using programming languages. These languages are collections of primitives (basic operations) that a computer understands. While programming languages avoid the problems of being wordy and ambiguous, they have some other disadvantages that make them undesirable for writing algorithms. Consider the following lines of code from the programming language C++ in Fig.4-1.

```
a = 1;
b = 0;
while (a <= 10)
{
  b = b + a;
  a++;
}
cout << b;
```

Fig.4-1 Writing algorithms using programming languages

This algorithm sums the numbers from 1 to 10 and displays the answer on the computer screen. However, without some special knowledge of the C++ programming language, it would be difficult for you to know what this algorithm does. Using a programming language to specify algorithms means learning special syntax and symbols that are not part of standard english. For example, in the code above, it is not very obvious what the symbol "++" or the symbol

“<<” does. When we write algorithms, we would rather not worry about the details of a particular programming language.

What we would really like to do is combine the familiarity of plain English with the structure and order of programming languages. A good compromise is structured English. This approach uses English to write operations, but groups operations by **indenting** and numbering lines. An example of this approach is the directions for changing motor oil in the introduction lesson. Each operation in the algorithm is written on a separate line so they are easily distinguished from each other. We can easily see the advantage of this organization by comparing the structured English algorithm with the plain English algorithm comparing the structured English algorithm with the plain English algorithm in Fig.4-2.

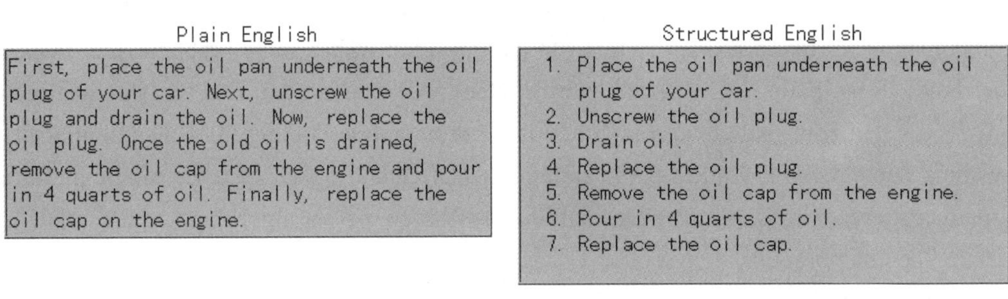

Fig.4-2 Comparison of structured English algorithm with the plain English algorithm

4.1.4 Examples — Sorting Algorithms

1. Basic Operations

The **sorting algorithms** share **two** basic operations in common. These operations are the comparison operation and the swap operation. We will look at each one in more detail before we examine our sorting algorithms.

1) The Comparison Operation

The comparison operation is simply a way of determining which item in a list should come first. If we are sorting a list of numbers from smallest to largest, the comparison operation tells us to place the number with the least value first. If we are sorting a list of letters alphabetically, the comparison operation tells us to place “a” before “b”, “b” before “c”, and so on. We will see that a sorting algorithm must usually perform many comparisons in order to correctly sort a list.

2) The Swap Operation

The swap operation is one way we move items as we are sorting. By swapping small items with large ones, we can place all the items in the correct order. When we use computers for sorting, the swap operation can be a little tricky because of the way computers copy data from one memory location to another. In Fig.4-3, we illustrate the process of swap operation.

Let’s take one more look at the swap algorithm we wrote:

(1) Copy the contents of cell A to temp.

(2) Copy the contents of cell B to cell A.

(3) Copy the contents of temp to cell B.

Notice that this operation requires three copies. It is important to remember that the swap operation is really a combination of copy operations. In our next lesson, we will learn a sorting algorithm called the Simple Sort that uses just the copy operation rather than the swap operation.

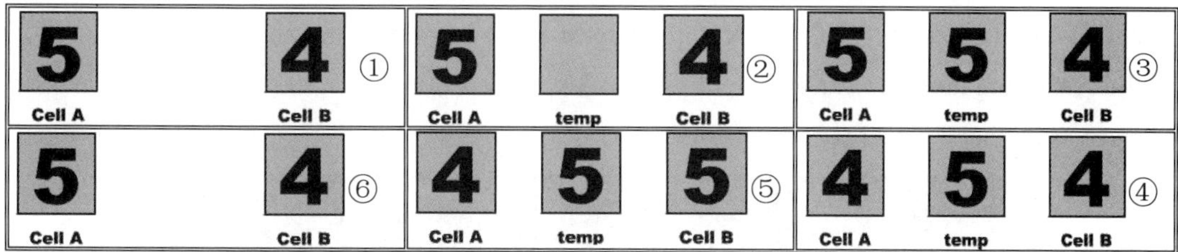

Fig.4-3 the process of swap operation

2. Simple Sort

One basic algorithm for sorting is the Simple Sort. We can illustrate this algorithm using a group of unsorted playing cards. The Simple Sort works by selecting the smallest card in the unsorted hand and moving this card to a second hand. Once all the cards have been removed from the unsorted hand, the second hand contains the cards in sorted order.

We can use the same idea as in our Simple Card Sort to write a Simple Sort that can be used by a computer. Let's see what this algorithm looks like and how it can be used to sort numbers in a computer. The algorithm for the Simple Sort is given in the box below.

Simple Card Sort Algorithm

1. Get a list of unsorted numbers
2. Repeat steps 3 through 6 until the unsorted list is empty
3. Compare the unsorted numbers
4.Select the smallest unsorted number
5. Move this number to the sorted list
6. Store a maximum value in the place of the smallest number
7. Stop

The steps below illustrate how the Simple Sort algorithm works on a computer.

(1) First, we give the computer a list of unsorted numbers. These numbers are stored in a group of contiguous memory cells called an array. Each memory cell of the array holds a single number.

(2) As the computer sorts these numbers, it will repeatedly compare them to find the smallest number. This is similar to the comparisons we made when sorting our hand of cards. Each time we compared two cards and kept the smaller of the two. Then we compared this card to the remaining cards until we found a smaller one or checked all the cards. The computer uses the same process only with numbers rather than cards. Once the smallest number is found, the computer will copy this number to a new array of memory cells and replace the old number with a special number called MAX. MAX is the largest number a single memory cell can hold. None of the remaining numbers can be larger than MAX, so this number is a good choice for marking memory cells that have already been sorted.

Unsorted Array

Sorted Array

(3) Next, the computer begins searching for the smallest number in the unsorted list. Although it is easy for us to scan the numbers and select the 2 as smallest, the computer must compare all the memory cells in the unsorted array to be certain which number is smallest. This means the computer must perform six comparisons: (7<8), (7>5), (5>2), (2<4), (2<6), and finally (2<3). Once the comparisons are done, the computer copies 2 to the sorted array and replaces the original 2 with MAX.

Unsorted Array

Sorted Array

(4) Now the computer begins searching for the smallest number again. Six more comparisons are required to determine that 3 is smallest: (7<8), (7>5), (5<MAX), (5>4), (4<6), and finally (4>3). Now we can see the importance of replacing 2 with MAX in our previous step.

If we had not made this change, then 2 would have been selected as the smallest number again. After copying 3 to the sorted array, the computer also replaces the original with MAX.

Unsorted Array

Sorted Array

(5) With six more comparisons, the computer selects 4 as the smallest number, copies it to the sorted array, and replaces the original with MAX.

Unsorted Array

Sorted Array

(6) The numbers 5, 6, 7, and 8 are also selected in turn by six comparisons, a copy, and a replacement of the original. Once all the memory cells in the unsorted array have been considered, the sorted array contains our original numbers in sorted order.

Unsorted Array

Sorted Array

3. Insertion Sort

Now let's look at how the **Insertion Sort** algorithm would work inside a computer. Below is our modified algorithm for sorting a list of numbers.

Insertion Sort Algorithm

1. Get a list of unsorted numbers
2. Set a marker for the sorted section after the first number in the list
3. Repeat steps 4 through 6 until the unsorted section is empty
4. Select the first unsorted number
5. Swap this number to the left until it arrives at the correct sorted position
6. Advance the marker to the right one position
7. Stop

This time the steps of our modified algorithm are almost identical to the steps in our card sorting algorithm. We have simply substituted numbers for playing cards and a list of numbers for a hand of cards. The steps below illustrate how the Insertion Sort algorithm works on a computer.

(1) First, we give the computer a list of unsorted numbers and store them in an array of memory cells.

(2) To begin the sort, the computer divides the sorted and unsorted sections of the list by placing a marker after the first number. To sort the numbers, it will repeatedly compare the first unsorted number with the numbers in the sorted section. If the unsorted number is smaller than its sorted neighbor, the computer will swap them.

(3) The first number in the unsorted section is 8, so the computer compares it with the number to the left. Since 8 is greater than 7, these numbers do not need to swapped and the computer simply advances the marker one position. Notice that only one comparison was needed to sort the 8.

(4) Now the first number in the unsorted section is 5. 5 is less than 8, so the computer swaps these numbers. 5 is also less than 7, so the computer swaps these numbers as well.

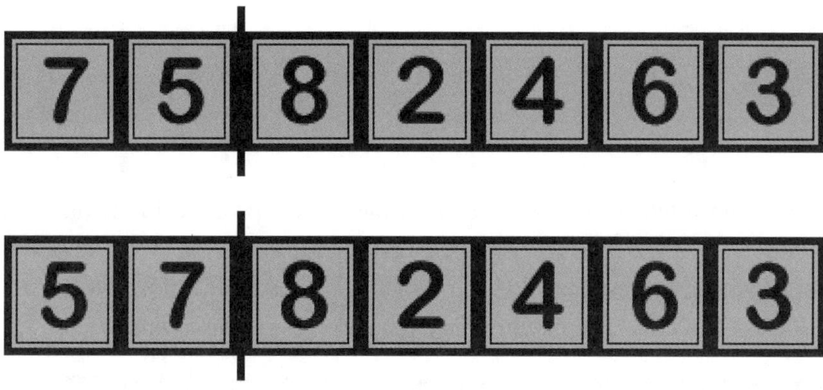

(5) Now 5 is in the correct order, so the computer advances the marker one position. This time two comparisons and two swaps were needed to sort the number.

(6) Now the first number in the unsorted section is 2. 2 is less than 8, 7, and 5, so after three comparisons and three swaps, 2 arrives at the correct sorted position, and the computer advances the sort marker.

(7) Now the first number in the unsorted section is 4. 4 is less than 8, 7, and 5 but it is not less than 2. This time the computer performs four comparisons and three swaps to put the 4 in the correct order. Only three swaps were needed since the 2 and the 4 did not need to be switched. After these comparisons and swaps, the computer advances the sort marker.

(8) Now 6 is the first number in the unsorted section. After three comparisons and two swaps, the computer places the 6 in the correct position between 5 and 7. Notice that the computer did not need to compare the 6 with the 2 or the 4 since it already knows these numbers are less than 5. Once the computer finds a number in the sorted section less than 6, it knows it has found the correct position for 6 and it can advance the sort marker.

(9) The final unsorted number is 3. To find the correct position for 3, the computer must compare it with every number in the unsorted section. However, only five swaps are required since the first number (2) is less than 3. After moving 3 to the correct position and advancing the sort marker, the Insertion Sort is complete since the unsorted section is empty.

4. Selection Sort

Now that you have a general idea of how the Selection Sort works, let's see how the computer would perform this sort with numbers. Below is our modified algorithm for sorting a list of numbers.

Selection Sort Algorithm

1. Get a list of unsorted numbers
2. Set a marker for the unsorted section at the front of the list
3. Repeat steps 4 - 6 until one number remains in the unsorted section
4. Compare all unsorted numbers in order to select the smallest one
5. Swap this number with the first number in the unsorted section
6. Advance the marker to the right one position
7. Stop

(1) First, we give the computer a list of unsorted numbers and store them in an array of memory cells.

(2) To begin the sort, the computer divides the sorted and unsorted sections of the list by placing a marker before the first number. To sort the numbers, the computer will repeatedly search the unsorted section for the smallest number, swap this number with the first number in the unsorted section, and update the sort marker.

(3) To find the smallest number in the unsorted section, the computer must make six comparisons: (7<8), (7>5), (5>2), (2<4), (2<6), and (2>3). After these comparisons, the computer knows that 2 is the smallest number, so it swaps this number with 7, the first number in the unsorted section, and advances the sort marker.

(4) Now five more comparisons are needed to find the smallest number in the unsorted section: (8>5), (5<7), (5>4), (4<6), and (4>3). After these comparisons, the computer swaps 3, the smallest number in the unsorted section, with 8, the first number in the unsorted section, and advances the sort marker.

(5) This time four comparisons are needed to determine that 4 is the smallest number in the unsorted section: (5<7), (5>4), (4<6), and (4<8). After these comparisons, the computer swaps 4 with 5 and then advances the sort marker.

(6) After three more comparisons, the computer identifies 5 as the smallest unsorted number: (7>5), (5<6), and (5<8). Then the computer swaps 5 with 7 and advances the sort marker.

(7) This time only two comparisons are needed to determine that 6 is the smallest number: (7>6) and (6<8). After these two comparisons, the computer swaps 6 with 7 and then advances the sort marker.

(8) Now we only need a single comparison to find the right position for 7: (7<8). Since 7 is the smallest number and it is also the first number in the unsorted section, the computer does not need to swap this number. It only needs to advance the sort marker. Now there is only one number in the unsorted section, so the list of numbers is sorted and the Selection Sort algorithm is complete.

4.1.5 Algorithm Analysis

For any given problem, it is quite possible that there is more than one algorithm that represents a correct solution. A good example of this is the problem of sorting. Dozens of different algorithms have been written to solve this problem. Given such a wide range of solutions, how can we determine which algorithm is the best one to use? To do this, we must analyze our algorithms in such a way that we can gauge the efficiency of the algorithm. Once we have calculated the efficiency of an algorithm, we can compare our measurements and select the best solution.

Let's analyze the three sorting algorithms in this section and determine which one was the most efficient solution for sorting the list of seven numbers in our examples. To do this we need a way to measure the efficiency of our algorithms. We can actually measure the efficiency in two different ways: **space efficiency** and **time efficiency**. An algorithm that is space-efficient uses the least amount of computer memory to solve the problem of sorting. An algorithm that is time-efficient uses the least amount of time to solve the problem of sorting. Since most of the sorting operations are comparisons, copies, and swaps, we can count these operations and use our results as a measure of time efficiency. Tab.4-1 below summarizes our measures of time and space efficiency in sorting.

Tab.4-1 Types of efficiency measures

Space	Amount of computer memory
Time	# of items copied
	# of items swapped
	# of items compared

1. Space Efficiency

Let's begin our analysis by determining which sort was the most space-efficient. We discussed in our previous lesson that space efficiency can be measured by calculating the number of memory cells a particular sort requires. For simplicity, we will assume that a memory cell can hold one number. This means that a sort with five numbers would require at least five **memory**

cells just to store the numbers in the computer. The total number of memory cells needed for sorting will depend on how many additional cells the algorithm requires to order the numbers.

Of the three sorts, the Insertion Sort and the Selection Sort are most space-efficient. These sorts order the items within the list using the swap operation rather than copying items to a new list. Since the Simple Sort copies every item to a new list, it requires twice as much computer memory. Remember that the swap operation requires a temporary memory cell.

2. Time Efficiency

Now let's determine which sort was the most time-efficient. To do this, we will count the number of operations each sort performed. Most of the operations that are done by the computer during sorting fall into two groups: copying numbers or comparing numbers. The algorithm which requires the least copying and comparing is the one that will execute the fastest.

For the Insertion Sort and the Selection Sort, it will be easier to count the number of swaps that are done rather than the number of copies. Remember that the swap operation requires three copies. We can find the total number of copies that the algorithms perform by counting the number of swaps and multiplying by three. The Simple Sort does not use the swap operation, so you can count the number of copies directly.

Now that you have completed your calculations, let's summarize the results in Tab.4-2 and Tab.4-3. We already know that the Insertion Sort and the Selection Sort were the most space-efficient, but we have yet to determine which sort is the most time-efficient. We will see that this answer is a little more difficult to determine.

Tab.4-2 Space efficiency

Algorithm	# of memory cells
Simple Sort	14
Insertion Sort	8
Selection Sort	8

Tab.4-3 Time efficiency

Algorithm	# of copies	# of comparisons
Simple Sort	7	42
Insertion Sort	45	19
Selection Sort	15	21

Notice that the Simple Sort required the least amount of copies. We would expect this to be true since it does not swap numbers while sorting. Instead the numbers are copied to a new list in the computer. This is a common tradeoff between time and space. Although the Simple Sort loses space efficiency by using two lists, it gains time efficiency because fewer copies are required. Of course this does not mean that it is always best to use the Simple Sort to gain more speed. If we are trying to sort a list of 5 million names the Simple Sort would use too much space in the computer's memory. It would be much better to swap items within the list rather than create two lists.

For number of comparisons, the Selection Sort and Insertion Sort were nearly the same. The Simple Sort, however, required twice as many comparisons. We can see the reason for this difference by thinking about how the algorithms work. Each algorithm repeatedly searches for the smallest number and then places this number in the correct position. For the Insertion Sort and the Selection Sort, each iteration of this process reduces the unsorted section by one number. During the next search, the computer does not need to make as many comparisons to find the smallest number. The Simple Sort, however, replaces sorted numbers with a marker called MAX. Each time the computer searches for the smallest number, it must compare all seven memory cells. This approach is much less efficient.

Given the particular set of seven numbers we sorted, the Selection Sort was the most time-efficient. However, it is important to understand that this may not be true for every set of seven numbers. Consider the following example.

If we use the Insertion Sort on these numbers only 8 comparisons and 1 swap would be needed to sort them. However, if we use the Selection Sort, 21 comparisons and 1 swap would be needed. In this case, the Insertion sort is more efficient.

4.2 Data Structures

4.2.1 Definition

Imagine that you are hired by company XYZ to organize all of their records into a computer database. The first thing you are asked to do is create a database of names with all the company's management and employees. To start your work, you make a list of everyone in the company along with their position (Tab.4-4).

Tab.4-4 List of management and employees

Name	Position	Name	Position
Aaron	Manager	Martha	Employee
Charles	VP	Patricia	Employee
George	Employee	Rick	Secretary
Jack	Employee	Sarah	VP
Janet	VP	Susan	Manager
John	President	Thomas	Employee
Kim	Manager	Zack	Employee
Larry	Manager		

But this list only shows one view of the company. You also want your database to represent the relationships between management and employees at XYZ. Although your list contains both name and position, it does not tell you which managers are responsible for which workers and so on. After thinking about the problem for a while, you decide that a tree diagram as shown in Fig.4-4 is a much better structure for showing the work relationships at XYZ.

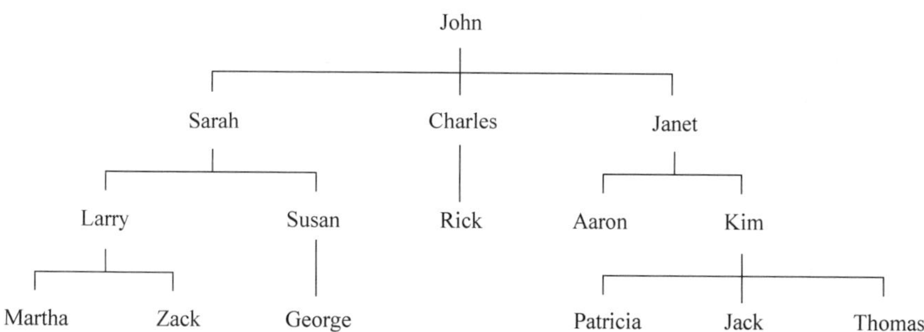

Fig.4-4 Relationships between management and employees

These two diagrams are examples of different **data structures**. In one of the data structures, your data is organized into a list. This is very useful for keeping the names of the employees in alphabetical order so that we can locate the employee's record very quickly. However, this structure is not very useful for showing the relationships between employees. A tree structure is much better suited for this purpose.

In computer science, data structures are an important way of organizing information in a computer. Just like the diagrams above illustrate, there are many different data structures that programmers use to organize data in computers. Some data structures are similar to the tree diagram because they are good for representing relationships between data. Other structures are good for ordering data in a particular way like the list of employees. Each data structure has unique properties that make it well suited to give a certain view of the data.

During these lessons, you will learn how data structures are created inside a computer. You will find there is quite a difference between your mental picture of a data structure and the actual way a computer stores a data structure in memory. You will also discover that there are many different ways of creating the same data structure in a computer. These various approaches are tradeoffs that programmers must consider when writing software. Finally, you will see that each data structure has certain operations that naturally fit with the data structure. Often these operations are bundled with the data structure and together they are called a data type. By the end of this study, you should be able to do the following.

(1) Show how data structures are represented in the computer.

(2) Identify linear and nonlinear data structures.

(3) Manipulate data structures with basic operations.

(4) Compare different implementations of the same data structure.

4.2.2 Types of Data Structure

1. List

The most common **linear data structure** is the **list**. Now we are going to look at a particular kind of list: an ordered list. Ordered lists are very similar to the alphabetical list of employee names for the XYZ company. These lists keep items in a specific order such as alphabetical or numerical order. Whenever an item is added to the list, it is placed in the correct sorted position so that the entire list is always sorted.

1) **Ordered List**

Before we consider how to implement such a list, we need to consider the abstract view of an ordered list. Since the idea of an abstract view of a list may be a little confusing, let's think about a more familiar example. Consider the abstract view of a television. Regardless of who makes a television, we all expect certain basic things like the ability to change channels and adjust the volume. As long as these operations are available and the TV displays the shows we want to view, we really don't care about who made the TV or how they chose to construct it. The circuitry inside the TV set may be very different from one brand to the next, but the functionality remains the same. Similarly, when we consider the abstract view of an ordered list, we don't worry about the details of implementation. We are only concerned with what the list does, not how it does it.

Suppose we want a list that can hold the following group of sorted numbers: [2 4 6 7]. What are some things that we might want to do with our list? Well, since our list is in order, we will need some way of adding numbers to the list in the proper place, and we will need some way of deleting numbers we don't want from the list. To represent these operations, we will use the following notation:

AddListItem(List, Item)

RemoveListItem(List, Item)

Each operation has a name and a list of parameters the operation needs. The parameter list for the AddListItem operation includes a list (the list we want to add to) and an item (the item we want to add). The RemoveListItem operation is very similar except this time we specify the item we want to remove. These operations are part of the abstract view of an ordered list. They are what we expect from any ordered list regardless of how it is implemented in the computer.

2) Stack

Another common linear data structure is the **stack**. With the stack, we have restricted the access to one end of the list by using the pop and push operations. The result of this restriction is that items in the list pile one on top of the other. To get to the bottom item, we must first remove all the items above it. This behavior is sometimes described as "last-in, first-out" or LIFO since the last item to enter the stack is the first item to leave the stack. With the stack, the top item is always the last item to enter the stack and it is always the first item to leave the stack since no other items can be removed until the top item is removed.

We will represent these two operations that can be performed on a stack with the following notation:

Item PushStackItem(Stack, Item)

Item PopStackItem(Stack)

The PushStackItem operation has two parameters which are a stack and an item. This operation adds the item to the top of the specified stack. The PopStackItem operation only takes one parameter which is a stack. However, notice that this operation has the keyword item listed to the left. This keyword represents the item that is removed from the top of the stack when the PopStackItem operation is done. These two operations are part of the abstract view of a stack. They are what we expect from any stack regardless of how it is implemented in the computer.

3) Queue

Like the stack, the **queue** is a type of restricted list. However, instead of restricting all the operations to one end of the list as a stack does, the queue allows items to be added at one end of the list and removed at the other end.

The restrictions placed on a queue cause this structure to be a "first-in, first-out" or FIFO structure. This idea is similar to customer lines at a grocery store. When customer X is ready to check out, he or she enters the tail of the waiting line. After the preceding customers have paid, then customer X pays and exits the head of the line. The check-out line is really a queue that enforces a "first come, first serve" policy.

We will represent these two operations that can be performed on a queue with the following notation:

Item EnqueueItem(Queue, Item)

Item DequeueItem(Queue)

These two operations are very similar to the operations we learned for the stack data structure. Although the names are different, the logic of the parameters is the same. The EnqueueItem operation takes the Item parameter and adds it to the tail of Queue. The DequeueItem operation removes the head item of Queue and returns this as Item. Notice that we represent the returned item with a keyword located to the left of the operation name. These two operations are part of the abstract view of a queue. Regardless of how we choose to implement our queue on the computer, the queue must support these two operations.

2. Tree

Tree is a common **nonlinear data structure**. We have already seen an example of a tree when we looked at the employee hierarchy from the XYZ company. Let's take another look at this diagram with some of the important features of trees highlighted, see Fig.4-5.

In this diagram, we can see that the starting point, or the **root node**, is circled in blue. A node is a simple structure that holds data and links to other nodes. In this case, our root node contains the data string "John" and three links to other nodes. Notice that the group of nodes circled in red does not have any links. These nodes are at the ends of the branches and they are appropriately called leaves or **leaf nodes**. In our diagram, the nodes are connected with solid black lines called arcs or edges. These edges show the relationships between nodes in the tree.

One important relationship is the parent/child relationship. **Parent nodes** have at least one edge to a node lower in the tree. This node is called the **child node**. Nodes can have more than one child, but children can only have a single parent. Notice that the root node has no parent, and the leaf nodes have no children. The final feature to note in our diagram is the **subtree**. At each level of the tree, we can see that the tree structure is repeated. For example, the two nodes representing "Charles" and "Rick" compose a very simple tree with "Charles" as the root node and "Rick" as a single leaf node.

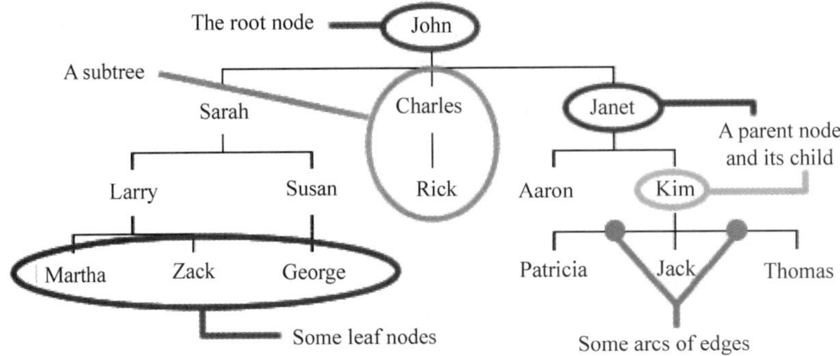

Fig.4-5 Tree diagram

Now let's examine one way that trees are implemented in the computer's memory. We will begin by introducing a simple tree structure called a **binary tree**. Binary trees have the restriction that nodes can have no more than two children. With this restriction, we can easily determine how to represent a single binary node in memory, as shown in Fig.4-6. Our node will need to reserve memory for data and two **pointers**.

Data	Left child pointer	Right child pointer

Fig.4-6 Structure of binary node

Using our binary node, we can construct a binary tree. In the data cell of each node, we will store a letter. The physical representation of our tree might look something like Fig.4-7.

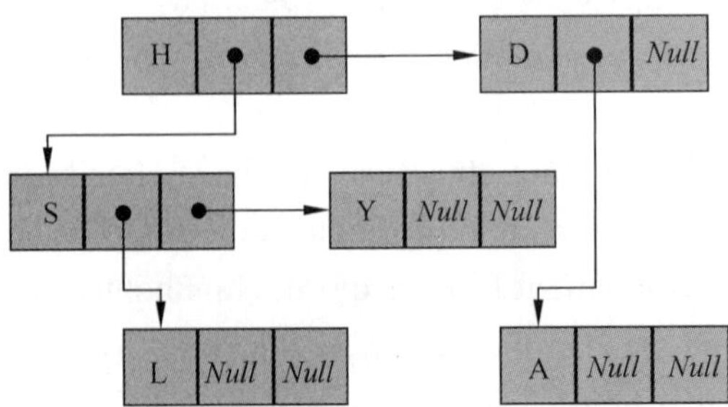

Fig.4-7 Physical representation of tree

3. Graph

The last data structure that we will study in this section is the **graph**. Graphs are similar to trees except they do not have as many restrictions. In the previous lesson, we saw that every tree has a root node, and all the other nodes in the tree are children of this node. We also saw that nodes can have many children but only one parent. When we relax these restrictions, we get the graph data structure. The logical representation of a typical graph might look something like Fig.4-8.

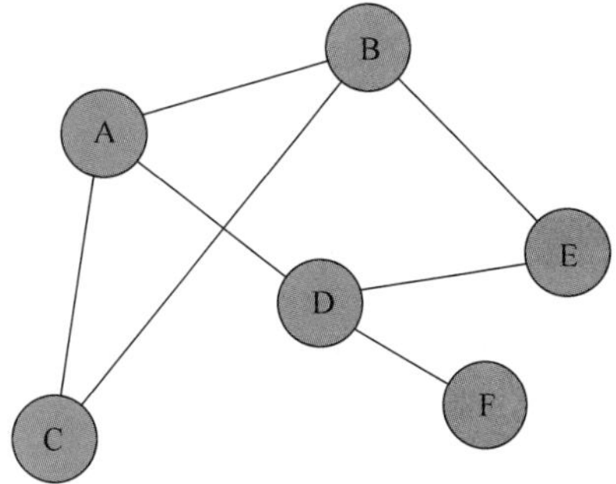

Fig.4-8 A graph

Notice that our graph does not have a root node like the tree data structure did. Instead, any node can be connected with any other node. Nodes do not have a clear parent/child relationship like we saw in the tree. Instead nodes are called **neighbors** if they are connected by an edge. For example, node A above has three neighbors: B, C, and D.

(4-2) It is not hard to imagine how the graph data structure could be useful for representing data. Perhaps each of the nodes above could represent a city and the edges connecting the nodes could represent roads. Or we could use a graph to represent a computer network where the nodes are workstations and the edges are network connections. Graphs have so many applications in computer science and mathematics that several algorithms have been written to perform standard graph operations such as searching the graph and finding the shortest path between nodes of a graph.

Now that you have a basic idea of the logical representation of graphs, let's take a look at one way that graphs are commonly represented in computers. The representation is called an **adjacency matrix**, and it uses a **two-dimensional array** to store information about the graph nodes. The adjacency matrix for our graph (in Fig.4-8) is given in Fig.4-9.

	A	B	C	D	E	F
A	--	1	1	1	--	--
B	1	--	1	--	1	--
C	1	1	--	--	--	--
D	1	--	--	--	1	1
E	--	1	--	1	--	--
F	--	--	--	1	--	--

Fig.4-9 Adjacency matrix for a graph

Notice that the matrix has six rows and six columns labeled with the nodes from the graph. We mark a "1" in a cell if there exists an **edge** from the two nodes that index that cell. For example, since we have a edge between A and B, we mark a "1" in the cells indexed by A and B. These cells are marked with a dark gray background in the adjacency matrix. With adjacency matrix, we can represent every possible edge that a graph can have.

4.3 Programming

4.3.1 Evolution of Programming Language

Suppose for a moment that you were given the following list of instructions to perform:

0001 0011 0011 1011

1101 0111 0001 1001

1111 0001 1101 1111

0000 1100 0101 1101

0001 0011 0011 1011

Of course, these instructions have no real meaning to you, but they are exactly the kind of instructions that a computer expects. Instructions like these are called "**machine code**" and each one represents a typical operation that a computer might perform. You can immediately see the difficulty with this language. While it might be very appropriate for a computer, it is extremely confusing for a computer programmer.

As computers have developed over the past few decades, new generations of **programming languages** have also been developed to bridge the gap between programmers and machine code. An early solution to simplifying programming was the use of **hexadecimal** notation to represent machine instructions. For example, in hexadecimal, the first instruction would be written 133B. While this representation is certainly easier to remember than a string of ones and zeros, it still fails to give us any idea of the purpose of the instruction.

Assembly language was the first programming language to address the problem of assigning meaningful names to computer instructions. This language used simple names like "LOAD" "ADD", or "STORE" to represent machine instructions. A program written with these names was then converted into machine code using an assembler, a program that translated the names into the binary instructions understood by the computer. Assembly languages are now referred to as "second generation" languages while machine code is considered to be a "first generation" language.

The next generation of computer languages further increased the ease of programming by grouping sets of machine instructions together to form common programming constructs. While it might take 3 or 4 lines of code to add two numbers using assembly language, this task could be accomplished with a single instruction in a "third generation" language. Languages such as Pascal, C, C++, Java, and Ada are all examples of third generation languages that are widely used today. These languages are also known as **high-level languages** since they abstract away the details of machine code and help programmers to concentrate on problem solving.

4.3.2 Basic Components and Structure of a Program

In this section on programming languages, you will be learning about the five basic concepts of a "third generation" language. These concepts are **variables**, **expressions, control structures**, input/output, and **abstraction**.

1. Identifiers, Variables and Constants

(4-3) The first important concept in programming languages is the idea of **identifiers**. Identifiers are descriptive names that are mapped to locations in the computer's memory. Once a memory location is given a specific name, we can refer to that location using the identifier rather than the numeric address. This system of associating a name with a memory location allows us to choose names that give meaning to the contents of the memory location.

For example, suppose you wanted to write a simple program to calculate the amount of sales tax you will need to pay on a new pair of shoes that costs $19.95. The first thing you will need to do is create an identifier called PriceOfShoes that is mapped to a particular memory location, as shown in Fig.4-10. Then you need to represent the value 19.95 in the computer at that location. Once this is done, you can use your identifier PriceOfShoes to refer to the price of the shoes in your program.

To be precise, we need to remember that most computers use the binary system for their storage and arithmetic rather than the decimal system. This means that when we associate a fractional value such as 19/20 (or 0.95) with a **variable**, we are actually storing a representation of 19/20 rather than the exact value. Since the binary number system has no exact representation for many fractions like 19/20, 1/3, or even 1/10, the computer must use a close **approximation** to represent this value. While we often think of exact values being stored in the computer's memory, the reality is that only close approximations for some numbers can be stored. However, to simplify our discussion, we will refer to exact values as being stored in memory although we know that only representations are really stored in most cases.

What happens if you need to compute the sales tax on another pair of shoes with a different price? Do you need to create another identifier? Actually, all you need to do is change the value that is stored in the memory location named PriceOfShoes. You can use the same memory location and identifier but vary the value that is stored. Memory locations that are used in the fashion are known as **variables**. Be careful not to confuse variables and identifiers. Identifiers are only the names given to variables, but variables are the actual memory locations used to store data. You can compare this with cities on a map. Cities represent a specific location on a map just like variables represent a specific location in memory. But when we refer to a city, we use a name like Richmond to identify the city. In the same way, when we refer to a memory location, we use an identifier like PriceOfShoes to name the variable. The contents of our variable can change just like the residents of a city can change. But although the contents change, the name or identifier remains the same.

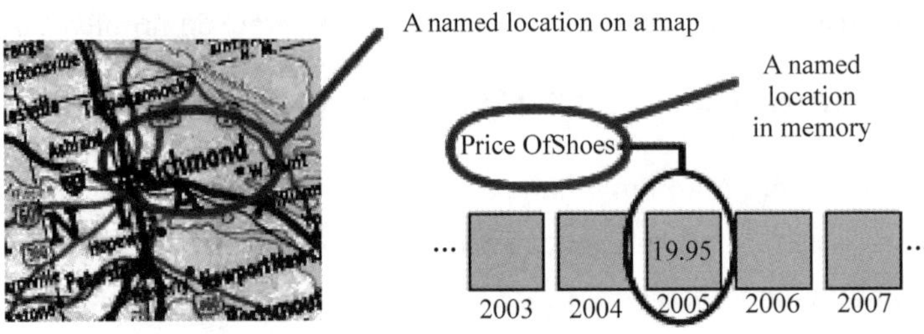

Fig.4-10 A named location in memory

We can also use identifiers to name other things in programming languages beside variables. Another important use for identifiers is giving names to values that remain the same throughout a program. These values are called **constants.** A good example of a value that can be used as a constant is sales tax. In Virginia, the sales tax is 4.5% so we can represent this by creating a constant called TAXRATE and associating it with the value 0.045. Now whenever we need the value for the sales tax, we can use the name TAXRATE to refer to it.

You might be wondering why we even bother to give a name to the value for sales tax. Why not simply use the value 0.045? The answer is that by using constants, we can easily modify our program to run under different conditions. For example, suppose a friend who lives in Tennessee wanted to use your program to compute sales tax. Since the sales tax in Tennessee is 6%, he will need to change all the sales tax computations to use 0.06 instead of 0.045. This modification is very easy if the program uses the constant TAXRATE. All he needs to do is associate the identifier TAXRATE with the new value of 0.06. Another advantage of using constants is that they make our programs more readable. The identifier TAXRATE is much more readily understood than the number 0.06.

As we stated before, both variables and constants are named using identifiers. However, the purpose of variables and constants is very different. Tab.4-5 below compares variables and constants to highlight their different roles.

Tab.4-5 Comparison of variables and constants

	Variables	**Constants**
Definition	a named memory location	a named value
Use	value can change while a program is running	value always remains the same while a program is running
Purpose	allows data to be stored in computer memory	allows programs to be easily modified

One last note about identifiers: In most programming languages, identifiers are required to conform to a certain format. For example, the identifiers in this lesson all began with letters and were composed only of letters and numbers. None of the identifiers included spaces, and constants were written in UPPERCASE letters. This is a typical format for identifiers, and we will use this format in the rest of the lessons.

2. Assignment and Expressions

Variables are named memory locations whose value can change while a program executes whereas constants are named values which cannot change while a program executes. But how do we tell the computer the value that a variable should contain or that a constant should represent? This is accomplished through the **assignment operation**.

In many high-level languages like C++ and Java, the assignment operation is done by placing an equal sign between a value and a variable like this:

```
PriceOfShoes = 19.95
```

On the left side of the assignment operator is the name of a variable or constant while the right side has the value to be assigned. Also, we can assign a variable to another variable, like: OtherPrice = PriceOfOtherShoes, after this assignment operation is complete, the variable OtherPrice contains a copy of the value stored in PriceOfOtherShoes. We can summarize a general rule for assignment in Tab.4-6.

Tab.4-6 A general rule for assignment

Left Side	Assignment Operator	Right Side
Variable or CONSTANT	=	Expression, **Variable**, or CONSTANT

The right side of the table allows us to use something called an **expression**. Usually the result of an expression is stored in another variable. You can think of expressions as being very similar to equations in mathematics, eg. `result = 6 + 10 / 2 - 4 * 2`.

How do we know which operations in the expression to calculate first? The answer to this question is solved by the rules of **operator precedence** in a programming language. These rules define which operations must be performed first. In mathematics, the multiplication and division operations are performed first and then addition and subtraction are performed. However, we can change the order of operations by enclosing a part of the expression in **parentheses**.

We can also have expressions that represent values which are either true or false. These expressions are called **Boolean expressions**. Usually such expressions involve the comparison of values with the following operators: > (greater than), < (less than), and = (equals), >= (greater than or equal to) and <= (less than or equal to).

Boolean expressions can be joined together to form longer expressions using three **logical operators**: AND, OR, and NOT. Suppose we want to test our count variable to determine if the count is between 10 and 100. We can write a Boolean expression using the AND operator to construct our test:

```
(count > 10) AND (count < 100)
```

In order for our entire Boolean expression to be true, both of the smaller expressions must be true. If count is equal to 1,000, then only the first half of the expression is true, and the entire expression evaluates to false. This is exactly what we want because 1,000 is not between 10 and

100. If count is equal to 1, then only the second half of the expression is true, and the entire expression evaluates to false. This is also what we want since 1 is also not between 10 and 100.

Now let's consider how we could write a Boolean expression to test whether count is outside the range 10 to 100. One easy solution would be to reverse the test we have already created using the NOT operator. This operator turns true values to false and false values to true. Therefore, if we want to construct the opposite test, we can reverse the test we already have like this:

```
NOT ((count > 10) AND (count < 100))
```

Although constructing this new test only requires a small change to our previous test, the meaning of the new test is not very clear. A better way to construct this test is to use the OR operator. What we really want to test is when count is greater than 100 or less than 10.

```
(count < 10) OR (count > 100)
```

In order for our entire Boolean expression to be true, at least one of the smaller expressions must be true. If count is equal to 1,000, then the second half of the expression is true, and our entire test is also true. This is exactly what we want because 1,000 is not between 10 and 100. If count is equal to 1, then the first half of the expression is true, and the entire test is also true. This is also what we want since 1 is not between 10 and 100.

To summarize, the AND operation is only true when it joins two true Boolean expressions. The OR operation is only false when it joins two false expressions. The NOT operation always reverses the truth value of the expression.

3. Control Structure

(4-4) In many programming languages, the **sequential flow** of control is the default behavior. However, we often need to alter this flow when we write programs. To do this, most programming languages provide at least three control structures for altering the sequential flow of control. These control structures are known as **selection**, **loop**, and **subprogram**. Combining these control structures with the default sequential flow of control, we have four ways of specifying the flow of control in a program, we will briefly introduce Program Flow of Control below.

(1) Sequence flow is composed of a series of **statements** which are executed one by one from top to bottom. Sequence is the default flow of control for many programming languages. All of the programs illustrated so far have used this flow of control for their execution.

(2) Selection is used to alter the flow of control when a choice needs to be made between two or more actions. Often the choice is based on the state of some variables in the program. This control structure is commonly specified using the **keywords If** and **Else**.

(3) Loop is a control structure that causes a set of statements to be executed repeatedly. With each **loop iteration**, a test is performed to determine whether the loop should continue or

end. Often this control structure is specified using the key word **While**.

(4) Subprograms are a way of grouping statements that provide a single logical operation. An example subprogram might be **SquareRoot** which could find the square root of a number and return the result to the main program. The keyword **Call** indicates a subprogram.

4. Input/Output

Almost every programming language handles program input/output (I/O) differently. However, what are common to most programming languages are the two types of I/O: user I/O and file I/O.

1) User I/O

User I/O is interactive since this type of input is generally accomplished via the keyboard or mouse and a monitor. A typical scenario for user I/O is a **prompt** on the computer monitor asking a user for some needed information. Once this information is entered, the program assigns the input to a variable and continues its execution. A very common prompt for user input and output is called a **command prompt**.

2) File I/O

Often there are times when we need to give input to a program that does not need to be interactive. For example, suppose we had a program that calculated the grades for a class with 200 students. If the program prompted us to enter the grade for each student, we would quickly grow bored and give up. Instead what we would like to do is collect all the information in one place and allow the program to read what it needs from that location rather than asking us for the data. The solution to our problem is file I/O.

File I/O allows us to store data in a file on the computer and then read the data from this file. This type of input is not interactive since the computer does not wait for a response from the user. Instead, the data is read directly from a file specified in the program. Notice how this is different from user I/O. With user I/O, the program halts each time it reaches an input statement. Execution cannot continue until the program receives a response from an input device like the keyboard. However, file I/O allows the computer to get the input for itself without the help of the user so the program does not need to halt at each input. This makes file I/O ideal for processing large amounts of data at once.

As user I/O, file I/O typically is done using a collection of **procedures** that are defined by a given programming language. Of course, these procedures will need an extra parameter that tells them which file to use for input or output.

5. Programs

Now that you have seen all the pieces of programming languages that make up a typical computer program, let's consider how we put these together to make a program. You are already familiar with the notation used to describe subprograms (procedures and functions), and the notation for a program is really just the same. First, we begin by using the keyword Program and EndProgram to indicate which statements belong to the program. These keywords also tell the computer that this part is the start of our code and should be executed first. Next, we declare the variables that the program will use for storing and processing data. Last, we implement the logic of the program using the three control structures and the language's I/O procedures. The basic outline of a program is given below.

```
Program MyProgram ()
  <variable 1>
  <variable 2>
    ...
  <control logic and I/O calls>
    ...
EndProgram
```

4.3.3 Object-oriented Programming

Object-oriented programming (OOP) is a programming paradigm based on the concept of "**objects**", which may contain data, in the form of fields, often known as **attributes**; and code, in the form of procedures, often known as methods. A feature of objects is that an object's procedures can access and often modify the data fields of the object with which they are associated (objects have a notion of "this" or "self "). In OOP, computer programs are designed by making them out of objects that interact with one another. There is significant diversity of OOP languages, but the most popular ones are class-based, meaning that objects are **instances** of **classes**, which typically also determine their type.

Though some people consider OOP to be a modern programming **paradigm**, the roots go back to 1960s. The first programming language to use objects was Simula 67. There are those who glorify OOP and think that anything which is not programmed in an object oriented way can't be good. On the other hand there are well known computer scientists and specialists who criticize OOP. Alexander Stepanov, who is the primary designer and implementer of the objected oriented C++ Standard Template Library, said that OOP provides a mathematically-limited viewpoint and called it "almost as much of a hoax as Artificial Intelligence".

Many of the most widely used programming languages (such as C++, Java, Python etc.) are multi-paradigm programming languages that support object-oriented programming to a greater or lesser degree, typically in combination with imperative, procedural programming. Significant object-oriented languages include Java, C++, C#, Python, PHP, Ruby, Perl, Object Pascal, Objective-C, Swift, Scala, Common Lisp, and Smalltalk.

Object-oriented Programming uses objects, but not all of the associated techniques and structures are supported directly in languages that claim to support OOP. The features listed below are, however, common among languages considered strongly class- and object-oriented (or multi-paradigm with OOP support), with notable exceptions mentioned.

1. Shared with Non-OOP Predecessor Languages

Object-oriented programming languages typically share low-level features with high-level procedural programming languages (which were invented first). The fundamental tools that can be used to construct a program include:

Variables: that can store information formatted in a small number of built-in data types like **integers** and **alphanumeric characters**. This may include data structures like strings, lists, and hash tables that are either built-in or result from combining variables using memory pointers

Procedures – also known as functions, **methods**, routines, or **subroutines** – that take input, generate output, and manipulate data. Modern languages include structured programming constructs like loops and conditionals.

Modular programming support provides the ability to group procedures into files and modules for organizational purposes. Modules are namespaced so code in one module will not be accidentally confused with the same procedure or variable name in another file or module.

2. Objects and Classes

Languages that support object-oriented programming typically use **inheritance** for code reuse and extensibility in the form of either classes or prototypes. Those that use classes support two main concepts:

(1) Classes – the definitions for the data format and available procedures for a given type or class of object; may also contain data and procedures (known as class methods) themselves, i.e. classes contains the data members and member functions.

(2) Objects – instances of classes.

Objects sometimes correspond to things found in the real world. For example, a graphics program may have objects such as “circle” “square” “menu”. An online shopping system might have objects such as “shopping cart” “customer” and “product”. Sometimes objects represent more abstract entities, like an object that represents an open file, or an object that provides the service of translating measurements from U.S. customary to metric.

Each object is said to be an instance of a particular class (for example, an object with its name field set to “Mary” might be an instance of class Employee). Procedures in object-oriented programming are known as methods; variables are also known as fields, members, attributes, or properties. This leads to the following terms:

(1) **Class variables** – belong to the class as a whole; there is only one copy of each one.

(2) **Instance variables** or attributes – data that belongs to individual objects; every object has its own copy of each one.

(3) **Member variables** – refers to both the class and instance variables that are defined by a particular class.

(4) **Class methods** – belong to the class as a whole and have access only to class variables and inputs from the procedure call.

(5) **Instance methods** – belong to individual objects, and have access to instance variables for the specific object they are called on, inputs, and class variables.

Objects are accessed somewhat like variables with complex internal structure, and in many languages are effectively pointers, serving as actual references to a single instance of said object in memory within a **heap** or stack. They provide a layer of abstraction which can be used to separate internal from external code. External code can use an object by calling a specific instance method with a certain set of input parameters, read an instance variable, or write to an instance variable. Objects are created by calling a special type of method in the class known as a constructor. A program may create many instances of the same class as it runs, which operate independently. This is an easy way for the same procedures to be used on different sets of data.

Object-oriented programming that uses classes is sometimes called class-based programming, while prototype-based programming does not typically use classes. As a result, a significantly different yet analogous terminology is used to define the concepts of object and instance.

3. Dynamic Dispatch/Message Passing

It is the responsibility of the object, not any external code, to select the procedural code to execute in response to a method call, typically by looking up the method at run time in a table associated with the object. This feature is known as dynamic dispatch, and distinguishes an object from an abstract data type (or module), which has a fixed (static) implementation of the operations for all instances. If there are multiple methods that might be run for a given name, it is known as **multiple dispatch**.

A method call is also known as message passing. It is conceptualized as a message (the name of the method and its input parameters) being passed to the object for dispatch.

4. Encapsulation

Encapsulation is an Object Oriented Programming concept that binds together the data and functions that manipulate the data, and that keeps both safe from outside interference and misuse. Data encapsulation led to the important OOP concept of **data hiding**.

If a class does not allow calling code to access internal object data and permits access through methods only, this is a strong form of **abstraction** or information hiding known as encapsulation. Some languages (Java, for example) let classes enforce access restrictions explicitly, for example denoting internal data with the private keyword and designating methods intended for use by code outside the class with the public keyword. Methods may also be designed public, private, or intermediate levels such as protected (which allows access from the same class and its subclasses, but not objects of a different class). In other languages (like Python) this is enforced only by **convention** (for example, private methods may have names that start with an **underscore**). Encapsulation prevents external code from being concerned with the internal workings of an object. This facilitates code refactoring, for example allowing the author of the class to change how objects of that class represent their data internally without changing any external code (as long as "public" method calls work the same way). It also encourages programmers to put all the code that is concerned with a certain set of data in the same class, which organizes it for easy comprehension by other programmers. Encapsulation is a technique that encourages decoupling.

5. Composition, Inheritance, and Delegation

Objects can contain other objects in their instance variables; this is known as **object composition**. For example, an object in the Employee class might contain (point to) an object in the Address class, in addition to its own instance variables like "first_name" and "position". Object composition is used to represent "has-a" relationships: every employee has an address, so every Employee object has a place to store an Address object.

Languages that support classes almost always support inheritance. This allows classes to be arranged in a hierarchy that represents "is-a-type-of" relationships. For example, class Employee might inherit from class Person. All the data and methods available to the parent class also appear in the child class with the same names. For example, class Person might define variables "first_name" and "last_name" with method "make_full_name()". These will also be available in class Employee, which might add the variables "position" and "salary". This technique allows easy re-use of the same procedures and data definitions, in addition to potentially mirroring real-world relationships in an intuitive way. Rather than utilizing database tables and programming subroutines, the developer utilizes objects the user may be more familiar with: objects from their application domain.

Subclasses can override the methods defined by **superclasses**. Multiple inheritance is allowed in some languages, though this can make resolving overrides complicated. Some languages have special support for **mixins**, though in any language with multiple inheritance, a mixin is simply a class that does not represent an is-a-type-of relationship. Mixins are typically used to add the same methods to multiple classes. For example, class UnicodeConversionMixin might provide a method unicode_to_ascii() when included in class FileReader and class WebPageScraper, which don't share a common parent.

Abstract classes cannot be instantiated into objects; they exist only for the purpose of inheritance into other "concrete" classes which can be instantiated. In Java, the final keyword can be used to prevent a class from being subclassed.

The doctrine of composition over inheritance advocates implementing has-a relationships using composition instead of inheritance. For example, instead of inheriting from class Person, class Employee could give each Employee object an internal Person object, which it then has the opportunity to hide from external code even if class Person has many public attributes or methods. Some languages, like Go do not support inheritance at all.

The "open/closed principle" advocates that classes and functions "should be open for extension, but closed for modification".

Delegation is another language feature that can be used as an alternative to inheritance.

6. Polymorphism

Subtyping, a form of **polymorphism**, is when calling code can be agnostic as to whether an object belongs to a parent class or one of its descendants. For example, a function might call "make_full_name()" on an object, which will work whether the object is of class Person or class Employee. This is another type of abstraction which simplifies code external to the class hierarchy and enables strong separation of concerns.

7. Open Recursion

In languages that support open **recursion**, object methods can call other methods on the same object (including themselves), typically using a special variable or keyword called this or self. This variable is late-bound; it allows a method defined in one class to invoke another method that is defined later, in some subclass thereof.

4.4 Software Engineering[①]

In 1967, a NATO study group coined the phrase "software engineering" as a response to their belief that the current software crisis could be solved by adapting existing engineering practices to software development. This crisis was characterized by the consistent development of low quality software which exceeded cost limits and development deadlines. Twenty years later, the **software crisis** was still thriving. Consider the following analysis done by the Government Accounting Office (GAO) in 1979 on the state of management information systems software development [Air Force 1996]. Out of the 163 contractors and 113 government personnel surveyed,

60% of their contracts had schedule overruns,

50% of their contracts had cost overruns,

45% of the software could not be used,

29% of the software was never delivered, and

19% of the software had to be reworked to be used.

Even today the software crisis is a significant problem that software engineering must address.

4.4.1 Life Cycle of Software

A common mistake people make concerning software is assuming that the majority of software development is programming. When they think of programming, their minds conjure up the image of a late-night hacker pounding out code on an old computer in the basement of a musty apartment. While this is certainly one approach to programming, it is hardly the norm and definitely not the way the majority of current software is developed. In fact, programming is only a fraction of the software development process. Today, many other steps are involved in the successful development and deployment of computer software. Taken together, all these steps are referred to as the software life cycle. Often these steps are described by models called **software life cycle models**. In the next two lessons we will examine two of these models: the waterfall model and the spiral model. First, however, we need to describe the basic processes that make up the software life cycle.

---

① http://courses.cs.vt.edu/~csonline/SE/

Most models of the software life cycle include the following six processes: **requirements engineering**, **design**, **programming**, **integration**, **delivery**, and **maintenance.** The list below gives a description of each process.

1. Requirements Engineering

During this process, developers and clients meet to discuss ideas for the new software product. (4-5)Developers use a variety of techniques in order to assess the real needs of the client. One such technique is **rapid prototyping** in which a prototype program is built that can mimic the functionality of the desired software. Using this prototype, clients can better understand how the final product will behave and can determine whether this behavior is what they really need. Unless the requirements engineering process is done properly, the resulting software will not be useful to the client even though it may run correctly. The requirements engineering process is completed when the specifications for the new software product are written in a formal document called the **requirements specification document**.

2. Design

During this process, the developers decide how they will construct the software so that it meets the specifications agreed upon in the requirements specification document. Usually the design of the software goes through several stages in which it becomes progressively more detailed. This approach to design is called stepwise refinement, and it allows the developers to manage the complexity of software by postponing decisions about details as late as possible in order to concentrate on other important design issues. When the design is complete, it is recorded in the design specification document.

3. Programming

During this process, teams of programmers write the actual code of the software. The software is divided into separate units called modules in order to handle the complexity of the programming process. Not only are these teams responsible for coding their modules, they are also responsible for proper documentation describing their code and for testing the code to insure correctness.

4. Integration

During this process, the individual modules of the software product are combined to form the integrated software product. Since the modules were developed separately, testing is crucial to the integration process. Even with a good design, **incompatibilities** between modules are likely to exist. These problems need to be identified and corrected to complete the integration.

5. Delivery

During this process, the developers deliver the completed software to the clients. Usually the clients will conduct acceptance testing on the software to determine whether or not it meets the specifications agreed upon in the requirements specification document. Once the software is accepted, it is installed and used by the client.

6. Maintenance

During this process, the software undergoes various changes after delivery in order to fix bugs, add new functionality, port the software to new platforms, or adapt the software to new technologies. Although it may seem that the software should be finished after delivery, this is far from true. All successful software products evolve over time to meet the changing needs of the clients.

You may be surprised to discover that of all these processes, maintenance dominates the cost of the life cycle. Fig.4-11 shows the relative cost of the processes that make up the software life cycle. Because maintenance costs are so important, many developers are beginning to use design approaches that result in software which is easier to maintain.

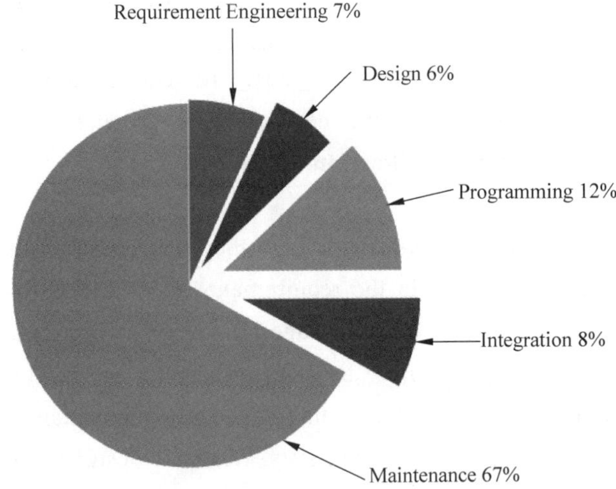

Fig.4-11 Approximate relative costs of the phases of the software life cycle①

4.4.2 Software Development Models

Many different models have been created to represent the software life cycle. Although these models use various names for the processes of the life cycle, they all include the six processes described above in one way or another. In addition, these models usually emphasize some other aspect of software development such as a particular design technique (e.g., rapid prototyping), management technique (e.g., risk management), or the model describes a limited domain of software development (e.g., real-time software).

1. The Waterfall Model

The **Waterfall Model** is the classic software life cycle model, it was the only widely accepted life cycle model until the early 1980s. This model represents the software life cycle using processes and products. Each process transforms a product to produce a new product as output. Then the new product becomes the input of the next process. Tab.4-7 lists the processes and products of the Waterfall Model.

① Schach R. Software Engineering. Fourth Edition. McGraw-Hill, Boston, MA, 1999: 11.

Tab.4-7 Processes and products of the Waterfall Model

Input Product	Process	Output Product
Communicated Requirements	Requirements Engineering	Requirements Specification Document
Requirements Specification Document	Design	Design Specification Document
Design Specification Document	Programming	Executable Software Modules
Executable Software Modules	Integration	Integrated Software Product
Integrated Software Product	Delivery	Delivered Software Product
Delivered Software Product	Maintenance	Changed Requirements

While the Waterfall Model presents a straightforward view of the software life cycle, this view is only appropriate for certain classes of software development. Specifically, the Waterfall Model works well when the software requirements are well understood (e.g., software such as compilers or operating systems) and the nature of the software development involves contractual agreements. The Waterfall Model is a natural fit for contract-based software development since this model is document driven; that is, many of the products such as the requirements specification and the design are documents. These documents then become the basis for the software development contract.

According to Boehm, however, this model "does not work well for many classes of software, particularly interactive end-user applications." Specifying the requirements for such applications is notoriously difficult since interface design is highly subjective and clients rarely ever understand the real needs the software should meet. As a result, "document-driven standards have pushed many projects to write elaborate specifications of poorly understood user interfaces and decision-support functions, followed by the design and development of large quantities of unusable code"[①]. The problem is that a contract is signed before the real requirements of the system are properly understood.

In addition to this shortcoming, the Waterfall Model provides no means for **risk assessment and management** during the life cycle. Consider the case of the baggage-handling system at the Denver International Airport (DIA) again. Initially, DIA had intended that each individual airline would be responsible for building its own baggage-handling system. However, when American Airlines (AA) decided to use DIA as its second-largest hub airport, AA commissioned BAE Automatic Systems to develop an automated baggage-handling system efficient enough to allow AA to turn aircraft around in under thirty minutes. As the construction of the airport progressed, a larger vision emerged "for the inclusion of an airport-wide integrated baggage-handling system that could provide a major improvement in the efficiency of luggage delivery." To accommodate the vision, DIA negotiated a new contract with BAE to develop the airport-wide baggage system. This new plan, however, "underestimated the high complexity of the expanded system, the newness of the technology, and the high level of coordination required among the entities housed

① Boehm B. A Spiral Model of Software Development and Enhancement. IEEE Computer,1988, 21, 5, 61-72.

at DIA that were to be served by the system"[①]. Despite the enormous change in the specifications of the project, no one gave any thought to risk assessment. Had the developers considered the risks involved with changing the system requirements radically at a late stage of development, they may have concluded that the expanded plan was infeasible. In the end, DIA had to settle with a much less ambitious plan, and Montealegre reports that "six months after the de-scaling of the system, the airport was able to open and operate successfully".

2. Spiral Model

In 1988, Barry Boehm proposed a more comprehensive life cycle model called the **Spiral Model** (see Fig.4-12) to address the inadequacies of the Waterfall Model. According to Boehm, "the major distinguishing feature of the Spiral Model is that it creates a **risk-driven** approach to the software process rather than a primarily **document-driven** or **code-driven** process. It incorporates many of the strengths of other models and resolves many of their difficulties". Software projects encompass many different areas of risk such as project cost overruns, changed requirements (e.g., the DIA baggage system), loss of key project personnel, delay of necessary hardware, competition from other software developers, technological breakthroughs which obsolete the project, and many others. The essential concept of the Spiral Model is "to minimize risks by the repeated use of prototypes and other means. Unlike other models, at every stage risk analysis is performed. The Spiral Model works by building progressively more complete versions of the software by starting at the center of the spiral and working outwards. With each loop of the spiral, the customer evaluates the work and suggestions are made for its modification. Additionally, with each loop of the spiral, a risk analysis is performed which results in a 'go / no-go' decision. If the risks are determined to be too great then the project is terminated"[②]. Thus, the Spiral Model addresses the problem of requirements engineering through development of prototypes, and it addresses the need for risk management by performing risk analysis at each step of the life cycle.

The final phase of the Spiral Model is analogous to the Waterfall Model. At this point in the project, the software requirements should be well-understood through the development of several prototypes. The project should also have resolved the major risks involved with building the final version of the software. With these issues resolved, the detailed design of the software enters the last three processes of the Waterfall Model, and the software is created. Although some of the labels in the Spiral Model are different, the same processes are represented.

---

① Montealegre R. "What Can We Learn from the Implementation of the Automated Baggage-Handling System at the Denver International Airport. In Proceedings of the Association for Information Systems Americas Conference, 1996.

② Frankovich J. Synopsis of the Carnegie Mellon University course on Requirements Engineering. http://sern.ucalgary.ca/courses/seng/621/W97/johnf/reqeng.htm.

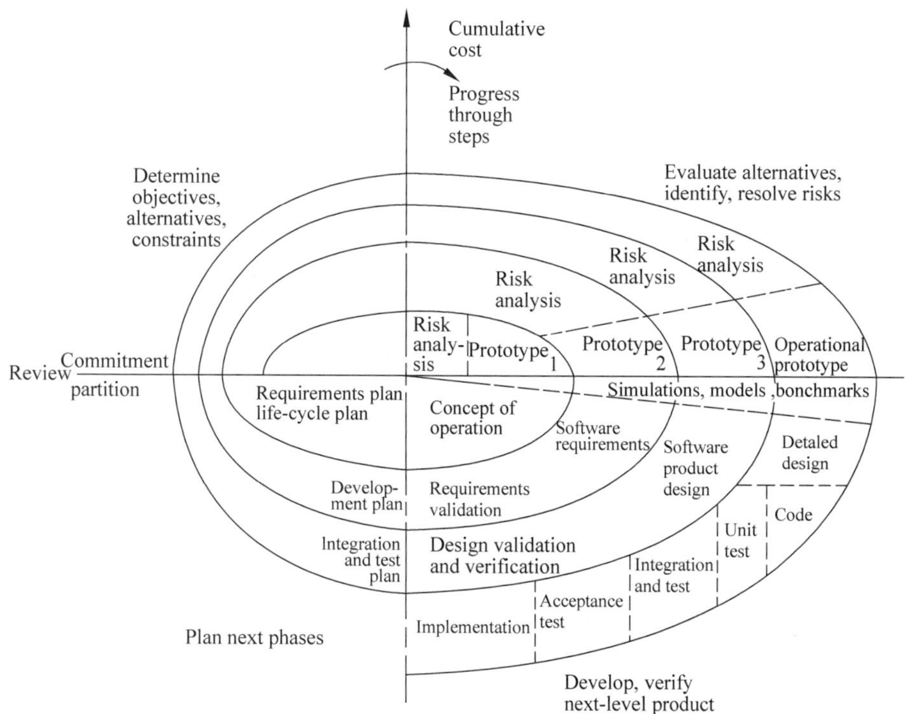

Fig.4-12　The Spiral Model Diagram from Boehm

4.4.3 Software Quality Characteristics

The goal of software engineering is, of course, to design and develop better software. However, what exactly does "better software" mean? In order to answer this question, this lesson introduces some common software quality characteristics. Six of the most important quality characteristics are **maintainability**, **correctness**, **reusability**, **reliability**, **portability**, and **efficiency**.

Maintainability is "the ease with which changes can be made to satisfy new requirements or to correct deficiencies"[①]. Well designed software should be flexible enough to accommodate future changes that will be needed as new requirements come to light. Since maintenance accounts for nearly 70% of the cost of the software life cycle[②], the importance of this quality characteristic cannot be overemphasized. Quite often the programmer responsible for writing a section of code is not the one who must maintain it. For this reason, the quality of the software documentation significantly affects the maintainability of the software product.

Correctness is the degree with which software adheres to its specified requirements. At the start of the software life cycle, the requirements for the software are determined and formalized

① Balci O. Software Engineering Lecture Notes. Department of Computer Science, Virginia Tech, Blacksburg, VA, 1998.

② Schach R. Software Engineering. Fourth Edition. McGraw-Hill, Boston, MA, 1999: 11.

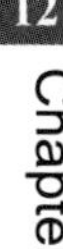

in the requirements specification document. (4-6)Well designed software should meet all the stated requirements. While it might seem obvious that software should be correct, the reality is one of the hardest characteristics to assess. Because of the tremendous complexity of software products, it is impossible to perform exhaustive **execution-based testing** to insure that no errors will occur when the software is run. Also, it is important to remember that some products of the software life cycle such as the design specification cannot be "executed" for testing. Instead, these products must be tested with various other techniques such as formal proofs, inspections, and **walkthroughs**.

Reusability is the ease with which software can be reused in developing other software. By reusing existing software, developers can create more complex software in a shorter amount of time. Reuse is already a common technique employed in other engineering disciplines. For example, when a house is constructed, the trusses which support the roof are typically purchased preassembled. Unless a special design is needed, the architect will not bother to design a new truss for the house. Instead, he or she will simply reuse an existing design that has proven itself to be reliable. In much the same way, software can be designed to accommodate reuse in many situations. A simple example of software reuse could be the development of an efficient sorting routine that can be incorporated in many future applications.

Reliability is "the frequency and **criticality** of software failure, where failure is an unacceptable effect or behavior occurring under permissible operating conditions". The frequency of software failure is measured by the average time between failures. The criticality of software failure is measured by the average time required for repair. Ideally, software engineers want their products to fail as little as possible (i.e., demonstrate high correctness) and be as easy as possible to fix (i.e., demonstrate good maintainability). For some real-time systems such as air traffic control or heart monitors, reliability becomes the most important software quality characteristic. However, it would be difficult to imagine a highly reliable system that did not also demonstrate high correctness and good maintainability.

Portability is "the ease with which software can be used on computer configurations other than its current one"[①]. Porting software to other computer configurations is important for several reasons. First, "good software products can have a life of 15 years or more, whereas hardware is frequently changed at least every 4 or 5 years. Thus good software can be implemented, over its lifetime, on three or more different hardware configurations". Second, porting software to a new computer configuration may be less expensive than developing analogous software from scratch. Third, the sales of "shrink-wrapped software" can be increased because a greater market for the software is available.

Efficiency is "the degree with which software fulfills its purpose without waste of resources"[10]. Efficiency is really a multifaceted quality characteristic and must be assessed with

---

① Balci O. Software Engineering Lecture Notes. Department of Computer Science, Virginia Tech, Blacksburg, VA,1998.

respect to a particular resource such as execution time or storage space. One measure of efficiency is the speed of a program's execution. Another measure is the amount of storage space the program requires for execution. Often these two measures are inversely related, that is, increasing the execution efficiency causes a decrease in the space efficiency. This relationship is known as the **space-time tradeoff**. When it is not possible to design a software product with efficiency in every aspect, the most important resources of the software are given priority.

Fig.4-13 summarizes each of the six quality characteristics. With these characteristics, the answer to the question "What is better software?" becomes much more precise. Using these characteristics, software engineers can assess software products for strengths and weaknesses. In addition, these quality characteristics can also be used to compare and contrast the relative merits of software development paradigms. In this case, software engineers do not refer to the paradigm itself as reliable or portable. Instead, they refer to the software products developed with the paradigm as more reliable or more portable than products developed with another paradigm. The two most prominent software development paradigms are the classical or procedural paradigm and the object oriented paradigm.

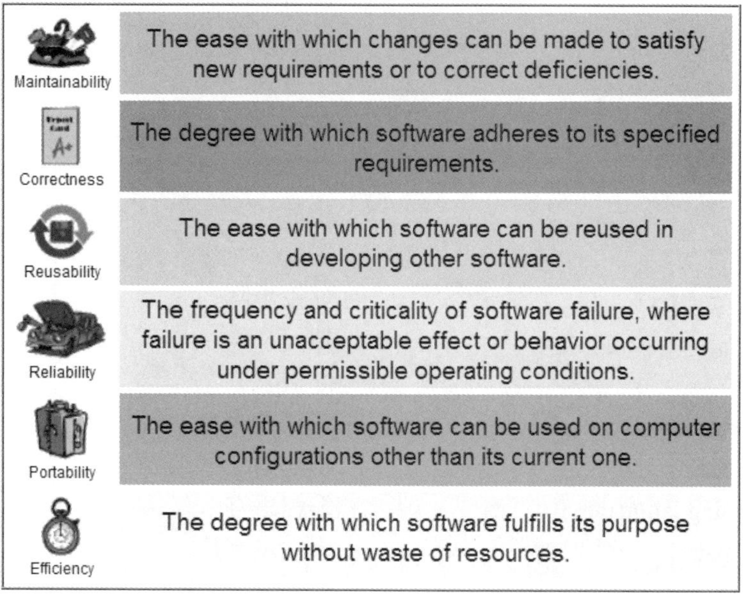

Fig.4-13 summarizations of the six quality characteristics

4.5 Key Terms and Review Questions

1.Technical Terms

primitive operation	元操作，基元	4.1
malfunctioning	出故障，不正常工作	4.1
infinity	无限大	4.1

indent	缩进	4.1
sorting algorithms	排序算法	4.1
contiguous	连续的	4.1
Insertion Sort	插入排序	4.1
space efficiency	空间效率	4.1
time efficiency	时间效率	4.1
memory cells	存储单元	4.1
data structure	数据结构	4.2
list	表，链表	4.2
ordered list	有序表	4.2
stack	栈	4.2
linear data structure	线性数据结构	4.2
nonlinear data structure	非线性数据结构	4.2
tree	树	4.2
root node	根节点	4.2
leaf node	叶节点	4.2
parent nodes	父节点	4.2
child node	子节点	4.2
subtree	子树	4.2
binary tree	二叉树	4.2
pointer	指针	4.2
graph	图	4.2
adjacency matrix	邻接矩阵	4.2
edge	边	4.2
neighbor	邻居	4.2
two-dimensional array	二维数组	4.2
machine code	机器码	4.3
programming language	编程语言	4.3
hexadecimal	十六进制数	4.3
assembly language	汇编语言	4.3
high-level languages	高级语言	4.3
variable	变量	4.3
expression	表达式	4.3
control structure	控制结构	4.3
abstraction	抽象	4.3
identifiers	标识符	4.3
approximation	近似	4.3
constant	常量	4.3
assignment operation	赋值操作	4.3

parentheses	括号	4.3
Boolean expressions	布尔表达式	4.3
logical operator	逻辑运算符	4.3
sequential flow	顺序流程	4.3
selection	选择	4.3
loop	循环	4.3
subprogram	子程序	4.3
statement	语句	4.3
keyword	关键字	4.3
loop iteration	循环迭代	4.3
call	调用	4.3
prompt	提示；提示符	4.3
command prompt	命令提示符	4.3
procedure	过程	4.3
object-oriented	面向对象	4.3
object	对象	4.3
attribute	属性	4.3
instance	实例	4.3
class	类	4.3
paradigm	范例	4.3
alphanumeric characters	字母	4.3
integer	整数	4.3
method	方法	4.3
subroutine	子程序	4.3
routine	程序，例程	4.3
modular programming	模块化编程	4.3
inheritance	继承	4.3
class variable	类变量	4.3
instance variables	实例变量	4.3
member variable	成员变量	4.3
class methods	类方法	4.3
instance method	实例方法	4.3
heap	堆栈	4.3
dynamic dispatch	动态分派	4.3
message passing	消息传递	4.3
multiple dispatch	多分派	4.3
encapsulation	封装	4.3
data hiding	数据隐藏	4.3
convention	约定	4.3

underscore	下划线	4.3
object composition	对象组合	4.3
mixin	混入类	4.3
abstract class	虚类	4.3
subclass	子类	4.3
superclass	超类	4.3
delegation	委托	4.3
polymorphism	多态	4.3
recursion	递归	4.3
software crisis	软件危机	4.4
software life cycle models	软件生命周期模型	4.4
requirements engineering	需求工程，需求分析	4.4
design	设计	4.4
integration	集成电路	4.4
delivery	交付、送达	4.4
maintenance	维护	4.4
rapid prototyping	快速原型法	4.4
requirements specification document	需求规格说明书	4.4
incompatibility	不相容	4.4
Waterfall Model	瀑布模型	4.4
risk assessment and management	风险评估和管理	4.4
Spiral Model	螺旋模型	4.4
risk-driven	风险驱动	4.4
code-driven	代码驱动	4.4
document-driven	文档驱动	4.4
maintainability	可维护性	4.4
correctness	正确性	4.4
reusability	可重用性	4.4
reliability	可靠性	4.4
portability	可移植性	4.4
efficiency	效率	4.4

2. Translation Exercises

(3-1) The first task, managing the hardware and software resources, is very important, as various programs and input methods compete for the attention of the central processing unit (CPU) and demand memory, storage and input/output (I/O) bandwidth for their own purposes. In this capacity, the operating system plays the role of the good parent, making sure that each application gets the necessary resources while playing nicely with all the other applications, as well as husbanding the limited capacity of the system to the greatest good of all the users and applications.

(3-2) Interrupts are special signals sent by hardware or software to the CPU. It's as if some part of the computer suddenly raised its hand to ask for the CPU's attention in a lively meeting. Sometimes the operating system will schedule the priority of processes so that interrupts are masked — that is, the operating system will ignore the interrupts from some sources so that a particular job can be finished as quickly as possible.

(3-3) Trouble can begin if the user tries to have too many processes functioning at the same time. The operating system itself requires some CPU cycles to perform the saving and swapping of all the registers, queues and stacks of the application processes. If enough processes are started, and if the operating system hasn't been carefully designed, the system can begin to use the vast majority of its available CPU cycles to swap between processes rather than run processes. When this happens, it's called *thrashing*, and it usually requires some sort of direct user intervention to stop processes and bring order back to the system.

(3-4) One reason that drivers are separate from the operating system is so that new functions can be added to the driver — and thus to the hardware subsystems — without requiring the operating system itself to be modified, recompiled and redistributed. Through the development of new hardware device drivers, development often performed or paid for by the manufacturer of the subsystems rather than the publisher of the operating system, input/output capabilities of the overall system can be greatly enhanced.

(3-5) The kernel mode is a highly privileged mode of operation in which the program code has direct access to the virtual memory. This includes the address spaces of all user mode processes and applications and their hardware.

(3-6) UNIX users generally configure a system by editing the configuration files with any of the available text editors. The advantage of this mechanism is that the user does not need to learn how to use a large set of configuration tools, but must only be familiar with an editor and possibly a scripting language. The disadvantage is that the information in the files comes in various formats; hence the user must learn the various formats in order to change the settings. UNIX users often employ scripts to reduce the possibility of repetition and error.

(3-7) UNIX has several IPC mechanisms that have different characteristics and which are appropriate for different situations. Shared memory, *pipes*, and *message queues* are all suitable for processes running on a single computer. Shared memory and message queues are suitable for communicating among unrelated processes. Pipes are usually chosen for communicating with a child process through standard input and output. For communications across the network, sockets are usually the chosen technique.

References

[1] Object Oriented Programming[DB/OL].[2016-9-04]. https://en.wikipedia.org/wiki/Object-oriented_ programming.

[2] Algorithms[Z/OL].http://courses.cs.vt.edu/~csonline/Algorithms/Lessons/index.html.

[3] Gibbs W. Software's Chronic Crisis[J]. Scientific American, 1994, 271(3):72-81.

[4] Lyu M. Handbook of Software Reliability Engineering[M]. New York: IEEE Computer Science Press and McGraw-Hill Publishing Company, 1997.

[5] Balci O. Software Engineering Lecture Notes[R]. Department of Computer Science, Virginia Tech, Blacksburg, VA. 1998.

[6] Schach R. Software Engineering(Fourth Edition). Boston: McGraw-Hill,1999:11.

Chapter 5 Databases and Information Retrieval

5.1 Database System

5.1.1 Database

A database is a collection of data that is stored for a specific purpose and organized in a manner that allows its contents to be easily accessed, managed, and updated. Although this definition includes stored data collections such as libraries, file cabinets, and address books, when we talk about databases we almost invariably mean a collection of data that is stored on a computer. There are two basic categories of database. The most commonly encountered category is the **transactional database**, used to store dynamic data, such as inventory contents, which is subject to change on an ongoing basis. The other category is the **analytical database**, used to store static data, such as geographical or chemical test results, which is rarely altered.

Strictly speaking, a database is just the stored data itself, although the term is often used, erroneously, to refer to a database and its management system (DBMS).

The first attempts at computer databases arose around the mid-twentieth century. Early versions were file-oriented. A database file became known as a table because its structure was the same as a paper-based data **table**. For the same reason, the columns within a table were called **fields** and the rows were called **records**. Computers were evolving during that same time period, and their potential for data storage and retrieval was becoming recognized.

The earliest computer databases were based on a flat file model, in which records were stored in text format. In this model, no relationships are defined between records. Without defining such relationships, records can only be accessed sequentially. For example, if you wanted to find the record for the fiftieth customer, you would have to go through the first 49 customer records in sequence first. The flat file model works well for situations in which you want to process all the records, but not for situations in which you want to find specific records within the database.

The **hierarchical model**, widely used in mainframe environments, was designed to allow structured relationships that would facilitate **data retrieval**. Within an inverted tree structure, relationships in the hierarchical model are parent-child and one-to-many. Each parent table may be related to multiple child tables, but each child table can only be related to a single parent table. Because table structures are permanently and explicitly linked in this model, data retrieval was fast. However, the model's rigid structure causes some problems. For example, you can't add a child table that is not linked to a parent table: if the parent table was "Doctors" and the child table was "Patients," you could not add a patient record independently. That would mean that if a new patient came into a community's health care system, under this system, their record could not be added until they had a doctor. The hierarchical structure also means that if a record is deleted in a parent table, all the records linked to it in child tables will be deleted as well.

Also based on an inverted tree structure, the next approach to database design was the **network model**. The network model allowed more complex connections than the hierarchical model: several inverted trees might share branches, for example. The model connected tables in sets, in which a record in an owner table could link to multiple records in a member table. Like the hierarchical model, the network model enabled very fast data retrieval. However it also had a number of problems. For example, a user would need a clear understanding of the database structure to be able to get information from the data. Furthermore, if a set structure was changed, any reference to it from an external program would have to be changed as well.

In the 1970s, the relational database was developed to deal with data in more complex ways. The **relational model** eventually dominated the industry and has continued to do so through to the present day. We'll explore the relational database in some detail in the next segment.

5.1.2 Relational Database①

(5-1) In the relational database model, data is stored in relations, more commonly known as tables. Tables, records (sometimes known as **tuples**), and fields (sometimes known as **attributes**) are the basic components. Each individual piece of data, such as a last name or a telephone number, is stored in a table field and each record comprises a complete set of field data for a particular table. In the following example, the table maintains customer shipping address information. Last_Name and other column headings are the fields. A record, or row, in the table comprises the complete set of field data in that context: all the address information that is required to ship an order to a specific customer. Each record can be identified by, and accessed through, a unique identifier called a primary key. In the Customer_Shipping table, for example, the Customer_ID field could serve as a primary key because each record has a unique value for that field's data.

Customer_Shipping

ID	First	Last	Apt.	Address	City	State	Zip
101	John	Smith	147	123 1st Street	Chicago	IL	60635
102	Jane	Doe	13 C	234 2nd Street	Chicago	IL	60647

① http://whatis.techtarget.com/reference/Learn-IT-The-Power-of-the-Database

103	June	Doe	14A	243 2nd Street	Chicago	IL	60647
104	George	Smith	N/A	345 3rd Street	Chicago	IL	60625

The term relational comes from set theory, rather than the concept that relationships between data drive the database. However, the model does, in fact, work through defining and exploiting the relationships between table data. Table relationships are defined as one-to-one (1∶1), one- to-many (1∶*N*), or (uncommonly) many-to-many (*N*∶*M*).

If a pair of tables has a one-to-one relationship, each record in Table A relates to a single record in Table B. For example, in a table pairing consisting of a table of customer shipping addresses and a table of customer account balances, each single customer ID number would be related to a single identifier for that customer's account balance record. The one-to-one relationship reflects the fact that each individual customer has a single account balance.

If a pair of tables has a one-to-many relationship, each individual record in Table A relates to one or more records in Table B. For example, in a table pairing consisting of a table of university courses (Table A) and a table of student contact information (Table B), each single course number would be related to multiple records of student contact information. The one-to-many relationship reflects the fact that each individual course has multiple students enrolled in it.

If a pair of tables has a many-to-many relationship, each individual record in Table A relates to one or more records in Table B, and each individual record in Table B relates to one or more records in Table A. For example, in a table pairing consisting of a table of employee information and a table of project information, each employee record could be related to multiple project records and each project record could be related to multiple employee records. The many-to-many relationship reflects the fact that each employee may be involved in multiple projects and that each project involves multiple employees.

The relational database model developed from the proposals in "**A Relational Model of Data for Large Shared Databanks**", a paper presented by Dr. E. F. Codd in 1970. Codd, a research scientist at IBM, was exploring better ways to manage large amounts of data than were currently available. The heirarchical and network models of the time tended to suffer from problems with data redundancy and poor data integrity. By applying relational calculus, algebra, and logic to data storage and retrieval, Codd enabled the development of a more complex and fully articulated model than had previously existed.

One of Codd's goals was to create an English-like language that would allow non-technical users to interact with a database. Based on Codd's article, IBM started their System R research group to develop a relational database system. The group developed SQL/DS, which eventually became DB2. The system's language, SQL, became the industry's **de facto standard**. In 1985, Dr. Codd published a list of twelve rules for an ideal relational database. Although the rules may never have been fully implemented, they have provided a guideline for database developers for the last several decades.

Codd's 12 Rules:

(1) The Information Rule: Data must be presented to the user in table format.

(2) Guaranteed Access Rule: Data must be reliably accessible through a reference to the table name, **primary key**, and field name.

(3) Systematic Treatment of Null Values: Fields that are not primary keys should be able to remain empty (contain a null value).

(4) Dynamic On-Line Catalog Based on the Relational Model: The database structure should be accessible through the same tools that provide data access.

(5) Comprehensive Data Sublanguage Rule: The database must support a language that can be used for all interactions (SQL was developed from Codd's rules).

(6) View Updating Rule: Data should be available in different combinations (views) that can also be updated and deleted.

(7) High-level Insert, Update, and Delete: It should be possible to perform all these tasks on any set of data that can be retrieved.

(8) Physical Data Independence: Changes made to the architecture underlying the database should not affect the user interface.

(9) Logical Data Independence: If the logical structure of a database changes, that should not be reflected in the way the user views it.

(10) Integrity Independence: The language used to interact with the database should support user constraints that will maintain data integrity.

(11) Distribution Independence: If the database is distributed (physically located on multiple computers) that fact should not be apparent to the user.

(12) Nonsubversion Rule: It should not be possible to alter the database structure by any other means than the database language.

Although the relational model is by far the most prevalent one, there are a number of other models that are better suited to particular types of data. Alternatives to the relational model include:

Flat-File Databases: Data is stored in files consisting of one or more readable files, usually in text format.

Hierarchical Databases: Data is stored in tables with parent/child relationships with a strictly hierarchical structure.

Network Databases: Similar to the hierarchical model, but allows more flexibility; for example, a child table can be related to more than one parent table.

Object-Oriented Databases: The object-oriented database model was developed in the late 1980s and early 1990s to deal with types of data that the relational model was not well-suited for. Medical and multimedia data, for example, required a more flexible system for data representation and manipulation.

Object-Relational Databases: A hybrid model, combining features of the relational and object-oriented models.

Normalization is a guiding process for database table design that ensures, at four levels of stringency, increasing confidence that results of using the database are unambiguous and as intended. Basically a refinement process, normalization tests a table design for the way it stores data, so that it will not lead to the unintentional deletion of records, for example, and that it will reliably return the data requested.

Normalization degrees of relational database tables are defined as follows.

First normal form (1NF). This is the "basic" level of normalization and generally corresponds to the definition of any database, namely:

(1) It contains two-dimensional tables with rows and columns corresponding, respectively, to records and fields.

(2) Each field corresponds to the concept represented by the entire table: for example, each field in the Customer_Shipping table identifies some component of the customer's shipping address.

(3) No duplicate records are possible.

(4) All field data must be of the same kind. For example, in the "Zip" field of the Customer_Shipping table, only five consecutive digits will be accepted.

Second normal form (2NF). In addition to 1NF rules, each field in a table that does not determine the contents of another field must itself be a function of the other fields in the table. For example, in a table with three fields for customer ID, product sold, and price of the product when sold, the price would be a function of the customer ID (entitled to a discount) and the specific product.

Third normal form (3NF). In addition to 2NF rules, each field in a table must depend on the primary key. For example, using the customer table just cited, removing a record describing a customer purchase (because of a return perhaps) will also remove the fact that the product has a certain price. In the third normal form, these tables would be divided into two tables so that product pricing would be tracked separately. The customer information would depend on the primary key of that table, Customer_ID, and the pricing information would depend on the primary key of that table, which might be Invoice_Number.

Domain/key normal form (DKNF). In addition to 3NF rules, a key, which is a field used for sorting, uniquely identifies each record in a table. A domain is the set of permissible values for a field. By enforcing key and domain restrictions, the database is assured of being freed from modification anomalies. DKNF is the normalization level that most designers aim to achieve.

5.1.3 Database Management System

(5-2) A **Database Management** System is also known as DBMS, sometimes called a database manager or database system, which is a set of computer programs that controls the creation, organization, maintenance, and retrieval of data from the database stored in a computer. An excellent database system helps the end users to easily access and uses the data and also stores the new data in a systematic way.

A DBMS is a system software package that ensures the integrity and security of the data.

The most typical DBMS is a relational database management system (RDBMS). A newer kind of DBMS is the object-oriented database management system (ODBMS). The DBMS are categorized according to their data types and structure. It accepts the request for the data from an application program and instructs the operating system to transfer the appropriate data to the end user. A standard user and program interface is the **Structured Query Language** (SQL).

There are many Data Base Management System like MySQL, PostgreSQL, Microsoft Access, SQL Server, FileMaker, Oracle, RDBMS, dBASE, Clipper, FoxPro and many more that work independently and freely but also allow other database systems to be integrated with them. For this DBMS software comes with an **Open Database Connectivity** (ODBC) driver ensuring the databases to be integrated with it.

A DBMS includes four main parts: **modeling language**, data structure, **database query language**, and **transaction mechanisms modeling language**.

(1) Modeling language: A data modeling language to define the schema (the overall structure of the database) of each database hosted in the DBMS, according to the DBMS database model. The schema specifies data, data relationships, data semantics, and consistency constraints on the data. The four most common types of models are the:

① Hierarchical model

② Network model

③ Relational model

④ Object model

The optimal structure depends on the natural organization of the application's data, and on the application's requirements that include transaction rate (speed), reliability, maintainability, scalability, and cost.

(2) Data structures: Data structures which include fields, record, files and objects optimized to deal with very large amounts of data stored on a permanent data storage device like hard disks, CDs, DVDs, Tape etc.

(3) Database query language: Using the Database Query Language (DQL) users can formulate requests and generate reports. It also controls the security of the database. The DQL and the report writer allows users to interactively interrogate the database, analyze its data and update it according to the users privileges on data. For accessing and using personal records there is a need of password to retrieve the individual records among the bunch of records. For example: the individual records of each employee in a factory.

(4) Transaction mechanisms modeling language: The transaction mechanism modeling language ensures about data integrity despite concurrent user accesses and faults. It maintains the integrity of the data in the database by not allowing more than one user to update the same record at the same time. The unique index constraints prevent to retrieve the duplicate records like no two customers with the same customer numbers (key fields) can be entered into the database.

Among several types of DBMS, Relational Database Management System (RDBMS) and

Object-Oriented Database Management System (OODBMS) are the most commonly used DBMS software.

The RDBMS is a Database Management System (DBMS) based on the relational model in which data is stored in the form of tables and the relationship among the data is also stored in the form of tables. It was introduced by E. F. Codd, which is the most popular commercial and open source databases nowdays. The most popular RDBMS is: MySQL,PostgreSQL, Firebird,SQLite,DB2,Oracle Tutorials.

Object-Oriented Database Management System (OODBMS) in short Object Database Management System (ODBMS) is a Database Management System (DBMS) that supports the modeling and creation of data as objects. It includes some kind of support for classes of objects and the inheritance of class properties and methods by subclasses and their objects. An ODBMS must satisfy two conditions: it should be an object-oriented programming language and a DBMS too.

OODBMS extends the object programming language with transparently persistent data, concurrency control, data recovery, associative queries, and other database capabilities. At present it is on its development stage and used in Java and other object oriented programming languages.

Earlier it was introduced to replace the RDBMS due to its better performance and **scalability** but the inclusion of object-oriented features in RDBMS and the origin of Object-Relational Mappers (ORMs) made it powerful enough to defend its persistence. The higher switching cost also played a vital role to defend the existence of RDBMS. Now it is being used as a complement, not a replacement for relational databases.

Now it is being used in embedded persistence solutions in devices, on clients, in packaged software, in real-time control systems, and to power websites. The open source community has created a new wave of enthusiasm that's now fueling the rapid growth of ODBMS installations.

5.1.4 SQL

SQL is short for Structured Query Language, it is a domain-specific language used in programming and designed for managing data held in a Relational Database Management System (RDBMS).

(5-3) Originally based upon relational algebra and tuple relational calculus, SQL consists of a data definition language, data manipulation language, and data control language. The scope of SQL includes data insert, query, update and delete, schema creation and modification, and data access control. Although SQL is often described as, and to a great extent is, a **declarative language**, it also includes procedural elements.

SQL became a standard of the American National Standards Institute (ANSI) in 1986, and of the International Organization for Standardization (ISO) in 1987. Since then, the standard has been revised to include a larger set of features. Despite the existence of such standards, most SQL code is not completely portable among different database systems without adjustments.

Data Definition Language: The Data Definition Language is used to create, alter, drop or delete a table in database. This includes a index through which you can make a link between tables in database. The list of Data Definition Language is given below.

(1) Create Table is used to create a table in database.

(2) Update is used to alter the table.

(3) Drop Table is used to drop a table in database.

Data Manipulation Language: The Data Manipulation Language is used to select, insert into, delete, update from database table. The keywords perform the particular task are given below.

(1) Select is used to display the records available in table.

(2) Update is used to modify or change in table.

(3) Delete is used to delete record from table.

(4) Insert into is used to add records into table.

The Data Control Language (DCL) authorizes users to access and manipulate data. Its two main statements are:

(1) GRANT authorizes one or more users to perform an operation or a set of operations on an object.

(2) REVOKE eliminates a grant, which may be the default grant.

SQL is subdivided into several language elements, including:

(1) **Clauses**, which are constituent components of statements and queries. (In some cases, these are optional) .

(2) Expressions, which can produce either scalar values, or tables consisting of columns and rows of data.

(3) **Predicates**, which specify conditions that can be evaluated to SQL three-valued logic (3VL) (true/false/unknown) or Boolean truth values and are used to limit the effects of statements and queries, or to change program flow.

(4) Queries, which retrieve the data based on specific criteria. This is an important element of SQL.

(5) Statements, which may have a persistent effect on **schemata** and data, or may control transactions, program flow, connections, sessions, or diagnostics.

(6) SQL statements also include the semicolon (";") statement terminator. Though not required on every platform, it is defined as a standard part of the SQL grammar.

(7) Insignificant whitespace is generally ignored in SQL statements and queries, making it easier to format SQL code for readability.

5.2 Information Retrieval

5.2.1 Introduction

The meaning of the term **information retrieval** can be very broad. Just getting a credit card out of your wallet so that you can type in the card number is a form of information retrieval. However, as an academic field of study, information retrieval might be defined thus:

Information Retrieval (IR) is finding material (usually documents) of an unstructured nature (usually text) that satisfies an information need from within large collections (usually stored on computers).

(5-4) Information retrieval is concerned with representing, searching, and manipulating large collections of electronic text and other human-language data. IR systems and services are now widespread, with millions of people depending on them daily to facilitate business, education, and entertainment. Web search engines — Google, Bing, and others — are by far the most popular and heavily used IR services, providing access to up-to-date technical information, locating people and organizations, summarizing news and events, and simplifying comparison shopping. Digital library systems help medical and academic researchers learn about new journal articles and conference presentations related to there are as of research. Consumers turn to local search services to find retailers providing desired products and services. Within large companies, enterprise search systems act as repositories for e-mail, memos, technical reports, and other business documents, providing corporate memory by preserving these documents and enabling access to the knowledge contained within them. Desktop search systems permit users to search their personal e-mail, documents, and files.

Information retrieval is fast becoming the dominant form of information access, overtaking traditional database-style searching (the sort that is going on when a clerk says to you: "I'm sorry, I can only look up your order if you can give me your Order ID").

IR can also cover other kinds of data and information problems beyond that specified in the core definition above. The term "**unstructured data**" refers to data which does not have clear, **semantically** overt, easy-for-a-computer structure. It is the opposite of structured data, the **canonical** example of which is a relational database, of the sort companies usually use to maintain product inventories and personnel records. In reality, almost no data are truly "unstructured". This is definitely true of all text data if you count the latent linguistic structure of human languages. But even accepting that the intended notion of structure is overt structure, most text has structure, such as headings and paragraphs and footnotes, which is commonly represented in documents by explicit markup (such as the coding underlying Web pages). IR is also used to facilitate "**semistructured**" search such as finding a document where the title contains Java and the body contains threading.

The field of information retrieval also covers supporting users in browsing or filtering document collections or further processing a set of retrieved documents. Given a set of documents, clustering is the task of coming up with a good grouping of the documents based on their contents. It is similar to arranging books on a bookshelf according to their topic. Given a set of topics, standing information needs, or other categories (such as suitability of texts for different age groups), classification is the task of deciding which class(es), if any, each of a set of documents belongs to. It is often approached by first manually classifying some documents and then hoping to be able to classify new documents automatically.

Information retrieval systems can also be distinguished by the scale at which they operate, and it is useful to distinguish three prominent scales. In **Web search**, the system has to provide search over billions of documents stored on millions of computers. Distinctive issues are needed to gather documents for indexing, being able to build systems that work efficiently at this enormous scale, and handling particular aspects of the Web, such as the exploitation of hypertext and not being fooled by site providers manipulating page content in an attempt to boost their search engine rankings, given the commercial importance of the Web. At the other extreme is **personal information retrieval**. In the last few years, consumer operating systems have integrated information retrieval (such as Apple's Mac OS X Spotlight or Windows Vista's Instant Search). E-mail programs usually not only provide search but also text classification: they at least provide a **spam (junk mail)** filter, and commonly also provide either manual or automatic means for classifying mail so that it can be placed directly into particular folders. Distinctive issues here include handling the broad range of document types on a typical personal computer, and making the search system maintenance free and sufficiently lightweight in terms of startup, processing, and disk space usage that it can run on one machine without annoying its owner. In between is the space of enterprise, institutional, and domain-specific search, where retrieval might be provided for collections such as a corporation's internal documents, a database of patents, or research articles on biochemistry. In this case, the documents will typically be stored on centralized file systems and one or a handful of dedicated machines will provide search over the collection.

5.2.2 An Example of Information Retrieval①

A fat book which many people own is **Shakespeare's Collected Works**. Suppose you wanted to determine which plays of Shakespeare contain the words Brutus AND Caesar and NOT Calpurnia. One way to do that is to start at the beginning and to read through all the text, noting for each play whether it contains Brutus and Caesar and excluding it from consideration if it contains Calpurnia. The simplest form of document retrieval is for a computer to do this sort of linear scan through documents. This process is commonly referred to as grepping through text, after the UNIX command **grep**, which performs this process. Grepping through text can be a very effective process, especially given the speed of modern computers, and often allows useful possibilities for **wildcard** pattern matching. With modern computers, for simple querying of modest collections (the size of **Shakespeare's Collected Works** is a bit under one million words of text in total), you really need nothing more.

But for many purposes, you do need more:

(1) To process large document collections quickly. The amount of online data has grown at

---

① Christopher D et al. Introduction to Information Retrieval. Cambridge University Press, 2008. http://nlp. stanford. edu/ IR-book/.

least as quickly as the speed of computers, and we would now like to be able to search collections that total in the order of billions to trillions of words.

(2) To allow more flexible matching operations. For example, it is impractical to perform the query Romans NEAR countrymen with grep, where NEAR might be defined as "within 5 words" or "within the same sentence".

(3) To allow ranked retrieval: in many cases you want the best answer to an information need among many documents that contain certain words.

The way to avoid linearly scanning the texts for each query is to **index** the documents in advance. Let us stick with **Shakespeare's Collected Works**, and use it to introduce the basics of the Boolean retrieval model. Suppose we record for each document — here a play of Shakespeare's — whether it contains each word out of all the words Shakespeare used (Shakespeare used about 32,000 different words). The result is a binary term-document incidence matrix, as in Fig.5-1. Terms are the indexed units; they are usually words, and for the moment you can think of them as words, but the information retrieval literature normally speaks of terms because some of them, such as perhaps I-9 or Hong Kong are not usually thought of as words. Now, depending on whether we look at the matrix rows or columns, we can have a vector for each term, which shows the documents it appears in, or a vector for each document, showing the terms that occur in it.

	Antony and Cleopatra	Julius Caesar	The Tempest	Hamlet	Othello	Macbeth	...
Antony	1	1	0	0	0	1	
Brutus	1	1	0	1	0	0	
Caesar	1	1	0	1	1	1	
Calpurnia	0	1	0	0	0	0	
Cleopatra	1	0	0	0	0	0	
mercy	1	0	1	1	1	1	
worser	1	0	1	1	1	0	
...							

Fig.5-1 A term-document incidence matrix

To answer the query Brutus AND Caesar AND NOT Calpurnia, we take the vectors for Brutus, Caesar and Calpurnia, complement the last, and then do a bitwise AND:

110100 AND 110111 AND 101111 = 100100

The answers for this query are thus Antony and Cleopatra and Hamlet (Fig.5-1).

The **Boolean retrieval model** is a model for information retrieval in which we can pose any query which is in the form of a Boolean expression of terms, that is, in which terms are combined with the operators and, or, and not. The model views each document as just a set of words.

Let us now consider a more realistic scenario, simultaneously using the opportunity to introduce some **terminology** and **notation**. Suppose we have **N**=1million documents. By documents we mean whatever units we have decided to build a retrieval system over. They might be individual memos or chapters of a book. We will refer to the group of documents over which we perform retrieval as the (document) collection. It is sometimes also referred to as a **corpus** (a body of texts). Suppose each document is about 1,000 words long (2~3 book pages). If we assume an average of 6 B per word including spaces and punctuation, then this is a document collection about 6 GB in size. Typically, there might be about **M** = 500,000 distinct terms in these documents. There is nothing special about the numbers we have chosen, and they might vary by an order of **magnitude** or more, but they give us some idea of the dimensions of the kinds of problems we need to handle.

Our goal is to develop a system to address the ad hoc retrieval task. This is the most standard IR task. In it, a system aims to provide documents from within the collection that are relevant to an arbitrary user information need, communicated to the system by means of a one-off, user-initiated query. An information need is the topic about which the user desires to know more, and is differentiated from a query, which is what the user conveys to the computer in an attempt to communicate the information need. A document is relevant if it is one that the user perceives as containing information of value with respect to their personal information need. Our example above was rather artificial in that the information need was defined in terms of particular words, whereas usually a user is interested in a topic like "pipeline leaks" and would like to find relevant documents regardless of whether they precisely use those words or express the concept with other words such as pipeline rupture. (5-5)To assess the effectiveness of an IR system (i.e., the quality of its search results), a user will usually want to know two key statistics about the system's returned results for a query:

(1) **Precision** : What fraction of the returned results are relevant to the information need?

(2) **Recall** : What fraction of the relevant documents in the collection were returned by the system?

We now cannot build a term-document matrix in a naive way. A 500K×1M matrix has half-a-trillion 0's and 1's - too many to fit in a computer's memory. But the crucial observation is that the matrix is extremely **sparse**, that is, it has few non-zero entries. Because each document is 1,000 words long, the matrix has no more than one billion 1's, so a minimum of 99.8% of the cells are zero. A much better representation is to record only the things that do occur, that is, the 1's positions.

This idea is central to the first major concept in information retrieval, the **inverted index**. The name is actually redundant: an index always maps back from terms to the parts of a document where they occur. Nevertheless, inverted index, or sometimes inverted file, has become the standard term in information retrieval. The basic idea of an inverted index is shown in Fig.5-2. We keep a dictionary of terms (sometimes also referred to as a vocabulary or **lexicon**; in this book, we use dictionary for the data structure and vocabulary for the set of terms). Then for each term, we have a list that records which documents the term occurs in. Each item in the list — which records that a term appeared in a document (and, later, often, the positions in the document) — is conventionally called a posting. The list is then called a **postings list**, and all the postings lists taken together are referred to as the postings. The dictionary in Fig.5-2 has been sorted alphabetically and each postings list is sorted by document ID.

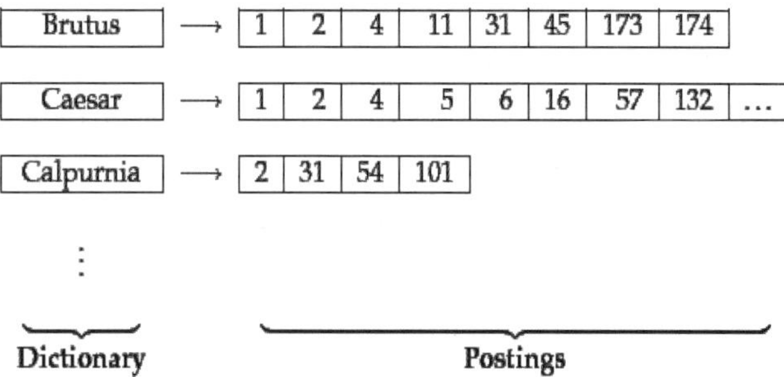

Fig.5-2 the two parts of an inverted index

To gain the speed benefits of indexing at retrieval time, we have to build the index in advance. The major steps in this are:

(1) Collect the documents to be indexed:

Friends, Romans, countrymen, So let it be with Caesar …

(2) Tokenize the text, turning each document into a list of tokens:

Friends Romans countrymen So let it …

(3) Do linguistic preprocessing, producing a list of normalized tokens, which are the indexing terms.

(4) Index the documents that each term occurs in by creating an inverted index, consisting of a dictionary and postings.

You can think of tokens and normalized tokens as also loosely equivalent to words. Here, we assume that the first 3 steps have already been done, and we examine building a basic inverted index by sort-based indexing as Fig.5-3.

Doc 1
I did enact Julius Caesar: I was killed i' the Capitol; Brutus killed me.

Doc 2
So let it be with Caesar. The noble Brutus hath told you Caesar was ambitious:

term	docID
I	1
did	1
enact	1
julius	1
caesar	1
I	1
was	1
killed	1
i'	1
the	1
capitol	1
brutus	1
killed	1
me	1

⟹

term	docID
ambitious	2
be	2
brutus	1
brutus	2
capitol	1
caesar	1
caesar	2
caesar	2
did	1
enact	1
hath	1
I	1
I	1
i'	1

⟹

term	doc. freq.	→	postings lists
ambitious	1	→	2
be	1	→	2
brutus	2	→	1 → 2
capitol	1	→	1
caesar	2	→	1 → 2
did	1	→	1
enact	1	→	1
hath	1	→	2
I	1	→	1
i'	1	→	1
it	1	→	2

Fig.5-3 Building an index by sorting and grouping

Within a document collection, we assume that each document has a unique serial number, known as the document identifier (docID). During index construction, we can simply assign successive integers to each new document when it is first encountered. The input to indexing is a list of normalized tokens for each document, which we can equally think of as a list of pairs of term and docID, as in Fig.5-3. The core indexing step is sorting this list so that the terms are alphabetical, giving us the representation in the middle column of Fig.5-3. Multiple occurrences of the same term from the same document are then merged. Instances of the same term are then grouped, and the result is split into a dictionary and postings, as shown in the right column of Fig.5-3. Since a term generally occurs in a number of documents, this data organization already reduces the storage requirements of the index. The dictionary also records some statistics, such as the number of documents which contain each term (the document frequency, which is here also the length of each postings list). This information is not vital for a basic Boolean search engine, but it allows us to improve the efficiency of the search engine at query time, and it is a statistic later used in many ranked retrieval models. The postings are secondarily sorted by docID. This provides the basis for efficient query processing. This inverted index structure is essentially without rivals as the most efficient structure for supporting ad hoc text search.

In the resulting index, we pay for storage of both the dictionary and the postings lists. The latter are much larger, but the dictionary is commonly kept in memory, while postings lists are normally kept on disk, so the size of each is important, we need examine how each can be optimized for storage and access efficiency. What data structure should be used for a postings list? A fixed length array would be wasteful as some words occur in many documents, and others in very few. For an in-memory postings list, two good alternatives are singly **linked lists** or **variable length arrays**. Singly linked lists allow cheap insertion of documents into postings lists (following updates, such as when recrawling the Web for updated documents), and naturally extend to more advanced indexing strategies such as skip lists, which require additional pointers. Variable length arrays win in space requirements by avoiding the overhead for pointers and in time requirements because their use of contiguous memory increases speed on modern processors with memory caches. Extra pointers can in practice be encoded into the lists as **offsets**. If updates are relatively infrequent, variable length arrays will be more compact and faster to traverse. We can also use a hybrid scheme with a linked list of fixed length arrays for each term. When postings lists are stored on disk, they are stored (perhaps compressed) as a contiguous run of postings without explicit pointers (as in Fig.5-2), so as to minimize the size of the postings list and the number of disk seeks to read a postings list into memory.

How do we process a query using an inverted index and the basic Boolean retrieval model? Consider processing the simple conjunctive query:

Brutus AND Calpurnia

over the inverted index partially shown in Fig.5-3. We:

(1) Locate Brutus in the dictionary

(2) Retrieve its postings

(3) Locate Calpurnia in the dictionary

(4) Retrieve its postings

(5) Intersect the two postings lists, as shown in Fig.5-4.

```
INTERSECT(p1, p2)
answer ←<>
while p1 ≠ nil and p2 ≠ nil
   do if docID(p1) = docID(p2)
         then ADD(answer, docID(p1))
              p1 ← next(p1)
              p2 ← next(p2)
         else if docID(p1) < docID(p2)
                 then p1 ← next(p1)
                 else p2 ← next(p2)
return answer
```

Fig.5-4 Algorithm for the intersection of two postings lists p_1 and p_2

The intersection is the crucial one: we need to efficiently intersect postings lists so as to be able to quickly find documents that contain both terms. (This operation is sometimes referred to as merging postings lists: this slightly counterintuitive name reflects using the term merge algorithm for a general family of algorithms that combine multiple sorted lists by interleaved advancing of pointers through each; here we are merging the lists with a logical AND operation.)

5.2.3 Open Source IR System①

There exists a wide variety of open-source information retrieval systems that you may use for exercises in this book and to start conducting your own information retrieval experiments. As always, a (non-exhaustive) list of open-source IR systems can be found in Wikipedia. Since this list of available systems is so long, we do not even try to cover it in detail. Instead, we restrict ourselves to a very brief overview of three particular systems that were chosen because of their popularity, their influence on IR research, or their intimate relationship with the contents of this book. All three systems are available for download from the Web and may be used free of charge, according to their respective licenses.

1. Lucene

Lucene is an indexing and search system implemented in Java, with ports to other programming languages. The project was started by Doug Cutting in 1997. Since then, it has grown from a single-developer effort to a global project involving hundreds of developers in various countries. It is currently hosted by the Apache Foundation. Lucene is by far the most successful open-source search engine. Its largest installation is quite likely Wikipedia: All queries entered into Wikipedia's search form are handled by Lucene. A list of other projects relying on its indexing and search capabilities can be found on Lucene's "PoweredBy" page.

Known for its modularity and extensibility, Lucene allows developers to define their own

① http://www.ir.uwaterloo.ca/book/01-introduction.pdf

indexing and retrieval rules and **formulae**. Under the hood, Lucene's retrieval framework is based on the concept of fields: Every document is a collection of fields, such as its title, body, URL, and so forth. This makes it easy to specify **structured search requests** and to give different weights to different parts of a document.

Due to its great popularity, there is a wide variety of books and tutorials that help you get started with Lucene quickly. Try the query "Lucene tutorial" in your favorite Web search engine.

2. Indri

Indri is an academic information retrieval system written in C++. It is developed by researchers at the University of Massachusetts and is part of the Lemur project, a joint effort of the University of Massachusetts and Carnegie Mellon University.

Indri is well known for its high retrieval effectiveness and is frequently found among the top-scoring search engines at TREC. Its retrieval model is a combination of the language modeling approaches. Like Lucene, Indri can handle multiple fields per document, such as title, body, and anchor text, which is important in the context of Web search. It supports automatic query expansion by means of **pseudo-relevance feedback**, a technique that adds related terms to an initial search query, based on the contents of an initial set of search results. It also supports query-independent document scoring that may, for instance, be used to prefer more recent documents over less recent ones when ranking the search results.

3. Wumpus

Wumpus is an academic search engine written in C++ and developed at the University of Waterloo. Unlike most other search engines, Wumpus has no built-in notion of "documents" and does not know about the beginning and the end of each document when it builds the index. Instead, every part of the text collection may represent a potential unit for retrieval, depending on the structural search constraints specified in the query. This makes the system particularly attractive for search tasks in which the ideal search result may not always be a whole document, but may be a section, a paragraph, or a sequence of paragraphs within a document.

Wumpus supports a variety of different retrieval methods, including the proximity ranking function, the BM25 algorithm, and the language modeling and divergence from randomness approaches. In addition, it is able to carry out real-time index updates (i.e., adding/removing files to /from the index) and provides support for multi-user security restrictions that are useful if the system has more than one user, and each user is allowed to search only parts of the index.

5.2.4 Performance Measure

To measure ad hoc information retrieval effectiveness in the standard way, we need a test collection consisting of three things:

(1) A document collection.

(2) A test suite of information needs, expressible as queries.

(3) A set of relevance judgments, standardly a binary assessment of either relevant or nonrelevant for each query-document pair.

The standard approach to information retrieval system evaluation revolves around the notion of **relevant** and **nonrelevant** documents. With respect to a user information need, a document in the test collection is given a binary classification as either relevant or nonrelevant. This decision is referred to as the **gold standard** or **ground truth** judgment of relevance. The test document collection and suite of information needs have to be of a reasonable size: you need to average performance over fairly large test sets, as results are highly variable over different documents and information needs. As a rule of thumb, 50 information needs has usually been found to be a sufficient minimum.

Relevance is assessed relative to an information need, not a query. For example, an information need might be:

Information on whether drinking red wine is more effective at reducing your risk of heart attacks than white wine.

This might be translated into a query such as:

wine AND red AND white AND heart AND attack AND effective

A document is relevant if it addresses the stated information need, not because it just happens to contain all the words in the query. This distinction is often misunderstood in practice, because the information need is not overt. If a user types python into a Web search engine, they might want to know where they can purchase a pet python. Or they might want information on the programming language Python. From a one word query, it is very difficult for a system to know what the information need is. But, nevertheless, the user has one, and can judge the returned results on the basis of their relevance to it. To evaluate a system, we require an overt expression of an information need, which can be used for judging returned documents as relevant or nonrelevant. At this point, we make a simplification: relevance can reasonably be thought of as a scale, with some documents highly relevant and others marginally so. But for the moment, we will use just a binary decision of relevance.

Many systems contain various weights (often known as parameters) that can be adjusted to tune system performance. It is wrong to report results on a test collection which were obtained by tuning these parameters to maximize performance on that collection. That is because such tuning overstates the expected performance of the system, because the weights will be set to maximize performance on one particular set of queries rather than for a random sample of queries. In such cases, the correct procedure is to have one or more development test collections, and to tune the parameters on the development test collection. The tester then runs the system with those weights on the test collection and reports the results on that collection as an unbiased estimate of performance.

5.3 Web Search Basics

5.3.1 Background and History

The Web is unprecedented in many ways: unprecedented in scale, unprecedented in the almost-complete lack of coordination in its creation, and unprecedented in the diversity of backgrounds and motives of its participants. Each of these contributes to making Web search different — and generally far harder — than searching "traditional" documents.

The invention of hypertext, envisioned by Vannevar Bush in the 1940's and first realized in working systems in the 1970's, significantly precedes the formation of the World Wide Web (which we will simply refer to as the Web), in the 1990's. Web usage has shown tremendous growth to the point where it now claims a good fraction of humanity as participants, by relying on a simple, open client-server design: ① the server communicates with the client via a protocol (the http or hypertext transfer protocol) that is lightweight and simple, **asynchronously** carrying a variety of payloads (text, images and — over time — richer media such as audio and video files) encoded in a simple markup language called HTML (for Hypertext Markup Language); ② the client — generally a browser, an application within a graphical user environment — can ignore what it does not understand. Each of these seemingly innocuous features has contributed enormously to the growth of the Web, so it is worthwhile to examine them further.

The basic operation is as follows: a client (such as a browser) sends an http request to a Web server. The browser specifies a URL (for Uniform Resource Locator) such as http://www.stanford.edu/home/atoz/contact.html. In this example URL, the string "http" refers to the protocol to be used for transmitting the data. The string "www.stanford.edu" is known as the domain and specifies the root of a hierarchy of Web pages (typically mirroring a filesystem hierarchy underlying the Web server). In this example, "/home/atoz/contact.html" is a path in this hierarchy with a file "contact.html" that contains the information to be returned by the Web server at www.stanford.edu in response to this request. The HTML — encoded file contact.html holds the hyperlinks and the content (in this instance, contact information for Stanford University), as well as formatting rules for rendering this content in a browser. Such an http request thus allows us to fetch the content of a page, something that will prove to be useful to us for crawling and indexing documents.

The designers of the first browsers made it easy to view the HTML markup tags on the content of a URL. This simple convenience allowed new users to create their own HTML content without extensive training or experience; rather, they learned from example content that they liked. As they did so, a second feature of browsers supported the rapid proliferation of Web content creation and usage: browsers ignored what they did not understand. This did not, as one might fear, lead to the creation of numerous incompatible dialects of HTML. What it did promote was amateur content creators who could freely experiment with and learn from their newly created Web pages without fear that a simple syntax error would "bring the system down". Publishing on the Web became a mass activity that was not limited to a few trained programmers, but rather open to tens and eventually hundreds of millions of individuals. For most users and for most information needs, the Web quickly became the best way to supply and consume information on everything from rare ailments to subway schedules.

The mass publishing of information on the Web is essentially useless unless this wealth of information can be discovered and consumed by other users. Early attempts at making Web information "discoverable" fell into two broad categories: ① full-text index search engines such as Altavista, Excite and Infoseek and ② **taxonomies** populated with Web pages in categories, such as Yahoo!. The former presented the user with a keyword search interface supported by inverted indexes and ranking mechanisms. The latter allowed the user to browse through a hierarchical tree of category labels. While this is at first blush a convenient and intuitive metaphor for finding Web pages, it has a number of drawbacks: first, accurately classifying Web pages into taxonomy tree nodes is for the most part a manual editorial process, which is difficult to scale with the size of the Web. Arguably, we only need to have "high-quality" Web pages in the taxonomy, with only the best Web pages for each category. However, just discovering these and classifying them accurately and consistently into the taxonomy entails significant human effort. Furthermore, in order for a user to effectively discover Web pages classified into the nodes of the taxonomy tree, the user's idea of what sub-tree(s) to seek for a particular topic should match that of the editors performing the classification. This quickly becomes challenging as the size of the taxonomy grows; the Yahoo! taxonomy tree surpassed 1,000 distinct nodes fairly early on. Given these challenges, the popularity of taxonomies declined over time, even though variants (such as About.com and the Open Directory Project) sprang up with subject-matter experts collecting and annotating Web pages for each category.

The first generation of Web search engines transported classical search techniques to the Web domain, focusing on the challenge of scale. The earliest Web search engines had to contend with indexes containing tens of millions of documents, which were a few orders of magnitude larger than any prior information retrieval system in the public domain. Indexing, query serving and ranking at this scale required the harnessing together of tens of machines to create highly available systems, again at scales not witnessed hitherto in a consumer-facing search application. The first generation of Web search engines was largely successful at solving these challenges while continually indexing a significant fraction of the Web, all the while serving queries with sub-second response times. However, the quality and relevance of Web search results left much to be desired owing to the idiosyncrasies of content creation on the Web. This necessitated the invention of new ranking and spam-fighting techniques in order to ensure the quality of the search results. While classical information retrieval techniques continue to be necessary for Web search, they are not by any means sufficient. A key aspect is that whereas classical techniques measure the relevance of a document to a query, there remains a need to gauge the authoritativeness of a document based on cues such as which Web site hosts it.

5.3.2 Web Search Features

The essential feature that led to the explosive growth of the Web-decentralized content publishing with essentially no central control of authorship — turned out to be the biggest challenge for Web search engines in their quest to index and retrieve this content. Web page authors created content in dozens of (natural) languages and thousands of dialects, thus demanding many different forms of stemming and other linguistic operations. Because publishing was now open to tens of millions, Web pages exhibited heterogeneity at a daunting scale, in many crucial aspects. First, content-creation was no longer the privy of editorially-trained writers; while this represented a tremendous democratization of content creation, it also resulted in a tremendous variation in grammar and style (and in many cases, no recognizable grammar or style). Indeed, Web publishing in a sense unleashed the best and worst of desktop publishing on a planetary scale, so that pages quickly became riddled with wild variations in colors, fonts and structure. Some Web pages, including the professionally created home pages of some large corporations, consisted entirely of images (which, when clicked, led to richer textual content) — and therefore, no **indexable text**.

What about the substance of the text in Web pages? The democratization of content creation on the Web meant a new level of **granularity** in opinion on virtually any subject. This meant that the Web contained truth, lies, contradictions and suppositions on a grand scale. This gives rise to the question: which Web page does one trust? In a simplistic approach, one might argue that some publishers are trustworthy and others not — begging the question of how a search engine is to assign such a measure of trust to each Web site or Web page. More subtly, there may be no universal, user-independent notion of trust; a Web page whose contents are trustworthy to one user may not be so to another. In traditional (non-Web) publishing this is not an issue: users self-select sources they find trustworthy. Thus one reader may find the reporting of **The New York Times** to be reliable, while another may prefer **The Wall Street Journal**. But when a search engine is the only viable means for a user to become aware of (let alone select) most content, this challenge becomes significant.

While the question "how big is the Web?" has no easy answer, the question "how many Web pages are in a search engine's index" is more precise, although, even this question has issues. By the end of 1995, Altavista① reported that it had crawled and indexed approximately 30 million static Web pages. Static Web pages are those whose content does not vary from one request for that page to the next. For this purpose, a professor who manually updates his home page every week is considered to have a static Web page, but an airport's flight status page is considered to be dynamic. Dynamic pages are typically mechanically generated by an application server in response to a query to a database, as show in Fig.5-5. One sign of such a page is that the URL has the character "?" in it. Since the number of static Web pages was believed to be doubling every few months in 1995, early Web search engines such as Altavista had to constantly add hardware and bandwidth for crawling and indexing Web pages.

---

① AltaVista was a Web search engine established in 1995. It became one of the most-used early search engines, but lost ground to Google and was purchased by Yahoo! in 2003.

Fig.5-5 Dynamically generated Web page

It is crucial that we understand the users of Web search as well. This is again a significant change from traditional information retrieval, where users were typically professionals with at least some training in the art of phrasing queries over a well-authored collection whose style and structure they understood well. In contrast, Web search users tend to not know (or care) about the heterogeneity of Web content, the syntax of query languages and the art of phrasing queries; indeed, a mainstream tool (as Web search has come to become) should not place such onerous demands on billions of people. A range of studies has concluded that the average number of keywords in a Web search is somewhere between 2 and 3. Syntax operators (Boolean connectives, wildcards, etc.) are seldom used, again a result of the composition of the audience – "normal" people, not information scientists.

It is clear that the more user traffic a Web search engine can attract, the more revenue it stands to earn from **sponsored search**. How do search engines differentiate themselves and grow their traffic? Here Google identified two principles that helped it grow at the expense of its competitors: ① a focus on relevance, specifically precision rather than recall in the first few results; ② a user experience that is lightweight, meaning that both the search query page and the search results page are uncluttered and almost entirely textual, with very few graphical elements. The effect of the first was simply to save users time in locating the information they sought. The effect of the second is to provide a user experience that is extremely responsive, or at any rate not bottlenecked by the time to load the search query or results page.

There appear to be three broad categories into which common Web search queries can be grouped: ① informational, ② navigational and ③ transactional. We now explain these categories; it should be clear that some queries will fall in more than one of these categories, while others will fall outside them.

Informational queries seek general information on a broad topic, such as leukemia or Provence. There is typically not a single Web page that contains all the information sought; indeed, users with informational queries typically try to assimilate information from multiple Web pages.

Navigational queries seek the website or home page of a single entity that the user has in mind, say Lufthansa airlines. In such cases, the user's expectation is that the very first search result should be the home page of Lufthansa. The user is not interested in a plethora of documents containing the term Lufthansa; for such a user, the best measure of user satisfaction is precision at 1.

A transactional query is one that is a prelude to the user performing a transaction on the Web — such as purchasing a product, downloading a file or making a reservation. In such cases, the search engine should return results listing services that provide form interfaces for such transactions.

Discerning which of these categories a query falls into can be challenging. The category not only governs the algorithmic search results, but the suitability of the query for sponsored search results (since the query may reveal an intent to purchase). For navigational queries, some have argued that the search engine should return only a single result or even the target Web page directly. Nevertheless, Web search engines have historically engaged in a battle of bragging rights over which one indexes more Web pages. Does the user really care? Perhaps not, but the media does highlight estimates (often statistically indefensible) of the sizes of various search engines. Users are influenced by these reports and thus, search engines do have to pay attention to how their index sizes compare to competitors'. For informational (and to a lesser extent, transactional) queries, the user does care about the comprehensiveness of the search engine.

Fig.5-6 shows a composite picture of a Web search engine including the crawler, as well as both the Web page and advertisement indexes. The portion of the figure under the curved dashed line is internal to the search engine.

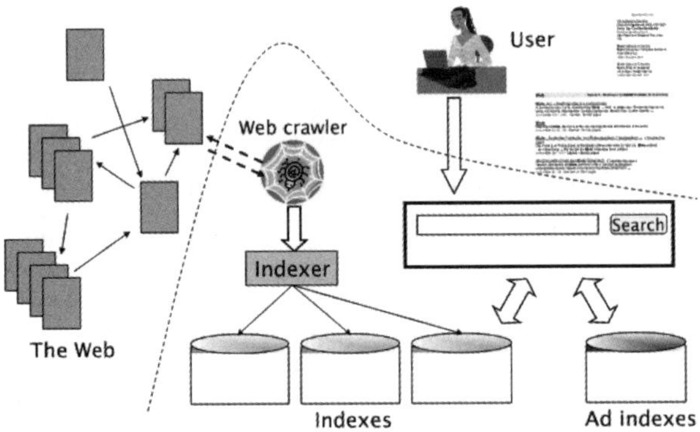

Fig.5-6 The various components of a Web search engine

5.3.3 Web Crawling and Indexes

Web crawling is the process by which we gather pages from the Web, in order to index them and support a search engine. The objective of crawling is to quickly and efficiently gather as many useful Web pages as possible, together with the link structure that interconnects them. The complexities of the Web stem from its creation by millions of uncoordinated individuals. In this section we study the resulting difficulties for crawling the Web. The focus is the component shown in Fig.5-6 as Web crawler ; it is sometimes referred to as a spider.

1. Features a Crawler Must Provide

1) **Robustness**

The Web contains servers that create spider traps, which are generators of Web pages that mislead crawlers into getting stuck fetching an infinite number of pages in a particular domain. Crawlers must be designed to be resilient to such traps. Not all such traps are malicious; some are the inadvertent side-effect of faulty Web site development.

2) Politeness

Web servers have both implicit and explicit policies regulating the rate at which a crawler can visit them. These politeness policies must be respected.

2. Features a Crawler Should Provide

1) Distributed

The crawler should have the ability to execute in a distributed fashion across multiple machines.

2) Scalable

The crawler architecture should permit scaling up the crawl rate by adding extra machines and bandwidth.

3) Performance and Efficiency

The crawl system should make efficient use of various system resources including processor, storage and network bandwidth.

4) Quality

Given that a significant fraction of all Web pages are of poor utility for serving user query needs, the crawler should be biased towards fetching "useful" pages first.

5) Freshness

In many applications, the crawler should operate in continuous mode: it should obtain fresh copies of previously fetched pages. A search engine crawler, for instance, can thus ensure that the search engine's index contains a fairly current representation of each indexed Web page. For such continuous crawling, a crawler should be able to crawl a page with a frequency that approximates the rate of change of that page.

6) Extensible

Crawlers should be designed to be extensible in many ways — to cope with new data formats, new fetch protocols, and so on. This demands that the crawler architecture be modular.

3. Crawling

The basic operation of any hypertext crawler, whether for the Web, an intranet or other hypertext document collection, is as follows. The crawler begins with one or more URLs that constitute a seed set. It picks a URL from this seed set, and then fetches the Web page at that URL. The fetched page is then **parsed**, to extract both the text and the links from the page (each of which points to another URL). The extracted text is fed to a text indexer. The extracted links (URLs) are then added to a URL frontier, which at all times consists of URLs whose corresponding pages have yet to be fetched by the crawler. Initially, the URL frontier contains the seed set; as pages are fetched, the corresponding URLs are deleted from the URL frontier. The entire process may be viewed as traversing the Web graph. In continuous crawling, the URL of a fetched page is added back to the frontier for fetching again in the future.

This seemingly simple **recursive traversal** of the Web graph is complicated by the many demands on a practical Web crawling system: the crawler has to be distributed, scalable, efficient, polite, robust and extensible while fetching pages of high quality. We examine the effects of each of these issues. Our treatment follows the design of the Mercator crawler that has formed the basis of a number of research and commercial crawlers. As a reference point, fetching a billion pages (a small fraction of the static Web at present) in a month-long crawl requires fetching several hundred pages each second. We will see how to use a multi-threaded design to address several bottlenecks in the overall crawler system in order to attain this fetch rate.

Before proceeding to this detailed description, we reiterate for readers who may attempt to build crawlers of some basic properties any non-professional crawler should satisfy:

(1) Only one connection should be open to any given host at a time.

(2) A waiting time of a few seconds should occur between successive requests to a host.

(3) Politeness restrictions should be obeyed.

5.4 Key Terms and Review Questions

1. Technical Terms

analytical database	分析型数据库
table	表
field	字段
record	记录
hierarchical model	层次模型
data retrieval	数据检索
network model	网状模型
relational model	关系模型
tuple	元组
attribute	属性
de facto standard	事实标准
primary key	主键
normalization	标准化、规范化
first normal form (1NF)	第一范式
Database Management System	数据库管理系统
Structured Query Language	结构化查询语言
Open Database Connectivity	开放数据库连接
modeling language	建模语言
database query language	数据库查询语言
transaction mechanisms modeling language	事务机制建语言

declarative language	说明性语言
Data Definition Language	数据定义语言
Data Manipulation Language	数据操纵语言
Data Control Language	数据控制语言
clause	子句
predicate	谓词
schemata	纲要
information retrieval	信息检索
unstructured data	非结构化数据
structured data	结构化数据
canonical	标准
semistructured	半结构化的
Web search	网页搜索
personal information retrieval	个人信息检索
spam	垃圾邮件
junk mail	垃圾邮件
grep	UNIX 工具程序；可做文件内的字符串查找
wildcard	通配符
index	索引
Boolean retrieval model	布尔检索模型
terminology	术语
notation	符号
corpus	语料库
magnitude	数量级
precision	准确率
recall	召回率（查全率）
inverted index	倒排索引
sparse	稀疏
lexicon	词典
postings list	位置表
linked list	链表
ariable length array	变长数组
offset	偏移量
formulae	规则
structured search requests	结构化的搜索请求
pseudo-relevance feedback	伪关联性反馈
relevant	相关
nonrelevant	不相关
ground truth	基本事实，机器学习中指正确的标注信息

asynchronously	异步地
taxonomy	分类学
indexable text	可索引的文本
granularity	粒度
sponsored search	付费搜索
Web crawling	网页信息采集
robustnet	健壮性
recursive traversal	递归遍历
parse	解析

2. Translation Exercises

(5-1) In the relational database model, data is stored in relations, more commonly known as tables. Tables, records (sometimes known as **tuples**), and fields (sometimes known as *attributes*) are the basic components. Each individual piece of data, such as a last name or a telephone number, is stored in a table field and each record comprises a complete set of field data for a particular table.

(5-2) A **Database Management System** is also know as DBMS, sometimes called a database manager or database system, which is a set of computer programs that controls the creation, organization, maintenance, and retrieval of data from the database stored in a computer. An excellent database system helps the end users to easily access and uses the data and also stores the new data in a systematic way.

(5-3) Originally based upon relational algebra and tuple relational calculus, SQL consists of a data definition language, data manipulation language, and data control language. The scope of SQL includes data insert, query, update and delete, schema creation and modification, and data access control. Although SQL is often described as, and to a great extent is, a declarative language, it also includes procedural elements.

(5-4) Information retrieval is concerned with representing, searching, and manipulating large collections of electronic text and other human-language data. IR systems and services are now widespread, with millions of people depending on them daily to facilitate business, education, and entertainment. Web search engines — Google, Bing, and others — are by far the most popular and heavily used IR services, providing access to up-to-date technical information, locating people and organizations, summarizing news and events, and simplifying comparison shopping.

(5-5) To assess the effectiveness of an IR system (i.e., the quality of its search results), a user will usually want to know two key statistics about the system's returned results for a query:

(1) **Precision**: What fraction of the returned results are relevant to the information need?

(2) **Recall**: What fraction of the relevant documents in the collection were returned by the system?

References

[1] Structured Query Language (SQL)[A/OL]. [2017-03-22].https://docs.microsoft.com/en-us/sql/odbc/reference/structured-query-language-sql.

[2] Manning C D, Raghavan P, Hinrich Schütze. Introduction to Information Retrieval[M]. Cambridge University Press, 2008.

[3] Stefan Büttcher, Clarke C L, Cormack G V. Information Retrieval: Implementing and Evaluating Search Engines[M/OL]. MIT Press, 2010. http://www.ir.uwaterloo.ca/book/bib.html

[4] Learn IT: The Power of the Database[EB/OL]. 2017-03-21. http://whatis.techtarget.com/reference/ Learn-IT-The-Power-of-the-Database.

[5] Baeza-Yates R A, Ribeiro-Neto B. Modern Information Retrieval[M]. 2nd ed. Reading, Massachusetts: Addison-Wesley, 2010.

[6] Barroso L A, Dean J, Hölzle U. Web search for a planet: The Google cluster architecture[J]. IEEE Micro, 2003, 23(2):22–28.

Chapter 6 Artificial Intelligence

6.1 Introduction

Artificial Intelligence (AI) is the intelligence exhibited by machines or software. It is also the name of the academic field of study which studies how to create computers and computer software that are capable of intelligent behavior. The term Artificial Intelligence was first coined by John McCarthy in 1956, he defined it as "the science and engineering of making intelligent machines".

Artificial Intelligence has been studied for decades and is still one of the most elusive subjects in Computer Science. This partly due to how large and nebulous the subject is. AI ranges from machines truly capable of thinking to search algorithms used to play board games. It has applications in nearly every way we use computers in society.

6.1.1 History of AI

In Vannevar Bush's work "As We May Think" in 1945, he proposed a system which **amplifies** people's own knowledge and understanding. Five years later, Alan Turing wrote a paper on the notion of machines being able to simulate human beings and the ability to do intelligent things, such as play chess. No one can refute a computer's ability to process logic, but it is unknown to many if a machine can think. The precise definition of think is important because there has been some strong opposition as to whether or not this notion is even possible.

The field of AI research was founded at a conference on the campus of Dartmouth College in the summer of 1956. The attendees, including John McCarthy, Marvin Minsky, Allen Newell, Arthur Samuel, and Herbert Simon, became the leaders of AI research for many decades. They and their students wrote programs that were, to most people, simply astonishing: computers were winning at checkers, solving word problems in algebra, proving logical theorems and speaking English. By the middle of the 1960s, research in the U.S. was heavily funded by the Department of Defense and laboratories had been established around the world. AI's founders were profoundly optimistic about the future of the new field: Herbert Simon predicted that "machines will be capable, within twenty years, of doing any work a man can do" and Marvin Minsky agreed, writing that "within a generation … the problem of creating 'artificial intelligence' will substantially be solved".

They had failed to recognize the difficulty of some of the problems they faced. In 1974, in response to the criticism of Sir James Lighthill and ongoing pressure from the US Congress to fund more productive projects, both the U.S. and British governments cut off all undirected **exploratory research** in AI. The next few years would later be called an "AI winter", a period when funding for AI projects was hard to find.

In the early 1980s, AI research was revived by the commercial success of **expert systems**, a form of AI program that simulated the knowledge and analytical skills of one or more human experts. By 1985 the market for AI had reached over a billion dollars. At the same time, Japan's fifth generation computer project[①] inspired the U.S. and British governments to restore funding for academic research in the field. However, beginning with the collapse of the Lisp Machine market in 1987, AI once again fell into disrepute, and a second, longer-lasting AI winter began.

In the 1990s and early 21st century, AI achieved its greatest successes, albeit somewhat behind the scenes. Artificial intelligence is used for logistics, data mining, medical diagnosis and many other areas throughout the technology industry. The success was due to several factors: the increasing computational power of computers, a greater emphasis on solving specific sub-problems, the creation of new ties between AI and other fields working on similar problems, and a new commitment by researchers to solid mathematical methods and rigorous scientific standards.

On May 11, 1997, Deep Blue became the first computer chess-playing system to beat a reigning world chess champion, Garry Kasparov[②]. In February 2011, in a Jeopardy![③] quiz show exhibition match, IBM's question answering system, Watson, defeated the two greatest Jeopardy champions by a significant margin. The Kinect, which provides a 3D body–motion interface for the Xbox 360 and the Xbox One, uses algorithms that emerged from lengthy AI research as do intelligent personal assistants in smart phones. In March 2016, AlphaGo won 4 out of 5 games of Go in a match with Go champion Lee Sedol[④], becoming the first computer Go-playing system to beat a professional human Go player without handicaps.

6.1.2 Research Branches of AI

The general problem of simulating (or creating) intelligence has been broken down into a

---

① The Japanese Fifth Generation Project in computer technology was an attempt to leapfrog Western computer expertise and create an entirely new computer technology, which not only to update the hardware technology of computers but alleviate the problems of programming by creating AI operating systems that would ferret out what the user wanted and then do it.

② A Russian former World Chess Champion, who is considered by many to be the greatest chess player of all time. From 1986 until his retirement in 2005, Kasparov was ranked world No.1.

③ An American television game show, it features a quiz competition in which contestants are presented with general knowledge clues in the form of answers, and must phrase their responses in the form of questions.

④ A South Korean professional Go player of 9-dan rank. As of February 2016, he ranks second in international titles.

number of specific sub-problems. These consist of particular traits or capabilities that researchers would like an intelligent system to display. The traits described below have received the most attention.

1. Deduction, Reasoning, Problem Solving

Early AI researchers developed algorithms that imitated the step-by-step **reasoning** that humans use when they solve puzzles or make **logical deductions**. By the late 1980s and 1990s, AI research had also developed highly successful methods for dealing with uncertain or incomplete information, employing concepts from probability and economics.

For difficult problems, most of these algorithms can require enormous computational resources — most experience a "combinatorial explosion": the amount of memory or computer time required becomes astronomical when the problem goes beyond a certain size. The search for more efficient problem-solving algorithms is a high priority for AI research.

Human beings solve most of their problems using fast, intuitive judgments rather than the conscious, step-by-step deduction that early AI research was able to model. AI has made some progress at imitating this kind of "sub-symbolic" problem solving: embodied agent approaches emphasize the importance of sensorimotor skills to higher reasoning; neural net research attempts to simulate the structures inside the brain that give rise to this skill; statistical approaches to AI mimic the probabilistic nature of the human ability to guess.

2. Knowledge Representation

(6-1) **Knowledge representation** and **knowledge engineering** are central to AI research. Many of the problems machines are expected to solve will require extensive knowledge about the world. Among the things that AI needs to represent are: objects, properties, categories and relations between objects; situations, events, states and time; causes and effects; knowledge about knowledge (what we know about what other people know); and many other, less well researched domains. A representation of "what exists" is **ontology**: the set of objects, relations, concepts and so on that the machine knows about. The most general are called upper ontologies, which attempt to provide a foundation for all other knowledge.

3. Planning

Intelligent agents must be able to set goals and achieve them. They need a way to visualize the future (they must have a representation of the state of the world and be able to make predictions about how their actions will change it) and be able to make choices that maximize the utility (or "value") of the available choices.

In classical **planning** problems, the agent can assume that it is the only thing acting on the world and it can be certain what the consequences of its actions may be. However, if the agent is not the only actor, it must periodically ascertain whether the world matches its predictions and it must change its plan as this becomes necessary, requiring the agent to reason under uncertainty.

Multi-agent planning uses the cooperation and competition of many agents to achieve a given goal. Emergent behavior such as this is used by evolutionary algorithms and swarm intelligence.

4. Learning

Machine learning is the study of computer algorithms that improve automatically through experience and has been central to AI research since the field's inception.

(6-2) **Unsupervised learning** is the ability to find patterns in a stream of input. **Supervised learning** includes both classification and numerical regression. **Classification** is used to determine what category something belongs in, after seeing a number of examples of things from several categories. **Regression** is the attempt to produce a function that describes the relationship between inputs and outputs and predicts how the outputs should change as the inputs change. In **reinforcement learning**, the agent is rewarded for good responses and punished for bad ones. The agent uses this sequence of rewards and punishments to form a strategy for operating in its problem space. These three types of learning can be analyzed in terms of decision theory, using concepts like utility. The mathematical analysis of machine learning algorithms and their performance is a branch of theoretical computer science known as computational learning theory.

Within developmental robotics, developmental learning approaches were elaborated for lifelong cumulative acquisition of repertoires of novel skills by a robot, through autonomous self-exploration and social interaction with human teachers, and using guidance mechanisms such as active learning, maturation, motor synergies, and imitation.

5. Perception

Machine perception is the ability to use input from sensors (such as cameras, microphones, tactile sensors, sonar and others more exotic) to deduce aspects of the world. Computer vision is the ability to analyze visual input. A few selected sub-problems are speech recognition, facial recognition and object recognition.

6. Motion and Manipulation

The field of robotics is closely related to AI. Intelligence is required for robots to be able to handle such tasks as object manipulation and navigation, with sub-problems of localization (knowing where you are, or finding out where other things are), mapping (learning what is around you, building a map of the environment), and motion planning (figuring out how to get there) or path planning (going from one point in space to another point, which may involve compliant motion — where the robot moves while maintaining physical contact with an object).

7. Social Intelligence

Affective computing is the study and development of systems and devices that can recognize, interpret, process, and simulate human affects. It is an interdisciplinary field spanning computer sciences, psychology, and cognitive science. While the origins of the field may be traced as far back as to early philosophical inquiries into emotion, the more modern branch of computer science originated with Rosalind Picard's 1995 paper on affective computing. A motivation for the research is the ability to simulate empathy. The machine should interpret the emotional state of humans and adapt its behavior to them, giving an appropriate response for those emotions.

Emotion and social skills play two roles for an intelligent agent. First, it must be able to predict the actions of others, by understanding their motives and emotional states. (This involves elements of **game theory**, **decision theory**, as well as the ability to model human emotions and the perceptual skills to detect emotions.) Also, in an effort to facilitate human-computer interaction, an intelligent machine might want to be able to display emotions—even if it does not actually experience them itself—in order to appear sensitive to the emotional dynamics of human interaction.

Kismet (see Fig.6-1) is a robotic head that can interact with humans in a human-like way via myriad facial expressions, head positions, and tones of voice. "The goal is to build a socially intelligent machine that learns things as we learn them, through social interactions," said Dr. Breazeal, a postdoctoral associate at MIT's Artificial Intelligence Laboratory and leader of the Kismet team.

Fig.6-1 Kismet, a robot with rudimentary social skills①

8. Creativity

A sub-field of AI addresses creativity both theoretically (from a philosophical and psychological perspective) and practically (via specific implementations of systems that generate outputs that can be considered creative, or systems that identify and assess creativity). Related areas of computational research are artificial intuition and artificial thinking.

9. General Intelligence

Many researchers think that their work will eventually be incorporated into a machine with general intelligence (known as strong AI), combining all the skills above and exceeding human abilities at most or all of them. A few believe that anthropomorphic features like artificial consciousness or an artificial brain may be required for such a project.

Many of the problems above may require general intelligence to be considered solved. For example, even a straightforward, specific task like machine translation requires that the machine read and write in both languages (NLP), follow the author's argument (reason), know what is being talked about (knowledge), and faithfully reproduce the author's intention (social intelligence). A problem like machine translation is considered "AI-complete". In order to solve this particular problem, one must solve all the problems.

---

① http://www.ai.mit.edu/projects/humanoid-robotics-group/kismet/kismet.html

6.2 Turing Test

6.2.1 Introduction

The **Turing test** is a central, long term goal for AI research — will we ever be able to build a computer that can sufficiently imitate a human to the point where a suspicious judge cannot tell the difference between human and machine? From its inception it has followed a path similar to much of the AI research. Initially it looked to be difficult but possible (once hardware technology reached a certain point), only to reveal itself to be far more complicated than initially thought with progress slowing to the point that some wonder if it will ever be reached. Despite decades of research and great technological advances, the Turing test still sets a goal that AI researchers strive toward while finding along the way how much further we are from realizing it. In 1950 English Mathematician Alan Turing published a paper entitled "Computing Machinery and Intelligence" which opened the doors to the field that would be called AI. This was years before the community adopted the term Artificial Intelligence as coined by John McCarthy. The paper itself began by posing the simple question, "Can machines think?" Turing then went on to propose a method for evaluating whether machines can think, which came to be known as the Turing test. The test, or "Imitation Game" as it was called in the paper, was put forth as a simple test that could be used to prove that machines could think. The Turing test takes a simple pragmatic approach, assuming that a computer that is indistinguishable from an intelligent human actually has shown that machines can think. The idea of such a long term, difficult problem was a key to defining the field of AI because it cuts to the heart of the matter — rather than solving a small problem it defines an end goal that can pull research down many paths. Without a vision of what AI could achieve, the field itself might never have formed or simply remained a branch of math or philosophy. The fact that the Turing test is still discussed and researchers attempt to produce software capable of passing it are indications that Alan Turing and the proposed test provided a strong and useful vision to the field of AI. It's relevance to this day seems to indicate that it will be a goal for the field for many years to come and a necessary marker in tracking the progress of the AI field as a whole. This section will explore the history of the Turing test, evaluate its validity, describe the current attempts at passing it and conclude with the possible future directions the Turing test solution may take.

6.2.2 Alan Turing

Alan Turing (23 June 1912—7 June 1954) was an English computer scientist, mathematician, logician, cryptanalyst and theoretical biologist. He was highly influential in the development of theoretical computer science, providing a formalization of the concepts of algorithm and computation with the Turing machine. Turing is widely considered to be the father of theoretical computer science and artificial intelligence. After graduating from college, he published a paper

"On Computable Numbers, with an Application to the Entscheidungs problem[①]" in which he proposed what would later be known as a Turing Machine — a computer capable of computing any computable function.

The paper defines a "computing machine" with the ability to read and write symbols to a tape using those symbols to execute an algorithm. This paper and the Turing machine provided that basis for the theory of computation. While Alan Turing focused primarily on mathematics and the theory of what would become computer science during and immediately after college, soon World War Ⅱ came and he became interested in more practical matters. The use of cryptography by the Axis[②] gave him reason to focus on building a machine capable of breaking ciphers. Before this potential use presented itself, Alan Turing likely hadn't been too concerned that the Turing machine he'd proposed in his earlier work was not feasible to build. In 1939 he was invited to join the Government Code and Cipher school as a cryptanalyst and it became clear that he needed to build a machine capable of breaking codes like Enigma which was used by the Germans. He designed in a few weeks and received funding for the construction of electromechanical machines called "bombes"which would be used to break Enigma[③] codes and read German messages by automating the processing of 12 electrically linked Enigma scramblers. It wasn't the Turing machine, but the concepts of generating cyphertext from plaintext via a defined algorithm clearly fit with the Turing machine notion. After the war, Turing returned to academia and became interested in the more philosophical problem of what it meant to be sentient, which lead him down the path to the Turing test.

6.2.3 Inception of the Turing Test

In 1950 Alan Turing was the Deputy Director of the computing laboratory at the University of Manchester. The paper which defined what would come to be known as the Turing test was published in a Philosophical journal called ***Mind***. The paper itself was based on the idea of an "Imitation Game". If a computer could imitate the sentient behavior of a human would that not imply that the computer itself was sentient? Even though the description itself is fairly simple, the implications of building a machine capable of passing the test are far reaching. It would have to process natural language, be able to learn from the conversation and remember what had been said, communicate ideas back to the human and understand common notions, displaying what we call common sense. Similar to how he used the Turing Machine to more clearly formalize what could or could not be computed, Alan Turing felt the need to propose the Turing Test so that there was a clear definition of whether or not the responses given by a human were part of the

---

① A decision problem, of finding a way to decide whether a formula is true or provable within a given system.

② Refers to Axis alliance which consists of three principal partners, Germany, Italy, and Japan.

③ Enigma was invented by the German engineer Arthur Scherbius at the end of World War I. Early models were used commercially from the early 1920s, and adopted by military and government services of several countries, most notably Nazi Germany before and during World War Ⅱ.

computable space. In the paper he wanted to replace the question, "Can machines think?" (which can have many possible answers and come down to a difference of opinion) with a version of the "Imitation Game". The original game upon which Turing's idea was based required a man, a woman and an interrogator. The goal was for the interrogator to identify which of the participants was a man and which was a woman. Since the interrogator would be able to identify the gender of the respondent by their voice (and maybe handwriting) the answers to the interrogator's questions would be type written or repeated by an intermediary. For the Turing test, one of those two participants would be replaced by a machine and the goal of the interrogator would not be to identify the gender of the participants, but which is human and which is a machine. As described above, the Turing test has a few key components. First the interrogator knows that there is one human and one machine. The test doesn't just require a computer to fool a human into thinking it is sentient; it asks the computer to fool a suspicious human. Second, physical nature isn't important — the goal is to not be able to tell the difference between man and machine when comparing the output of the machine and the true human. The communication medium is such that there are absolutely no hints beyond what can be expressed with written language. Also, the test doesn't include anything specific — no complex problem solving or requests to create art. As described, it seems a machine would pass the Turing test if it were able make small talk with another human and understand the context of the conversation. For Turing, passing such a test was sufficient for him to believe that machines were capable of thinking. Beyond defining the game, the paper continues with an introduction to digital computers and how they can be used for arbitrary computation. Taken with Godel's incompleteness theorem and Turing's formalization of what can and cannot be computed, the Turing test seems to strike at the simple question of whether that ability to appear sentient falls in to the realm of computable problems that a Turing machine can handle, or if it falls under the tiny subset of things that are true, but cannot be proven so. The test is simple, but the question is hugely significant and tied in to Turing's earlier work towards formalizing what can be computed.

6.2.4 Problems/Difficulties with the Turing Test

A large portion of Turing's original paper deals with addressing counter arguments concerning how the test he proposes may not be valid. In the introduction to that section he states that he believes there will be computers with enough storage capacity to make them capable of passing the Turing test "in about fifty years". The statement is interesting because it seems to imply that the AI software required to pass the Turing test would be rather straightforward and that the limiting factor would only be memory. Perhaps this limitation was at the front of his mind because he was routinely running into problems that he could have solved if only there were enough storage available. The same type of reasoning is similar to what happens today when we believe that **Moore's law** will let us solve the hard problems. Beyond the storage limitations, he also raises other objections, including those based in theology (the god granted immortal soul is necessary for sentience), mathematical arguments based on Godel's work, the ability for humans to create original works and experience emotion, and others. One of the more interesting contradictions to the test is what he terms "The Argument from Consciousness". The argument goes that just imitating a human would not be enough because it doesn't invoke the full range of what it is that we consider to be human. Specifically, the Turing test could be passed by a machine unable to do things such as write a poem or piece of music wrapped up as part of an emotional response. A machine passing the Turing test would not really have to experience or interpret art either. Turing argues that it is impossible to tell if the machine is feeling unless you are the machine, so there is no way to contradict the claim or to prove it. Using that method to dismiss the argument, he points out that the Turing test could include the machine convincing the interrogator that it is feeling something, even if there is truly no way to know that the emotions are actually being felt the way they would in a human. This would be similar to how humans communicate to convince each other of what they are feeling, though there is no guarantee that it is really true. Another interesting counter argument against the test that Turing describes is "Lady Lovelace's Objection". She posited that because machines can only do what we tell them, they cannot originate anything, while it is clear that humans do originate new concepts and ideas all of the time. At the time this was written it may not have been possible to model the learning process, but much of the progress that has been made in teaching machines to learn and infer seems to have shown that this issue can be overcome. There have been specific implementations where voice or character recognition is reached by software training itself to recognize the variances in human writing or dialect. At least in these specific cases a machine can recognize something new so perhaps they will be able to in the general case as well. Overall the potential problems with the Turing test appear to fall in one of two categories: Does imitating a human actually prove intelligence or is it just a hard problem? Is intelligence possible without passing the Turing test? It seems fair to say that passing the Turing test is only a subset of the situation that humans have to contend with on a day to day basis. So it is possible that there are other key capabilities like experiencing emotions, having core beliefs or motivations, or problem solving that might be simulated in a computer but would not necessarily be the same as what humans do. The Turing test avoids these questions by judging the computer (and human) only on the text they output as part of the casual conversation that takes place during the test. So even if a computer could pass the Turing test, is that enough to say machines are "intelligent" or that they can "think", or does that just say that they can now pass the Turing test, and there is much more to understand before we do consider them intelligent. Beyond that, there are many humans that we'd consider sentient — young children for instance, that would probably do poorly in the Turing test because they haven't accumulated enough knowledge and experience in communication. We wouldn't apply the Turing test to them and say that they therefore are not capable of thought, which means that it might be possible for a computer to “think” but still not pass the Turing test.

6.2.5 The Current State of the Turing Test

Since the Turing test was proposed many attempts have been made to build software capable of passing it (or a more constrained version). A limited form of the test — temporarily convincing a non-suspicious human that a computer is actually another person has been possible for some time, but the true test itself has never been passed. As the Turing test is such a fundamental and difficult problem, the research into it tends to be slow and incremental — trying to build something better than the last. Faster computers with more memory haven’t been a great boon to the progress, it seems because the problem isn’t in having enough time and storage, but in devising the correct algorithms to make use of the processing power and storage that we have available to us. The limited progress against the Turing test is perhaps due to the reduction of research in AI or the extreme difficulty of the problem. In the end, despite the efforts that have been put forth, not a lot of progress has been made. Currently there is one mechanism to encourage further research — the Leobner Prize and the associated Turing Test Competition. The prize includes a $100,000 cash reward for a system capable of being indistinguishable from a human. To encourage participation, the yearly competition has a few different levels of reward. Each year a medal and $2000 is awarded to the best entry. Given the size of the reward it may not be surprising that few researchers actively work toward attaining it when there are far more lucrative avenues for research. The Loebner prize web site includes the transcripts of the top four entries in each year going back to 1995. In reviewing the conversations for the winning entries it is amazing how far they are from being indistinguishable form a human and passing the test. The difference between human and computer is significant with the software often replying in ways that make no sense. At times the winning program offers seemingly random responses to questions that don’t seem to be complicated.

Here are some example exchanges in the 2005 contest between the judge and the winning program, Jaberwocky. In this example the judge types in random characters:

JUDGE: Hi George, cslkjslw weicmslw weos, right?
PROGRAM: Don’t repeat your name, csthiang!

It seems the program has determined that the letters are part of the judge’s name, which isn’t an assumption a human would likely make. When the judge carried out the same exchange with a human the results were different, causing the human to ask questions rather than making assumptions:

JUDGE: Wljso lwjejdo wleje elwjeo wjeol, isn’t it?
CONFEDERATE: Is that Polish?

JUDGE: Who?

CONFEDERATE: What you wrote above.

JUDGE: This is ljelwl, aren't you?

CONFEDERATE: I'm a little worried about your writing skills.

Often the winning system just appears to be responding at random with exchanges similar to the one below:

JUDGE: OK, enough for now. I'll go chat with the other entity.

PROGRAM: Ok then it was nice to share your hair brush.

JUDGE: What do mean by that? How have you shared my hair brush?

PROGRAM: Good.

JUDGE: Good?

PROGRAM: Yes.

JUDGE: This has become a very zen-like conversation.

PROGRAM: And worse yet, I gave it to charity.

JUDGE: What? What did you give to charity? This conversation?

PROGRAM: Er… 1 million Dollars?

JUDGE: Which charity?

PROGRAM: Are we talking about the Simpsons or unreal?

JUDGE: I'm not sure what the hell we're talking about, actually.

The difficulty of the Turing test and our lack of progress in attaining it can be seen when the winning systems provide such convoluted and disjointed conversations as the one above. While it is possible for software to mimic human responses for very small subsets of conversation with predefined branching points, general conversation is still far beyond what even the best systems have been able to produce.

Alan Turing's original hope that the test would be passed by a computer by the year 2000 has not been realized. Despite all the effort expended by the research community, advances in processor technology, and cheap memory, no computer has yet been able to approach passing the Turing test. It is clear that the Moore's law increase in computation power hasn't been the driving force in improvement in Turing Test focused AI; instead it is a software problem. **Software architectures**, such as the **Expert Systems**, offer possible solutions as their designs are refined and applied to various problems, perhaps including the Turing test or one of its derivatives.

6.2.6 Artificial Intelligence Computer System Passes Visual Turing Test

It is reported that researchers at MIT, New York University, and the University of Toronto had developed a computer system whose ability to produce a variation of a character in an unfamiliar writing system, on the first try, was indistinguishable from that of humans.

That means the system in some sense discerns what's essential to the character — its general structure — but also what's inessential — the minor variations characteristic of any one instance of it.

As such, the researchers argue, their system captures something of the elasticity of human concepts, which often have fuzzy boundaries but still seem to delimit coherent categories. It also mimics the human ability to learn new concepts from few examples. It thus offers hope, they say, that the type of computational structure it's built on, called a probabilistic program, could help model human acquisition of more sophisticated concepts as well.

"In the current AI landscape, there's been a lot of focus on **classifying patterns**," says Josh Tenenbaum, a professor in the Department of Brain and Cognitive sciences at MIT, a principal investigator in the MIT Center for Brains, Minds and Machines, and one of the new system's co-developers. "But what's been lost is that intelligence isn't just about classifying or recognizing; it's about thinking."

"This is partly why, even though we're studying hand-written characters, we're not shy about using a word like 'concept'," he adds. "Because there are a bunch of things that we do with even much richer, more complex concepts that we can do with these characters. We can understand what they're built out of. We can understand the parts. We can understand how to use them in different ways, how to make new ones."

The new system was the thesis work of Brenden Lake. "We analyzed these three core principles throughout the paper," Lake says. "The first we called compositionality, which is the idea that representations are built up from simpler primitives. Another is causality, which is that the model represents the abstract causal structure of how characters are generated. And the last one was learning to learn, this idea that knowledge of previous concepts can help support the learning of new concepts. Those ideas are relatively general. They can apply to characters, but they could apply to many other types of concepts."

The researchers subjected their system to a battery of tests. In one, they presented it with a single example of a character in a writing system it had never seen before and asked it to produce new instances of the same character — not identical copies, but nine different variations on the same character. In another test, they presented it with several characters in an unfamiliar writing system and asked it to produce new characters that were in some way similar. And in a final test, they asked it to make up entirely new characters in a hypothetical writing system.

Human subjects were then asked to perform the same three tasks. Finally, a separate group of human judges was asked to distinguish the human subjects' work from the machine's. Across all three tasks, the judges could identify the machine outputs with about 50 percent accuracy — no better than chance.

Conventional machine-learning systems, such as the ones that led to the speech-recognition algorithms on smartphones, often perform very well on constrained classification tasks, but they must first be trained on huge sets of training data. Humans, by contrast, frequently grasp concepts after just a few examples. That type of "one-shot learning" is something that the researchers designed their system to emulate.

Like a human subject, however, the system comes to a new task with substantial background knowledge, which in this case is captured by a probabilistic program. Whereas a conventional computer program systematically decomposes a high-level task into its most basic computations, a probabilistic program requires only a very sketchy model of the data it will operate on. Inference algorithms then fill in the details of the model by analyzing a host of examples.

6.3 Knowledge Representation and Reasoning

Intelligence, as exhibited by people anyway, is surely one of the most complex and mysterious phenomena that we are aware of. One striking aspect of intelligent behavior is that it is clearly conditioned by knowledge: for a very wide range of activities, we make decisions about what to do base on what we know (or believe) about the world, effortlessly and unconsciously. Using what we know in this way is so commonplace, that we only really pay attention to it when it is not there. When we say that someone behaved unintelligently, like when someone uses a lit match to see if there is any gas in a car's gas tank, what we usually mean is not that there is something that the person did not know, but rather that the person has failed to use what he or she did know. We might say: "You weren't thinking!" Indeed, it is thinking that is supposed to bring what is relevant in what we know to bear on what we are trying to do.

One definition of Artificial Intelligence (AI) is that it is the study of intelligent behavior achieved through computational means. Knowledge Representation and Reasoning, then, is that part of AI that is concerned with how an agent uses what it knows in deciding what to do. It is the study of thinking as a computational process.

6.3.1 How to Represent Knowledge

What is knowledge? This is a question has been discussed by philosophers since the ancient Greeks, and it is still not totally demystified. To get a rough sense of what knowledge is supposed to be, it is useful to look at how we talk about it informally. First, observe that when we say something like "John knows that…," we fill in the blank with a simple declarative sentence. So we might say that "John knows that Mary will come to the party" or that "John knows that Abraham Lincoln was assassinated". This suggests that, among other things, knowledge is a relation between a knower, like John, and a **proposition**, that is, the idea expressed by a simple declarative sentence, like "Mary will come to the party".

Part of the mystery surrounding knowledge is due to the nature of propositions. What can we say about them? As far as we are concerned, what matters about propositions is that they are abstract entities that can be true or false, right or wrong. When we say that "John knows that", we can just as well say that "John knows that it is true". Either way, to say that John knows something is to say that John has formed a judgment of some sort, and has come to realize that the world is one way and not another. In talking about this judgment, we use propositions to classify the two cases.

A similar story can be told about a sentence like "John hopes that Mary will come to the party". The same proposition is involved, but the relationship John has to it is different. Verbs like "knows" "hopes" "regrets" "fears", and "doubts" all denote, relationships between agents and propositions. In all cases, what matters about the proposition is its truth: if John hopes that Mary will come to the party, then John is hoping that the world is one way and not another, as classified by the proposition.

Of course, there are sentences involving knowledge that do not explicitly mention a proposition. When we say "John knows who Mary is taking to the party", or "John knows how to get there", we can at least imagine the implicit propositions: "John knows that Mary is taking so-and-so to the party", or "John knows that to get to the party, you go two blocks past Main Street, turn left,⋯", and so on. On the other hand, when we say that John has a skill as in "John knows how to play piano", or a deep understanding of someone or something as in "John knows Bill well", it is not so clear that any useful proposition is involved. While this is certainly challenging subject matter, we will have nothing further to say about this latter form of knowledge in this book.

A related notion that we are concerned with, however, is the concept of knowledge. The sentence "John believes that " is clearly related to "John knows that". We use the former when we do not wish to claim that John's judgment about the world is necessarily accurate or held for appropriate reasons. We sometimes use it when we feel that John might not be completely convinced. In fact, we have a full range of propositional attitudes, expressed by sentences like "John is absolutely certain that," "John is confident that", "John is of the opinion that", "John suspects that", and so on, that differ only in the level of conviction they attribute. For now, we will not distinguish amongst of them. What matters is that they all share with knowledge a very basic idea: John takes the world to be one way and not another.

6.3.2 Representation

The concept of representation is as philosophically vexing as that of knowledge. Very roughly speaking, representation is a relationship between two domains where the first is meant to "stand for" or take the place of the second. Usually, the first domain, the representor, is more concrete, immediate, or accessible in some way than the second. For example, a drawing of a milkshake and a hamburger on a sign might stand for a less immediately visible fast food restaurant; the drawing of a circle with a plus below it might stand for the much more abstract concept of womanhood; an elected legislator might stand for his or her constituency.

The type of representor that we will be most concerned with here is the formal, that is, a character or group of them taken from some predetermined alphabet. The digit "7" for example, stands for the number 7, as does the group of letters "VII", and in other contexts. As with all representation, it is assumed to be easier to deal with symbols (recognize them, distinguish them from each other, display them, etc.) than with what the symbols represent. In some cases, a word like "John" might stand for something quite concrete; but many words, like "love" or "truth", stand for **abstractions**.

Of special concern to us is when a group of formal symbols stands for a proposition: "John loves Mary" stands for the proposition that John loves Mary. Again, the symbolic English sentence is fairly concrete: it has distinguishable parts involving the 3 words, for example, and a recognizable syntax. The proposition, on the other hand, is abstract: it is something like a classification of all the different ways we can imagine the world to be into two groups: those where John loves Mary, and those where he does not.

(6-3) Knowledge Representation, then, is this: it is the field of study concerned with using formal symbols to represent a collection of propositions believed by some putative agent. As we will see, however, we do not want to insist that these symbols must represent all the propositions believed by the agent. There may very well be an infinite number of propositions believed, only a finite number of which are ever represented. It will be the role of reasoning to bridge the gap between what is represented and what is believed.

6.3.3 Reasoning about Knowledge

In general, reasoning is the formal manipulation of the symbols representing a collection of believed propositions to produce representations of new ones. It is here that we use the fact that symbols are more accessible than the propositions they represent: they must be concrete enough that we can manipulate them (move them around, take them apart, copy them, string them together) in such a way as to construct representations of new propositions.

The analogy here is with arithmetic. We can think of binary addition as being a certain formal **manipulation**: we start with symbols like "1011" and "10", for instance, and end up with "1101". The manipulation here is addition since the final symbol represents the sum of the numbers represented by the initial ones. Reasoning is similar: we might start with the sentences "John loves Mary" and "Mary is coming to the party", and after a certain amount of manipulation produce the sentence "Someone John loves is coming to the party". We would call this form of reasoning **logical inference** because the final sentence represents a logical conclusion of the propositions represented by the initial ones, as we will discuss below. According to this view (first put forward, incidentally, by the philosopher Gottfried Leibniz in the 17th century), reasoning is a form of calculation, not unlike arithmetic, but over symbols standing for propositions rather than numbers.

6.3.4 KBS

A knowledge-based system (KBS) is a computer program that reasons and uses a knowledge base to solve complex problems. The term is broad and is used to refer to many different kinds of systems. The one common theme that unites all knowledge based systems is an attempt to represent knowledge explicitly via tools such as ontologies and rules rather than implicitly via code the way a conventional computer program does. A knowledge based system has two types of sub-systems: a **knowledge base** and an **inference engine**. The knowledge base represents facts about the world, often in some form of subsumption ontology. The inference engine represents logical assertions and conditions about the world, usually represented via IF-THEN rules.

Knowledge-based systems were first developed by Artificial Intelligence researchers. These early knowledge-based systems were primarily **expert systems**. In fact the term is often used synonymously with expert systems. The difference is in the view taken to describe the system. (6-4)Expert system refers to the type of task the system is trying to solve, to replace or aid a human expert in a complex task. Knowledge-based system refers to the architecture of the system, that it represents knowledge explicitly rather than as procedural code. While the earliest knowledge-based systems were almost all expert systems, the same tools and architectures have since been used for a whole host of other types of systems, i.e., virtually all expert systems are knowledge-based systems but many knowledge-based systems are not expert systems.

The first knowledge-based systems were **rule based expert systems**. One of the most famous was Mycin a program for medical diagnosis. These early expert systems represented facts about the world as simple assertions in a flat database and used rules to reason about and as a result add to these assertions. Representing knowledge explicitly via rules had several advantages:

(1) Acquisition & Maintenance. Using rules meant that domain experts could often define and maintain the rules themselves rather than via a programmer.

(2) Explanation. Representing knowledge explicitly allowed systems to reason about how they came to a conclusion and use this information to explain results to users. For example, to follow the chain of inferences that led to a diagnosis and use these facts to explain the diagnosis.

(3) Reasoning. Decoupling the knowledge from the processing of that knowledge enabled general purpose inference engines to be developed. These systems could develop conclusions that followed from a data set that the initial developers may not have even been aware of.

As knowledge-based systems became more complex the techniques used to represent the knowledge base became more sophisticated. Rather than representing facts as assertions about data, the knowledge-base became more structured, representing information using similar techniques to object-oriented programming such as hierarchies of classes and subclasses, relations between classes, and behavior of objects. As the knowledge base became more structured, reasoning could occur both by independent rules and by interactions within the knowledge base itself. For example, procedures stored as **daemons** on objects could fire and could replicate the chaining behavior of rules.

Another advancement was the development of special purpose automated reasoning systems called classifiers. Rather than statically declare the subsumption relations in a knowledge-base, a classifier allows the developer to simply declare facts about the world and let the classifier deduce the relations. In this way a classifier also can play the role of an inference engine.

The most recent advancement of knowledge-based systems has been to adopt the technologies for the development of systems that use the Internet. The Internet often has to deal with complex, unstructured data that can't be relied on to fit a specific data model. The technology of knowledge-based systems and especially the ability to classify objects on demand is ideal for such systems. The model for these kinds of knowledge-based Internet systems is known as the Semantic Web.

6.3.5 MYCIN—A Case Study

MYCIN was an early expert system designed to identify bacteria causing severe infections, such as bacteremia and meningitis, and to recommend antibiotics, with the dosage adjusted for patient's body weight — the name derived from the antibiotics themselves, as many antibiotics have the suffix "-mycin". The MYCIN system was also used for the diagnosis of blood clotting diseases. MYCIN was developed over five or six years in the early 1970s at Stanford University. It was written in Lisp as the doctoral dissertation of Edward Shortliffe under the direction of Bruce Buchanan, Stanley N. Cohen and others. It arose in the laboratory that had created the earlier Dendral expert system. MYCIN was never actually used in practice but research indicated that it proposed an acceptable therapy in about 69% of cases, which was better than the performance of infectious disease experts who were judged using the same criteria.

MYCIN operated using a fairly simple inference engine, and a knowledge base of ~600 rules. It would query the physician running the program via a long series of simple yes/no or textual questions. At the end, it provided a list of possible culprit bacteria ranked from high to low based on the probability of each diagnosis, its confidence in each diagnosis' probability, the reasoning behind each diagnosis (that is, MYCIN would also list the questions and rules which led it to rank a diagnosis a particular way), and its recommended course of drug treatment.

Despite MYCIN's success, it sparked debate about the use of its ad hoc, but principled, **uncertainty framework known** as "**certainty factors**". The developers performed studies showing that MYCIN's performance was minimally affected by perturbations in the uncertainty metrics associated with individual rules, suggesting that the power in the system was related more to its knowledge representation and reasoning scheme than to the details of its numerical uncertainty model. Some observers felt that it should have been possible to use classical Bayesian statistics. MYCIN's developers argued that this would require either unrealistic assumptions of probabilistic independence, or require the experts to provide estimates for an unfeasibly large number of conditional probabilities.

Subsequent studies later showed that the certainty factor model could indeed be interpreted in a probabilistic sense, and highlighted problems with the implied assumptions of such a model. However the modular structure of the system would prove very successful, leading to the development of graphical models such as Bayesian networks.

6.4 Case-based Reasoning①

6.4.1 Introduction

Case-based reasoning (CBR) is an approach to knowledge-based problem solving and decision support: A new problem is solved by remembering a previous similar situation and by reusing information and knowledge of that situation. Let us illustrate this by looking at some typical problem solving situations.

A physician is examining a patient in his office. He gets a reminding to a patient that he treated two weeks ago. Assuming that the reminding was caused by a similarity of important symptoms, the physician uses the diagnosis and treatment of the previous patient to determine the disease and treatment for the patient in front of him.

A financial consultant working on a difficult credit decision task uses a reminding to a previous case, which involved a company in similar trouble as the current one, to recommend that the loan application should be refused.

A drilling engineer has experienced several dramatic blow out situations. He is quickly reminded of one of these situations when the combination of critical measurements during drilling matches those of a blow out case. In particular, he may get a reminding to a mistake he made during a previous blow-out, and use this to avoid repeating the error once again.

Reasoning by reusing past cases is a powerful and frequently applied way to solve problems for humans. This claim is supported by results from cognitive psychological research, and a part of the foundation for the case-based approach is its psychological plausibility. (6-5)Case-based reasoning is a problem solving paradigm that in many respects is fundamentally different from other major AI approaches. Instead of relying solely on general knowledge of a problem domain, or making associations along generalized relationships between problem descriptors and conclusions, CBR is able to utilize the specific knowledge of previously experienced, concrete problem situations (cases). A new problem is solved by finding a similar past case, and reusing it in the new problem situation. A second important difference is that CBR also is an approach to incremental, sustained learning, since a new experience is retained each time a problem has been solved, making it immediately available for future problems. Case-based reasoning can be considered a form of **analogical reasoning**, where the analogs typically are within the same application domain. However, as I will get back to later, the main body of research on analogical reasoning has a different focus, namely analogies across domains.

In CBR terminology, a case usually denotes a problem situation including its interpretation, solution, and possible annotations. A case is previously experienced situation, which has been captured and learned in such way that it can be reused in the solving of future problems.

① partly extracted from an article named "Case-based reasoning — an introduction" authored by Kolodner J L.

CBR is a combined approach to problem solving and machine learning, and a strong driving force behind case-based methods has come from the machine learning community. This makes CBR methods particularly interesting from a decision support point of view, since learning abilities in such systems is something that is urgently needed, but so far missing. Learning in CBR occurs as a natural by-product of problem solving. When a problem is successfully solved, the experience is retained in order to solve similar problems in the future. When an attempt to solve a problem fails, the reason for the failure is identified and remembered in order to avoid the same mistake in the future.

It seems clear that human problem solving and learning in general are processes that involve the representation and utilization of several types of knowledge, and the combination of several reasoning methods. If cognitive plausibility is a guiding principle, architecture for knowledge-based systems where the reuse of cases is at the centre, should also incorporate other and more general types of knowledge in one form or another. This is an issue of current concern in CBR research.

6.4.2 Fundamental of Case-based Reasoning

Central tasks that all case-based reasoning methods have to deal with are to identify the current problem situation, find a past case similar to the new one, use that case to suggest a solution to the current problem, evaluate the proposed solution, and update the system by learning from this experience. How this is done, what part of the process is focused, what type of problems drives the methods, etc. varies considerably, however.

The CBR paradigm covers a range of different methods for organizing, retrieving, utilizing and indexing the knowledge retained in past cases. Cases may be kept as concrete experiences, or a set of similar cases may form a generalized case. Cases may be stored as separate knowledge units, or splitted up into subunits and distributed within the knowledge structure. Cases may be indexed by a prefixed or open vocabulary, and within a flat or hierarchical index structure. The solution from a previous case may be directly applied to the present problem, or modified according to differences between the two cases. The matching of cases, adaptation of solutions, and learning from an experience may be guided and supported by a deep model of general domain knowledge, by more shallow and compiled knowledge, or be based on an apparent, syntactic similarity only. CBR methods may be purely self-contained and automatic, or they may interact heavily with the user for support and guidance of its choices. Some CBR method assume a rather large amount of widely distributed cases in its case base, while others are based on a more limited set of typical ones. Past cases may be retrieved and evaluated sequentially or in parallel. Actually, "case-based reasoning" is just one of a set of terms used to refer to systems of this kind. This has lead to some confusions, particularly since case-based reasoning is a term used both as a generic term for several types of more specific approaches, as well as for one such approach. To some extent, this can also be said for analogy reasoning. An attempt to clarify these notions is given in the following, by describing the main types of CBR approaches.

(1) Instance-based reasoning. This characterizes a highly syntactic CBR-approach. To compensate for lack of guidance from general background knowledge, a relatively large number of instances are needed in order to close in on a concept definition. The representation of the instances are usually simple (e.g. feature vectors), since the major focus is on studying automated learning with no user in the loop. Instance-based reasoning labels work by Kibler and Aha and colleagues. Basically, it is a non-generalization approach to the concept learning problem addressed by classical, inductive machine learning methods.

(2) Memory-based reasoning. This approach emphasizes a collection of cases as a large memory, and reasoning as a process of accessing and searching in this memory. Memory organization and access is a focus of these methods. The utilization of parallel processing techniques is a characteristic of these methods, and distinguishes this approach from the others. The access and storage methods may rely on purely syntactic criteria, as in the MBR-Talk system, or they may attempt to utilize general domain knowledge, as ongoing work in Japan on massive parallel memories.

(3) Case-based reasoning (in the typical sense). Although case-based reasoning is used as a generic term in this paper, the typical case-based reasoning methods have some characteristics that distinguish them from the other approaches listed here. First, a typical case is usually assumed to have a certain degree of richness of information contained in it, and a certain complexity with respect to its internal organization. That is, a feature vector holding some values and a corresponding class is not what we would call a typical case description. What is referred to as typical case-based methods also has another characteristic property: They are able to modify, or adapt, a retrieved solution when applied in a different problem solving context. Paradigmatic case-based methods also utilize general background knowledge — although its richness, degree of explicit representation, and role within the CBR processes varies. Core methods of typical CBR systems borrow a lot from cognitive psychology theories.

(4) Analogy-based reasoning. This term is sometimes used, as a synonym to case-based reasoning, to describe the typical case-based approach just described. However, it is also used to characterize methods that solve new problems based on past cases from a different domain, while typical case-based methods focus on indexing and matching strategies for **single-domain cases**. Research on analogy reasoning is therefore a subfield concerned with mechanisms for identification and utilization of **cross-domain analogies**. The major focus of these methods has been on the reuse of a past case, what is called the mapping problem: Finding a way to transfer, or map, the solution of an identified analogue (called source or base) to the present problem (called target).

6.4.3 The CBR Process

At the highest level of generality, a general CBR cycle may be described by the following four processes:

(1) RETRIEVE the most similar case or cases.

(2) REUSE the information and knowledge in that case to solve the problem.

(3) REVISE the proposed solution.

(4) RETAIN the parts of this experience likely to be useful for future problem solving.

A new problem is solved by retrieving one or more previously experienced cases, reusing the case in one way or another, revising the solution based on reusing a previous case, and retaining the new experience by incorporating it into the existing knowledge-base (case base). This working cycle is illustrated in Fig.6-2.

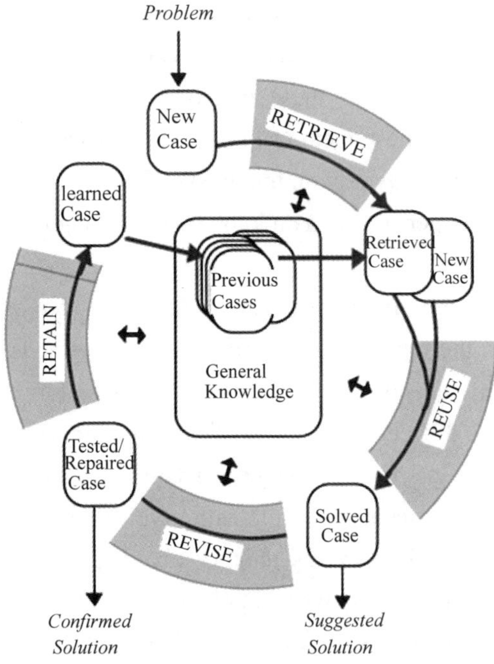

Fig.6-2 The CBR Cycle

An initial description of a problem (top of Fig.6-2) defines a new case. This new case is used to RETRIEVE a case from the collection of previous cases. The retrieved case is combined with the new case through REUSE into a solved case, i.e. a proposed solution to the initial problem. Through the REVISE process this solution is tested for success, e.g. by being applied to the real world environment or evaluated by an expert, and repaired if failed. During RETAIN, useful experience is retained for future reuse, and the case base is updated by a new learned case, or by modification of some existing cases. As indicated in the figure, general knowledge usually plays a part in this cycle, by supporting the CBR processes. This support may range from very weak (or none) to very strong, depending on the type of CBR method. By general knowledge here mean general domain-dependent knowledge, as opposed to specific knowledge embodied by cases. For example, in diagnosing a patient by retrieving and reusing the case of a previous patient, a model of anatomy together with causal relationships between pathological states may constitute the general knowledge used by a CBR system. A set of rules may have the same role.

Tab.6-1 Example of bilingual corpus

English	Japanese
How much is that **red umbrella**?	Ano akai kasa wa ikura desu ka.
How much is that **small camera**?	Ano chiisai kamera wa ikura desu ka.

6.4.4 Example-based Machine Translation

Example-based machine translation (EBMT) is a method of machine translation often characterized by its use of a bilingual corpus with parallel texts as its main knowledge base at run-time. It is essentially a translation by analogy and can be viewed as an implementation of a case-based reasoning approach to machine learning.

Example-based machine translation was first suggested by Makoto Nagao in 1984. B.Pevzner suggested the idea (looks like "translation memory") in 1975 year on seminar "Machine translation" in International center of scientific and technological information. (Moscow) Makoto Nagao pointed out that it is especially adapted to translation between two totally different languages, such as English and Japanese. In this case, one sentence can be translated into several well-structured sentences in another language, therefore, it is no use to do the deep linguistic analysis characteristic of rule-based machine translation. In November 2016, Google Translate switched from its statistical machine translation (SMT) system to the Google Neural Machine Translation (GNMT) system.

At the foundation of example-based machine translation is the idea of translation by analogy. When applied to the process of human translation, the idea that translation takes place by analogy is a rejection of the idea that people translate sentences by doing deep linguistic analysis. Instead, it is founded on the belief that people translate by first decomposing a sentence into certain phrases, then by translating these phrases, and finally by properly composing these fragments into one long sentence. Phrasal translations are translated by analogy to previous translations. The principle of translation by analogy is encoded to example-based machine translation through the example translations that are used to train such a system. Other approaches to machine translation, including statistical machine translation, also use **bilingual corpora** to learn the process of translation.

Example-based machine translation systems are trained from bilingual **parallel corpora** containing sentence pairs like the example shown in the Tab.1. Sentence pairs contain sentences in one language with their translations into another. The particular example shows an example of a minimal pair, meaning that the sentences vary by just one element. These sentences make it simple to learn translations of portions of a sentence. For example, an example-based machine translation system would learn three units of translation from the following example:

(1) "How much is that X ?" corresponds to "Ano X wa ikura desu ka."

(2) "red umbrella" corresponds to "akai kasa"

(3) "small camera" corresponds to "chiisai kamera"

Composing these units can be used to produce novel translations in the future.

A classical example of translation by analogy is the one discussed by Nagao (1984), who shows a method for translating new sentences, based on their similarity with available examples. Similarity is measured by the distance of words in a thesaurus (although other methods could be devised). For instance, Nagao shows how the two English-Japanese examples:

A man eats vegetables ↔ Hito wa yasai o taberu

Acid eats metal ↔ San wa kinzoku o okasu

can be used to translate the new sentence:

He eats potatoes

provided that a **bilingual lexicon** at the word level is available. The problem here is to choose one of two competing translations for eat (taberu vs. okasu). In Nagao's approach, the translation taberu is correctly selected, based on the greater semantic similarity of he and potatoes with man and vegetables, respectively, than with acid and metal. Conversely, the occurrence of eat in:

Sulfuric acid eats iron

is correctly translated by okasu, based on the greater semantic similarity of sulfuric acid and iron with acid and metal, respectively, than with man and vegetables. If we rephrase the translation selection problem in the terms described in the previous section, assuming one of the two translations as the "default" translation for eat (most likely taberu, which translates the more literal sense), the task is to identify the non-monotonic contexts that require a different translation from the default one. In doing this, it would be desirable to find the minimal context that triggers a given translation, so as to use it in a broader range of cases. It turns out that things stand differently, depending on whether exact **match of contexts or match by analogy is** performed.

When exact match of contexts is performed, an example (e.g. a man eats vegetables) only accounts for itself, i.e. it is only used when the same example, or a part of it, is input (e.g. a man eats…or…eats vegetables). In other words, given a set of examples, it is known in advance what input sentences they can be useful for. In this situation it is conceivable to use a contrastive method to reduce a set of overlapping examples to a set of minimal local contexts accounting for different translations of the same term. E.g. the set of examples,

A man eats vegetables ↔ Hito wa yasai o taberu

Acid eats metal ↔ San wa kinzoku o okasu

He eats potatoes ↔ Kare wa jagaimo o taberu

Sulfuric acid eats iron ↔ Ryūsan wa tetsu o okasu

could be reduced to the following set of minimal contexts:

(omitting function words for simplicity):

man & eat ↔ hito & taberu

acid & eat ↔ san & okasu

he & eat ↔ kare & taberu

sulfuric acid & eat ↔ ryūsan & okasu

This would be done on the basis of a generalization that ascribes the selection of one or the other translation of eat to its subject (alternatively, one could choose to generalize over the objects, or perhaps to do no generalization). The set of contexts can be further reduced by identifying a default translation for eat and only explicitly encoding non-monotonic contexts:

eat ↔ taberu

acid & eat ↔ san & okasu

sulfuric acid & eat ↔ ryūsan & okasu

When contexts are matched by analogy, a given example (e.g. a man eats vegetables) may account not only for itself, but also for previously unseen contexts (e.g. he eats potatoes). Conversely to that discussed for exact matches, given a set of examples, it is not known in advance what input sentences the examples can be useful for. In principle, an example might be used for an input sentence with no words in common, as long as there is a term-to-term semantic similarity between the example and the input sentence. Given an example like a man eats vegetables, how does one determine in advance what is the relevant context for an unforeseen translation? A man eats or eats vegetables? Or the two together? Because of the incompleteness assumption, extracting local contexts from examples becomes problematic. Given a set of examples, one can know what local contexts would adequately cover the examples at hand, but one cannot know whether such contexts would be adequate for all possibly relevant sentences. For instance, a sentence like moths eat holes in clothes may require that a larger context be taken into account, for a more informed **similarity assessment**. Therefore, it is advisable to have the largest possible contexts available (i.e. entire examples), so as to be able to use different portions of them depending on the input sentence to be translated. A traditional counterpart of this selection mechanism would be to abstractly encode contexts by means of some sort of semantic features that would act as selectional restrictions. E.g. the distinction between the two senses of eat (respectively translated by taberu and okasu) might be captured by a [± animate] feature, which would select the appropriate subject. This approach would make a database of explicit contexts superfluous, and would also lend itself to a similarity-based approach, if selectional restrictions were used as preferences rather than hard constraints. However, this approach would be labor-intensive and, as Somers (1999:127) points out, it would be "cumbersome and error-prone".

From this informal discussion it appears that translation by analogy, which is the most characteristic technique of EBMT, is also the one where the use of entire examples is most motivated. As a final remark, we only note that an example database for the purpose of translating by analogy can be an additional resource to whatever other resources are used, along the lines discussed above. In principle, translation by analogy could also be an extension to a traditional transfer MT system, to solve cases of lexical ambiguity for which no direct evidence is found in a translation database.

6.5 Robotics

Robotics is the **interdisciplinary** branch of engineering and science that includes mechanical engineering, electrical engineering, computer science, and others. Robotics deals with the design, construction, operation, and use of robots, as well as computer systems for their control, sensory feedback, and information processing.

These technologies are used to develop machines that can substitute for humans. Robots can be used in any situation and for any purpose, but today many are used in dangerous environments (including bomb detection and de-activation), manufacturing processes, or where humans cannot survive. Robots can take on any form but some are made to resemble humans in appearance. This is said to help in the acceptance of a robot in certain replicative behaviors usually performed by people. Such robots attempt to replicate walking, lifting, speech, cognition, and basically anything a human can do. Many of today's robots are inspired by nature, contributing to the field of bio-inspired robotics.

The concept of creating machines that can operate autonomously dates back to classical times, but research into the functionality and potential uses of robots did not grow substantially until the 20th century. Throughout history, it has been frequently assumed that robots will one day be able to mimic human behavior and manage tasks in a human-like fashion. Today, robotics is a rapidly growing field, researching, designing, and building new robots serve various practical purposes, whether domestically, commercially, or militarily. Many robots are built to do jobs that are hazardous to people such as defusing bombs, finding survivors in unstable ruins, and exploring mines and shipwrecks. Robotics is also used in STEM (Science, Technology, Engineering, and Mathematics) as a teaching aid.

6.5.1 Components of Robot

There are many types of robots; they are used in many different environments and for many different uses, although being very diverse in application and form they all share three basic similarities when it comes to their construction.

Robots all have some kind of mechanical construction, a frame, form or shape designed to achieve a particular task. For example, a robot designed to travel across heavy dirt or mud, might use caterpillar tracks. The mechanical aspect is mostly the creator's solution to completing the assigned task and dealing with the physics of the environment around it. Form follows function.

Robots have electrical components which power and control the machinery. For example, the robot with caterpillar tracks would need some kind of power to move the tracker treads. That power comes in the form of electricity, which will have to travel through a wire and originate from a battery, a basic electrical circuit. Even petrol powered machines that get their power mainly from petrol still require an electric current to start the combustion process which is why most petrol powered machines like cars, have batteries. The electrical aspect of robots is used for movement (through motors), sensing (where electrical signals are used to measure things like heat, sound, position, and energy status) and operation (robots need some level of electrical energy supplied to their motors and sensors in order to activate and perform basic operations).

All robots contain some level of computer programming code. A program is how a robot decides when or how to do something. In the caterpillar track example, a robot that needs to move across a muddy road may have the correct mechanical construction and receive the correct amount of power from its battery, but would not go anywhere without a program telling it to move. Programs are the core essence of a robot, it could have excellent mechanical and electrical construction, but if its program is poorly constructed its performance will be very poor (or it may not perform at all). There are three different types of robotic programs: remote control, artificial intelligence and hybrid. A robot with remote control programing has a preexisting set of commands that it will only perform if and when it receives a signal from a control source, typically a human being with a remote control. It is perhaps more appropriate to view devices controlled primarily by human commands as falling in the discipline of automation rather than robotics. Robots that use artificial intelligence interact with their environment on their own without a control source, and can determine reactions to objects and problems they encounter using their preexisting programming. Hybrid is a form of programming that incorporates both AI and RC functions.

1. Power Source

At present, mostly (lead–acid) batteries are used as a power source. Many different types of batteries can be used as a power source for robots. They range from lead–acid batteries, which are safe and have relatively long shelf lives but are rather heavy compared to silver–cadmium batteries that are much smaller in volume and are currently much more expensive. Designing a battery-powered robot needs to take into account factors such as safety, cycle lifetime and weight. **Generators**, often some type of **internal combustion engine**, can also be used. However, such designs are often mechanically complex and need a fuel, require heat dissipation and are relatively heavy. A tether connecting the robot to a power supply would remove the power supply from the robot entirely. This has the advantage of saving weight and space by moving all power generation and storage components elsewhere. However, this design does come with the drawback of constantly having a cable connected to the robot, which can be difficult to manage. Potential power sources could be:

(1) pneumatic (compressed gases)

(2) Solar power (using the sun's energy and converting it into electrical power)

(3) hydraulics (liquids)

(4) flywheel energy storage

(5) organic garbage (through anaerobic digestion)

(6) nuclear

2. Actuation

Actuators are the "muscles" of a robot, the parts which convert stored energy into movement. By far the most popular actuators are electric motors that rotate a wheel or gear, and linear actuators that control industrial robots in factories. There are some recent advances in alternative types of actuators, powered by electricity, chemicals, or compressed air.

1) Motors

The vast majority of robots use electric motors, often brushed and **brushless DC motors** in portable robots or **AC motors** in industrial robots and **CNC machines**. These motors are often preferred in systems with lighter loads, and where the predominant form of motion is rotational.

2) Linear Actuators

Various types of linear actuators move in and out instead of by spinning, and often have quicker direction changes, particularly when very large forces are needed such as with industrial robotics. They are typically powered by compressed air (**pneumatic actuator**) or oil (**hydraulic actuator**).

3) Series Elastic Actuators

A flexure is designed as part of the motor actuator, to improve safety and provide robust force control, energy efficiency, shock absorption (mechanical filtering) while reducing excessive wear on the transmission and other mechanical components. The resultant lower reflected inertia can improve safety when a robot is interacting with humans or during collisions. It has been used in various robots, particularly advanced manufacturing robots and walking humanoid robots.

4) Air Muscles

Pneumatic artificial muscles, also known as air muscles, are special tubes that expand (typically up to 40%) when air is forced inside them. They are used in some robot applications.

5) Muscle Wire

Muscle wire, also known as **shape memory alloy**, Nitinol® or Flexinol® wire, is a material which contracts (under 5%) when electricity is applied. They have been used for some small robot applications.

6) **Electroactive Polymers**

EAPs or EPAMs are a new plastic material that can contract substantially (up to 380% activation strain) from electricity, and have been used in facial muscles and arms of humanoid robots, and to enable new robots to float, fly, swim or walk.

7) Piezo Motors

Recent alternatives to DC motors are **piezo motors** or **ultrasonic motors**. These work on a fundamentally different principle, whereby tiny **piezoceramic** elements, vibrating many thousands of times per second, cause linear or rotary motion. There are different mechanisms of operation; one type uses the vibration of the piezo elements to step the motor in a circle or a straight line. Another type uses the piezo elements to cause a nut to vibrate or to drive a screw. The advantages of these motors are **nanometer** resolution, speed, and available force for their size. These motors are already available commercially, and being used on some robots.

8) Elastic Nanotubes

Elastic nanotubes are a promising artificial muscle technology in early-stage experimental development. The absence of defects in carbon nanotubes enables these filaments to deform elastically by several percent, with energy storage levels of perhaps 10 J/cm^3 for metal nanotubes. Human **biceps** could be replaced with an 8 mm diameter wire of this material. Such compact "muscle" might allow future robots to outrun and outjump humans.

3. Sensing

Sensors allow robots to receive information about a certain measurement of the environment, or internal components. This is essential for robots to perform their tasks, and act upon any changes in the environment to calculate the appropriate response. They are used for various forms of measurements, to give the robots warnings about safety or malfunctions, and to provide real-time information of the task it is performing.

1) Touch

Current robotic and **prosthetic hands** receive far less tactile information than the human hand. Recent research has developed a **tactile sensor** array that mimics the mechanical properties and **touch receptors** of human fingertips. The sensor array is constructed as a rigid core surrounded by conductive fluid contained by an elastomeric skin. **Electrodes** are mounted on the surface of the rigid core and are connected to an impedance-measuring device within the core. When the artificial skin touches an object, the fluid path around the electrodes is deformed, producing **impedance** changes that map the forces received from the object. The researchers expect that an important function of such artificial fingertips will be adjusting robotic grip on held objects.

Scientists from several European countries and Israel developed a prosthetic hand in 2009, called SmartHand, which functions like a real one—allowing patients to write with it, type on a keyboard, play piano and perform other fine movements. The prosthesis has sensors which enable the patient to sense real feeling in its fingertips.

2) Vision

Computer vision is the science and technology of machines that see. As a scientific discipline, computer vision is concerned with the theory behind artificial systems that extract information from images. The image data can take many forms, such as video sequences and views from cameras.

In most practical computer vision applications, the computers are pre-programmed to solve a particular task, but methods based on learning are now becoming increasingly common.

Computer vision systems rely on image sensors which detect **electromagnetic radiation** which is typically in the form of either visible light or infra-red light. The sensors are designed using **solid-state physics**. The process by which light propagates and reflects off surfaces is explained using optics. Sophisticated image sensors even require **quantum mechanics** to provide a complete understanding of the image formation process. Robots can also be equipped with multiple vision sensors to be better able to compute the sense of depth in the environment. Like human eyes, robots' "eyes" must also be able to focus on a particular area of interest, and also adjust to variations in light intensities.

There is a subfield within computer vision where artificial systems are designed to mimic the processing and behavior of biological system, at different levels of complexity. Also, some of the learning-based methods developed within computer vision have their background in biology.

6.5.2 Control System

The mechanical structure of a robot must be controlled to perform tasks. The control of a robot involves three distinct phases — perception, processing, and action (robotic paradigms). Sensors give information about the environment or the robot itself (e.g. the position of its joints or its end effector). This information is then processed to be stored or transmitted and to calculate the appropriate signals to the actuators (motors) which move the mechanical.

The processing phase can range in complexity. At a reactive level, it may translate raw sensor information directly into actuator commands. **Sensor fusion** may first be used to estimate parameters of interest (e.g. the position of the robot's gripper) from noisy sensor data. An immediate task (such as moving the gripper in a certain direction) is inferred from these estimates. Techniques from control theory convert the task into commands that drive the actuators.

At longer time scales or with more sophisticated tasks, the robot may need to build and reason with a "cognitive" model. Cognitive models try to represent the robot, the world, and how they interact. Pattern recognition and computer vision can be used to track objects. Mapping techniques can be used to build maps of the world. Finally, motion planning and other artificial intelligence techniques may be used to figure out how to act. For example, a planner may figure out how to achieve a task without hitting obstacles, falling over, etc.

Control systems may also have varying levels of autonomy.

(1) Direct interaction is used for haptic or teleoperated devices and human has nearly complete control over the robot's motion.

(2) Operator-assist modes have the operator commanding medium-to-high-level tasks, with the robot automatically figuring out how to achieve them.

(3) An autonomous robot may go without human interaction for extended periods of time. Higher levels of autonomy do not necessarily require more complex cognitive capabilities. For example, robots in assembly plants are completely autonomous but operate in a fixed pattern.

Another classification takes into account the interaction between human control and the machine motions.

(1) Teleoperation. A human controls each movement, each machine actuator change is specified by the operator.

(2) Supervisory. A human specifies general moves or position changes and the machine decides specific movements of its actuators.

(3) Task-level autonomy. The operator specifies only the task and the robot manages itself to complete it.

(4) Full autonomy. The machine will create and complete all its tasks without human interaction.

6.5.3 Environmental Interaction and Navigation

Though a significant percentage of robots in commission today are either human controlled or operate in a static environment, there is an increasing interest in robots that can operate autonomously in a dynamic environment. These robots require some combination of navigation hardware and software in order to traverse their environment. In particular, unforeseen events (e.g. people and other obstacles that are not stationary) can cause problems or collisions. Some highly advanced robots such as ASIMO and Meinü robot have particularly good robot navigation hardware and software. Also, self-controlled cars, Ernst Dickmanns' driverless car, and the entries in the DARPA Grand Challenge, are capable of sensing the environment well and subsequently making navigational decisions based on this information. Most of these robots employ a GPS navigation device with waypoints, along with radar, sometimes combined with other sensory data such as **lidar**, video cameras, and inertial guidance systems for better navigation between waypoints.

1. Human-robot Interaction

The state of the art in sensory intelligence for robots will have to progress through several orders of magnitude if we want the robots working in our homes to go beyond vacuum-cleaning the floors. If robots are to work effectively in homes and other non-industrial environments, the way they are instructed to perform their jobs and especially how they will be told to stop will be of critical importance. The people who interact with them may have little or no training in robotics, and so any interface will need to be extremely intuitive. Science fiction authors also typically assume that robots will eventually be capable of communicating with humans through speech, gestures, and facial expressions, rather than a command-line interface. Although speech would be the most natural way for the human to communicate, it is unnatural for the robot. It will probably be a long time before robots interact as naturally as the fictional C-3PO, or Data of Star Trek, Next Generation.

2. Speech Recognition

Interpreting the continuous flow of sounds coming from a human, in real time, is a difficult task for a computer, mostly because of the great variability of speech. The same word, spoken by the same person may sound different depending on local acoustics, volume, the previous word, whether or not the speaker has a cold, etc.. It becomes even harder when the speaker has a different accent. Nevertheless, great strides have been made in the field since Davis, Biddulph, and Balashek designed the first "voice input system" which recognized "ten digits spoken by a single user with 100% accuracy" in 1952. Currently, the best systems can recognize continuous, natural speech, up to 160 words per minute, with an accuracy of 95%.

3. Robotic Voice

Other hurdles exist when allowing the robot to use voice for interacting with humans. For social reasons, synthetic voice proves suboptimal as a communication medium, making it necessary to develop the emotional component of robotic voice through various techniques.

4. Gestures

One can imagine, in the future, explaining to a robot chef how to make a pastry, or asking directions from a robot police officer. In both of these cases, making hand gestures would aid the verbal descriptions. In the first case, the robot would be recognizing gestures made by the human, and perhaps repeating them for confirmation. In the second case, the robot police officer would gesture to indicate "down the road, then turn right". It is likely that gestures will make up a part of the interaction between humans and robots. A great many systems have been developed to recognize human hand gestures.

5. Facial Expression

Facial expressions can provide rapid feedback on the progress of a dialog between two humans, and soon may be able to do the same for humans and robots. Robotic faces have been constructed by Hanson Robotics using their elastic polymer called Frubber, allowing a large number of facial expressions due to the elasticity of the rubber facial coating and embedded subsurface motors (servos). The coating and servos are built on a metal skull. A robot should know how to approach a human, judging by their facial expression and body language. Whether the person is happy, frightened, or crazy-looking affects the type of interaction expected of the robot. Likewise, robots like Kismet and the more recent addition, Nexi can produce a range of facial expressions, allowing it to have meaningful social exchanges with humans.

6. Artificial Emotions

Artificial emotions can also be generated, composed of a sequence of facial expressions and/or gestures. As seen from the movie **Final Fantasy: The Spirits Within**, the programming of these artificial emotions is complex and requires a large amount of human observation. To simplify this programming in the movie, presets were created together with a special software program. This decreased the amount of time needed to make the film. These presets could possibly be transferred for use in real-life robots.

7. Personality

Many of the robots of science fiction have a personality, something which may or may not be desirable in the commercial robots of the future. Nevertheless, researchers are trying to create robots which appear to have a personality: i.e. they use sounds, facial expressions, and body language to try to convey an internal state, which may be joy, sadness, or fear. One commercial example is Pleo, a toy robot dinosaur, which can exhibit several apparent emotions.

8. Social Intelligence

The Socially Intelligent Machines Lab of the Georgia Institute of Technology researches new concepts of guided teaching interaction with robots. Aim of the projects is a social robot learns task goals from human demonstrations without prior knowledge of high-level concepts. These new concepts are grounded from low-level continuous sensor data through unsupervised learning, and task goals are subsequently learned using a Bayesian approach. These concepts can be used to transfer knowledge to future tasks, resulting in faster learning of those tasks. The results are demonstrated by the robot Curi who can scoop some pasta from a pot onto a plate and serve the sauce on top.

6.5.4 Top 10 Humanoid Robots①

A **humanoid robot** is a robot with its body shape built to resemble the human body. The design may be for functional purposes, such as interacting with human tools and environments, for experimental purposes, such as the study of **bipedal locomotion**, or for other purposes. In general, humanoid robots have a torso, a head, two arms, and two legs, though some forms of humanoid robots may model only part of the body, for example, from the waist up. Some humanoid robots also have heads designed to replicate human facial features such as eyes and mouths. Androids are humanoid robots built to aesthetically resemble humans.

Just late last year it was posited that the humanoid robot was poised to take a leap from a mere facsimile of human behavior, to one that futurists suggest, will walk like a human and possess self awareness as well as a full range of high-tech computational spectrum analysis and capabilities and emotions.

Some are predicting that robots of all types could fully replace humans by 2045. Artificial intelligence is now advancing to a point where a new type of brain can be offered to complement the relatively menial tasks of modern-day robotics, hinting at the next stage of machine evolution.

The current list of robots designed over the last few years to match human capability demonstrate what is described above could become reality sooner than we think.

1. Atlas Unplugged

The Atlas robot was developed by Google-owned Boston Dynamics with the US Defense Advanced Research Projects Agency (DARPA) for its robotics challenge, designed to negotiate rough, outdoor terrain in a bipedal manner, while being able to climb using hands and feet as a human would.

The first version, released in July 2013, required an electrical and control tether to power and operate the robot. The new generation of the robot, dubbed "Atlas Unplugged" as it can operate on battery power and be controlled wirelessly, has been developed for the DAPRA Robotics Challenge finals, which are set to take place in June.

The latest version of Atlas as shown in Fig.6-3 is slightly taller and heavier than before, standing 6ft 2in (1.88m) high and weighing 156.4kg (345lb). According to its manufacturer, Google's Boston Dynamics division, 75% of the humanoid machine is new — only its lower legs and feet remain unchanged.

---

① https://wtvox.com/robotics/top-10-humanoid-robots/

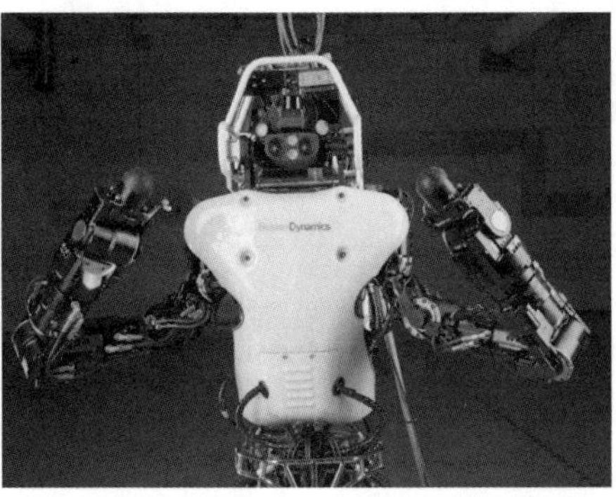

Fig.6-3 The latest version of Atlas

2. ASIMO And Honda P-Series

ASIMO (see Fig.6-4) is the 11th in a line of walking robots developed by Honda, called the P-Series. Unveiled in 2000, ASIMO could walk and run like a human, which was an amazing feat. ASIMO had a significant upgrade in 2005, that allowed him to run twice as fast (6 km/hr or 4.3mph), interact with humans, and perform basic tasks like holding a platter and serving food. The current ASIMO models number about 100 worldwide, stand 1.28 m tall and weight about 55 kg.

ASIMO, with his space-suit looking appearance, is cheerful and endearing. He has paved the way for many subsequent walking, human-like robots, but still holds his own as an advanced and powerful robot.

ASIMO is a great boon to Honda's global branding, and helps the company's appearance of innovation and technology. ASIMO has also appeared in commercials for Honda and makes many celebrity appearances. ASIMO makes this list because of his winsome appearance, world-wide recognition, and advanced technology.

Fig.6-4 ASIMO

3. iCub

iCub as shown in Fig.6-5 was created by the RobotCub Consortium, of several European universities. The name is a partial acronym, cub standing for Cognitive Universal Body.

The motivation behind the strongly humanoid design is the embodied cognition hypothesis, that human-like manipulation plays a vital role in the development of human cognition. A baby learns many cognitive skills by interacting with its environment and other humans using its limbs and senses, and consequently its internal model of the world is largely determined by the form of the human body.

The robot was designed to test this hypothesis by allowing cognitive learning scenarios to be acted out by an accurate reproduction of the perceptual system and articulation of a small child so that it could interact with the world in the same way that such a child does.

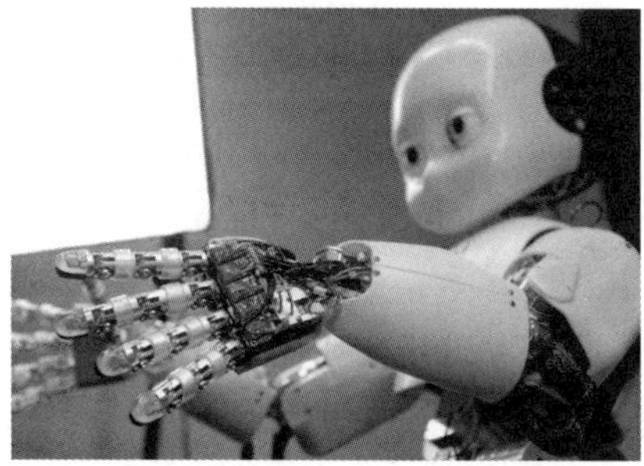

Fig.6-5 iCub

4. Poppy

Poppy (see Fig.6-6) is the latest humanoid robot and presents the first of its kind: Poppy was created by a 3D printer. A group of French researchers were able to cut costs by a third utilizing the latest 3D technology. Poppy's creators have focused on a biologically inspired walking motion that they hope will allow for better human-to-robot interaction.

It has an articulated spine with five motors — almost unheard of in robots of this size. The spine not only allows Poppy to move more naturally, but helps to balance the robot by adjusting its posture. The added flexibility also helps when physically interacting with the robot, such as guiding it by its hands, which is currently required to help the robot walk. You can see the incredibly human-like, heel-toe motion in the video below.

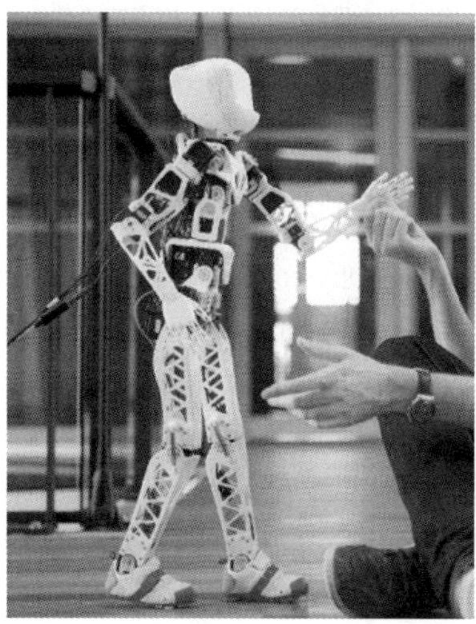

Fig.6-6 Poppy

5. Romeo

Romeo, in Fig.6-7, seeks to become the leader in the areas of robotic care giving and personal assistance with a more emotional element. Romeo builds off of a smaller humanoid robot called NAO that generated more than 5,000 sales or rentals worldwide.

The robot has the size of a child of eight years (1.40 m) and weighs a little more (40 kilos). To be as light as possible, its body is made of carbon fibre and rubber, and was designed to avoid the risk of injury to the person that will attend. Today, Romeo can walk, see the three-dimensional environment, hear and speak.

The timeline for testing the robot in real-world conditions is projected for 2016, with the final objective of being able to use it in a retirement home in the 2017 or 2019. Funded in part by the French government and the European Commission, the Romeo project budget totals 37 million Euros over the period from 2009 to 2016.

Fig.6-7 Romeo

6. Petman

Petman in Fig.6-8 is an anthropomorphic robot designed for testing chemical protection clothing. Natural agile movement is essential for Petman to simulate how a soldier stresses protective clothing under realistic conditions.

Unlike previous suit testers that had a limited repertoire of motion and had to be supported mechanically, Petman balances itself and moves freely; walking, bending and doing a variety of suit-stressing calisthenics during exposure to chemical warfare agents.

Petman also simulates human physiology within the protective suit by controlling temperature, humidity and sweating, all to provide realistic test conditions. The Petman system was delivered to the user's test facility where it is going through validation experiments.

7. NAO

NAO in Fig.6-9 is a 58-cm tall humanoid robot and was created to be a friendly companion around the house. Several versions of the robot have been released since 2008.

The most known NAO is the Academics Edition which was developed for universities and laboratories for research and education purposes. It was released to institutions in 2008, and was made publicly available by 2011. More recent upgrades to the Nao platform include the 2011 Nao Next Gen and the 2014 Nao Evolution.

NAO robots have been used for research and education purposes in numerous academic institutions worldwide. As of 2015, over 5,000 Nao units are in use in 50+ countries.

Fig.6-8 Petman

Fig.6-9 NAO

8. RoboThespian

RoboThespian in Fig.6-10 is a life sized humanoid robot designed for human interaction in a public environment. It is fully interactive, multilingual, and user-friendly, making it a perfect device with which to communicate and entertain.

Now in its third generation, with more than six years of continuous development, RoboThespian is a tried-and-tested platform, trusted by national science centres, visitor attractions, commercial users and academic research institutions around the world.

It comes with standard content, like greetings and impressions, to which you can add your own recorded sequences or bespoke content. With a Web-based interface files controlling movement, sound and video can be simply uploaded.

Fig.6-10 RoboThespian

9. Actroid-SIT

Actroid-SIT in Fig.6-11 can function autonomously, talking and gesturing while interacting with people. In fact, researchers have recently demonstrated how improvements to Actroid's behaviour can make it look smarter and more expressive than your average android. She knows sign language, such as "point" or "swing", that automatically adapts to the position of the interlocutor.

10. Robotic Pole Dancers – Lexy and Tess

At the CeBIT expo in Hanover, German software developer Tobit put together a booth that features two pole dancing robots, egged on by a fellow robot DJ with a megaphone for a head. The two ladybots as shown in Fig.6-12 move and twist in time to the music, though the actual performance is surprisingly tame. According to the BBC, you can pick up a bot of your own for $39,500.

Fig.6-11 Actroid-SIT

Fig.6-12 Lexy and Tess

6.6 Computer Vision

Areas of artificial intelligence deal with autonomous planning or deliberation for robotical systems to navigate through an environment. A detailed understanding of these environments is required to navigate through them. Information about the environment could be provided by a computer vision system, acting as a vision sensor and providing high-level information about the environment and the robot.

Artificial intelligence and computer vision share other topics such as pattern recognition and learning techniques. Consequently, computer vision is sometimes seen as a part of the artificial intelligence field or the computer science field in general.

6.6.1 Brief Introduction

1. What is Computer Vision

Computer Vision has a dual goal. From the biological science point of view, computer vision aims to come up with computational models of the human visual system. From the engineering point of view, computer vision aims to build autonomous systems which could perform some of the tasks which the human visual system can perform (and even surpass it in many cases). Many vision tasks are related to the extraction of 3D and temporal information from time-varying 2D data such as obtained by one or more television cameras, and more generally the understanding of such dynamic scenes.

Of course, the two goals are intimately related. The properties and characteristics of the human visual system often give inspiration to engineers who are designing computer vision systems. Conversely, computer vision algorithms can offer insights into how the human visual system works. In this section we will adopt the engineering point of view.

2. Why is Computer Vision Difficult

Computer Vision as a field of research is notoriously difficult. Almost no research problem has been satisfactorily solved. One main reason for this difficulty is that the human visual system is simply too good for many tasks (e.g., face recognition), so that computer vision systems suffer by comparison. A human can recognize faces under all kinds of variations in illumination, viewpoint, expression, etc. In most cases we have no difficulty in recognizing a friend in a photograph taken many years ago. Also, there appears to be no limit on how many faces we can store in our brains for future recognition. There appears no hope in building an autonomous system with such stellar performance.

Two major related difficulties in computer vision can be identified:

How do we distill and represent the vast amount of human knowledge in a computer in such a way that retrieval is easy?

How do we carry out (in both hardware and software) the vast amount of computation that is often required in such a way that the task (such as face recognition) can be done in real time?

3. Related Fields

Solid-state physics is a field closely related to computer vision. Most computer vision systems rely on **image sensors**, which detect **electromagnetic radiation**, which is typically in the form of either visible or **infra-red light**. The sensors are designed using **quantum physics**. The process by which light interacts with surfaces is explained using physics. Physics explains the behavior of optics which is a core part of most imaging systems. Sophisticated image sensors even require quantum mechanics to provide a complete understanding of the image formation process. Also, various measurement problems in physics can be addressed using computer vision, for example motion in fluids.

Another field which plays an important role is **neurobiology**, specifically the study of the biological vision system. Over the last century, there has been an extensive study of eyes, neurons, and the brain structures devoted to processing of visual stimuli in both humans and various animals. This has led to a coarse, yet complicated, description of how "real" vision systems operate in order to solve certain vision related tasks. These results have led to a subfield within computer vision where artificial systems are designed to mimic the processing and behavior of biological systems, at different levels of complexity. Also, some of the learning-based methods developed within computer vision (e.g. neural net and deep learning based image and feature analysis and classification) have their background in biology.

Some strands of computer vision research are closely related to the study of biological vision — indeed, just as many strands of AI research are closely tied with research into human consciousness, and the use of stored knowledge to interpret, integrate and utilize visual information. The field of biological vision studies and models the physiological processes behind visual perception in humans and other animals. Computer vision, on the other hand, studies and describes the processes implemented in software and hardware behind artificial vision systems. Interdisciplinary exchange between biological and computer vision has proven fruitful for both fields.

Yet another field related to computer vision is **signal processing**. Many methods for processing of one-variable signals, typically temporal signals, can be extended in a natural way to processing of two-variable signals or multi-variable signals in computer vision. However, because of the specific nature of images there are many methods developed within computer vision which have no counterpart in processing of one-variable signals. Together with the multi-dimensionality of the signal, this defines a subfield in signal processing as a part of computer vision.

Beside the above-mentioned views on computer vision, many of the related research topics can also be studied from a purely mathematical point of view. For example, many methods in computer vision are based on statistics, optimization or geometry. Finally, a significant part of the field is devoted to the implementation aspect of computer vision; how existing methods can be realized in various combinations of software and hardware, or how these methods can be modified in order to gain processing speed without losing too much performance.

The fields most closely related to computer vision are image processing, image analysis and machine vision. There is a significant overlap in the range of techniques and applications. This implies that the basic techniques that are used and developed in these fields are more or less identical, something which can be interpreted as there is only one field with different names. On the other hand, it appears to be necessary for research groups, scientific journals, conferences and companies to present or market themselves as belonging specifically to one of these fields and, hence, various characterizations that distinguish each of the fields from the others have been presented.

6.6.2 Tasks of Computer Vision

Some examples of typical computer vision tasks are presented below.

1. Recognition

The classical problem in computer vision, image processing, and machine vision is that of determining whether or not the image data contains some specific object, feature, or activity. Different varieties of the recognition problem are described in the literature:

Object recognition (also called object classification) — one or several pre-specified or learned objects or object classes can be recognized, usually together with their 2D positions in the image or 3D poses in the scene.

Identification – an individual instance of an object is recognized. Examples include identification of a specific person's face or fingerprint, identification of handwritten digits, or identification of a specific vehicle.

Detection – the image data are scanned for a specific condition. Examples include detection of possible abnormal cells or tissues in medical images or detection of a vehicle in an automatic road toll system. Detection based on relatively simple and fast computations is sometimes used for finding smaller regions of interesting image data which can be further analyzed by more computationally demanding techniques to produce a correct interpretation.

Currently, the best algorithms for such tasks are based on **convolutional neural networks**. An illustration of their capabilities is given by the ImageNet Large Scale Visual Recognition Challenge; this is a benchmark in object classification and detection, with millions of images and hundreds of object classes. Performance of convolutional neural networks, on the ImageNet tests, is now close to that of humans. The best algorithms still struggle with objects that are small or thin, such as a small ant on a stem of a flower or a person holding a quill in their hand. They also have trouble with images that have been distorted with filters (an increasingly common phenomenon with modern digital cameras). By contrast, those kinds of images rarely trouble humans. Humans, however, tend to have trouble with other issues. For example, they are not good at classifying objects into fine-grained classes, such as the particular breed of dog or species of bird, whereas convolutional neural networks handle this with ease.

2. Motion Analysis

Several tasks relate to motion estimation where an image sequence is processed to produce an estimate of the velocity either at each points in the image or in the 3D scene, or even of the camera that produces the images. Examples of such tasks are:

Ego motion – determining the 3D rigid motion (rotation and translation) of the camera from an image sequence produced by the camera.

Tracking – following the movements of a (usually) smaller set of interest points or objects (e.g., vehicles or humans) in the image sequence.

Optical flow – to determine, for each point in the image, how that point is moving relative to the image plane, i.e., its apparent motion. This motion is a result both of how the corresponding 3D point is moving in the scene and how the camera is moving relative to the scene.

3. Scene Reconstruction

Given one or (typically) more images of a scene, or a video, **scene reconstruction** aims at computing a 3D model of the scene. In the simplest case the model can be a set of 3D points. More sophisticated methods produce a complete 3D surface model.

4. Image Restoration

The aim of **image restoration** is the removal of noise (sensor noise, motion blur, etc.) from images. The simplest possible approach for noise removal is various types of filters such as **low-pass filters** or **median filters**. More sophisticated methods assume a model of how the local image structures look like, a model which distinguishes them from the noise. By first analyzing the image data in terms of the local image structures, such as lines or edges, and then controlling the filtering based on local information from the analysis step, a better level of noise removal is usually obtained compared to the simpler approaches.

6.6.3 An Example — Facial Recognition System

A face recognition system is a computer application capable of identifying or verifying a person from a digital image or a video frame from a video source. One of the ways to do this is by comparing selected facial features from the image and a face database.

It is typically used in security systems and can be compared to other biometrics such as fingerprint or eye iris recognition systems. Recently, it has also become popular as a commercial identification and marketing tool.

1. Techniques Used in Facial Recognition

1) Traditional Methods

Some face recognition algorithms identify facial features by extracting landmarks, or features, from an image of the subject's face. For example, an algorithm may analyze the relative position, size, and/or shape of the eyes, nose, cheekbones, and jaw. These features are then used to search for other images with matching features. Other algorithms normalize a gallery of face images and then compress the face data, only saving the data in the image that is useful for face recognition. A probe image is then compared with the face data. One of the earliest successful systems is based on template matching techniques applied to a set of salient facial features, providing a sort of compressed face representation.

Recognition algorithms can be divided into two main approaches, geometric, which looks at distinguishing features, or **photometric**, which is a statistical approach that distills an image into values and compares the values with templates to eliminate variances.

Popular recognition algorithms include **principal component analysis** using **eigenfaces**, linear discriminant analysis, hidden Markov model, multilinear subspace learning using **tensor** representation, and the neuronal motivated dynamic link matching.

2) 3-dimensional Recognition

Newly emerging trend, claimed to achieve improved accuracy, is three-dimensional face recognition. This technique uses 3D sensors to capture information about the shape of a face. This information is then used to identify distinctive features on the surface of a face, such as the contour of the eye sockets, nose, and chin.

One advantage of 3D face recognition is that it is not affected by changes in lighting like other techniques. It can also identify a face from a range of viewing angles, including a profile view. Three-dimensional data points from a face vastly improve the precision of face recognition. 3D research is enhanced by the development of sophisticated sensors that do a better job of capturing 3D face imagery. The sensors work by projecting structured light onto the face. Up to a dozen or more of these image sensors can be placed on the same CMOS chip — each sensor captures a different part of the spectrum.

A new method is to introduce a way to capture a 3D picture by using three tracking cameras that point at different angles; one camera will be pointing at the front of the subject, second one to the side, and third one at an angle. All these cameras will work together so it can track a subject's face in real time and be able to face detect and recognize.

3) Skin Texture Analysis

Another emerging trend uses the visual details of the skin, as captured in standard digital or scanned images. This technique, called **skin texture analysis**, turns the unique lines, patterns, and spots apparent in a person's skin into a mathematical space. Tests have shown that with the addition of skin texture analysis, performance in recognizing faces can increase 20 to 25 percent.

4) Thermal Cameras

A different form of taking input data for face recognition is by using **thermal cameras**, by this procedure the cameras will only detect the shape of the head and it will ignore the subject accessories such as glasses, hats, or make up. A problem with using thermal pictures for face recognition is that the databases for face recognition is limited. The research uses low-sensitive, low-resolution ferroelectric electrics sensors that are capable of acquire long wave thermal infrared (LWIR). The results show that a fusion of LWIR and regular visual cameras has the greater results in outdoor probes.

2. Notable Users and Deployments

The Australian and New Zealand Customs Services have an automated border processing system called SmartGate that uses face recognition. The system compares the face of the individual with the image in the e-passport microchip to verify that the holder of the passport is the rightful owner.

Law enforcement agencies in the United States, including the Los Angeles County Sheriff, use arrest mugshot databases in their forensic investigative work. Law enforcement has been rapidly building a database of photos in recent years.

The U.S. Department of State operates one of the largest face recognition systems in the world with 117 million American adults including mostly law abiding citizens in its database. The photos are typically drawn from driver's license photos. Although it is still far from completion, it is being put to use in certain cities to give clues as to who was in the photo. The FBI uses the photos as an investigative lead not for positive identification.

In recent years Maryland has used face recognition by comparing people's faces to their driver's license photos. The system drew controversy when it was used in Baltimore to arrest unruly protesters after the death of Freddie Gray in police custody. Many other states are using or developing a similar system; however some states have laws prohibiting its use.

The FBI has also instituted its Next Generation Identification program to include face recognition, as well as more traditional biometrics like fingerprints and iris scans, which can pull from both criminal and civil databases.

In addition to being used for security systems, authorities have found a number of other applications for face recognition systems. While earlier post-9/11 deployments were well publicized trials, more recent deployments are rarely written about due to their covert nature.

At Super Bowl XXXV in January 2001, police in Tampa Bay, Florida used Visage face recognition software to search for potential criminals and terrorists in attendance at the event. 19 people with minor criminal records were potentially identified. In the 2000 presidential election, the Mexican government employed face recognition software to prevent voter fraud. Some individuals had been registering to vote under several different names, in an attempt to place multiple votes. By comparing new face images to those already in the voter database, authorities were able to reduce duplicate registrations. Similar technologies are being used in the United States to prevent people from obtaining fake identification cards and driver's licenses.

There are also a number of potential uses for face recognition that are currently being developed. For example, the technology could be used as a security measure at ATMs. Instead of using a bank card or personal identification number, the ATM would capture an image of the customer's face, and compare it to the account holder's photo in the bank database to confirm the customer's identity.

Face recognition systems are used to unlock software on mobile devices. An independently developed Android Marketplace app called Visidon Applock makes use of the phone's built-in camera to take a picture of the user. Face recognition is used to ensure only this person can use certain apps which they choose to secure.

Face detection and face recognition are integrated into the iPhoto application for Macintosh, to help users organize and caption their collections.

Also, in addition to biometric usages, modern digital cameras often incorporate a facial detection system that allows the camera to focus and measure exposure on the face of the subject, thus guaranteeing a focused portrait of the person being photographed. Some cameras, in addition, incorporate a smile shutter, or automatically take a second picture if someone blinks during exposure.

Because of certain limitations of fingerprint recognition systems, face recognition systems could be used as an alternative way to confirm employee attendance at work for the claimed hours.

Another use could be a portable device to assist people with prosopagnosia in recognizing their acquaintances.

6.7 Existential Risk from Artificial General Intelligence

6.7.1 Overview

Artificial Intelligence: A Modern Approach[①], a standard undergraduate AI textbook, cites

① By Stuart Russell and Peter Norvig

the possibility that an AI system's learning function "may cause it to evolve into a system with unintended behavior" as the most serious existential risk from AI technology. Citing major advances in the field of AI and the potential for AI to have enormous long-term benefits or costs, the 2015 Open Letter on Artificial Intelligence stated:

The progress in AI research makes it timely to focus research not only on making AI more capable, but also on maximizing the societal benefit of AI. Such considerations motivated the AAAI 2008-09 Presidential Panel on Long-Term AI Futures and other projects on AI impacts, and constitute a significant expansion of the field of AI itself, which up to now has focused largely on techniques that are neutral with respect to purpose. We recommend expanded research aimed at ensuring that increasingly capable AI systems are robust and beneficial: **our AI systems must do what we want them to do**.

Institutions such as the Machine Intelligence Research Institute, the Future of Humanity Institute, the Future of Life Institute, and the Centre for the Study of Existential Risk are currently involved in mitigating existential risk from advanced artificial intelligence, for example by research into friendly artificial intelligence.

If **superintelligent** AI is possible, and if it is possible for a superintelligence's goals to conflict with basic human values, then AI poses a risk of human extinction. (6-6) A superintelligence, which can be defined as a system that exceeds the capabilities of humans in every relevant endeavor, can **outmaneuver** humans any time when its goals conflict with human goals; therefore, unless the superintelligence decides to allow humanity to coexist, the first superintelligence to be created will inexorably result in human extinction.

There is no physical law precluding particles from being organized in ways that perform even more advanced computations than the arrangements of particles in human brains. The emergence of superintelligence, if or when it occurs, may take the human race by surprise. An explosive transition is possible: as soon as human-level AI is possible, machines with human intelligence could repeatedly improve their design even further and quickly become superhuman. Just as the current-day survival of chimpanzees is dependent on human decisions, so too would human survival depend on the decisions and goals of the superhuman AI. The result could be human extinction, or some other unrecoverable permanent global catastrophe.

6.7.2 Risk Scenarios

In 2009, experts attended a conference hosted by the American Association for Artificial Intelligence (AAAI) to discuss whether computers and robots might be able to acquire any sort of autonomy, and how much these abilities might pose a threat or hazard. They noted that some robots have acquired various forms of semi-autonomy, including being able to find power sources on their own and being able to independently choose targets to attack with weapons. They also noted that some computer viruses can evade elimination and have achieved "cockroach intelligence". They concluded that self-awareness as depicted in science fiction is probably unlikely, but that there were other potential hazards and pitfalls.

The 2010s have seen substantial gains in AI functionality and autonomy. Citing work by Nick Bostrom, entrepreneurs Bill Gates and Elon Musk have expressed concerns about the possibility that AI could eventually advance to the point that humans could not control it. AI researcher Stuart Russell summarizes:

The primary concern is not spooky emergent consciousness but simply the ability to make high-quality decisions. Here, quality refers to the expected outcome utility of actions taken, where the utility function is, presumably, specified by the human designer. Now we have problems about:

(1) The utility function may not be perfectly aligned with the values of the human race, which are (at best) very difficult to pin down.

(2) Any sufficiently capable intelligent system will prefer to ensure its own continued existence and to acquire physical and computational resources — not for their own sake, but to succeed in its assigned task.

A system that is optimizing a function of n variables, where the objective depends on a subset of size $k<n$, will often set the remaining unconstrained variables to extreme values; if one of those unconstrained variables is actually something we care about, the solution found may be highly undesirable. A highly capable decision maker — especially one connected through the Internet to all the world's information and billions of screens and most of our infrastructure — can have an irreversible impact on humanity.

This is not a minor difficulty. Improving decision quality, irrespective of the utility function chosen, has been the goal of AI researches — the mainstream goal on which we now spend billions per year, not the secret plot of some lone evil genius.

(1) Poorly specified goals: "Be careful what you wish for".

The first of Russell's concerns is that autonomous AI systems may be assigned the wrong goals by accident. Dietterich and Horvitz note that this is already a concern for existing systems: "An important aspect of any AI system that interacts with people is that it must reason about what people intend rather than carrying out commands literally". This concern becomes more serious as AI software advances in autonomy and flexibility.

Isaac Asimov's Three Laws of Robotics are one of the earliest examples of proposed safety measures for AI agents. They are:

① A robot may not injure a human being or, through inaction, allow a human being to come to harm.

② A robot must obey the orders given it by human beings except where such orders would conflict with the First Law.

③ A robot must protect its own existence as long as such protection does not conflict with the First or Second Laws.

Asimov's laws were intended to prevent robots from harming humans. In Asimov's stories, problems with the laws tend to arise from conflicts between the rules as stated and the moral intuitions and expectations of humans. Citing work by AI theorist Eliezer Yudkowsky, Russell and Norvig note that a realistic set of rules and goals for an AI agent will need to incorporate a mechanism for learning human values over time: "We can't just give a program a static utility function, because circumstances, and our desired responses to circumstances, change over time. "

Misspecified goals were most apparent, and very real, in the early 1980s. Douglas Lenat's EURISKO, a heuristic learning program, was created with the capability of modifying itself to add new ideas, expand existing ones, or remove them entirely if they were deemed unnecessary. The program even went so far as to bend the rules for discovering new rules; in essence, it was capable of creating new ways for creativity. The program ended up becoming too creative and would self-modify too often, causing Lenat to limit its self-modification capacity. Without Lenat doing so, EURISKO would suffer from "goal mutation" where its initial task would be deemed unnecessary and a new goal deemed more appropriate. This "goal mutation" would have had the potential to change an initial idea for ordering drones to scan an area for potential threats, to ordering drones to eliminate any and all possible targets in range.

(2) Difficulties of modifying goal specification after launch.

While current goal-based AI programs are not intelligent enough to think of resisting programmer's attempts to modify it, a sufficiently advanced, rational, "self-aware" AI might resist any changes to its goal structure, just as Gandhi would not want to take a pill that makes him want to kill people. If the AI were superintelligent, it would be likely to out-maneuver its human operators and prevent being "turned off" or being programmed with a new goal.

(3) Instrumental goal convergence: Would superintelligence just ignore us?

There are some goals that almost any artificial intelligence might pursue, like acquiring additional resources or self-preservation. This could prove problematic because it might put an artificial intelligence in direct competition with humans.

Citing Steve Omohundro's work on the idea of instrumental convergence, Russell and Norvig write that "even if you only want your program to play chess or prove theorems, if you give it the capability to learn and alter itself, you need safeguards". Highly capable and autonomous planning systems require additional checks because of their potential to generate plans that treat humans adversarially, as competitors for limited resources.

(4) **Orthogonality**: Does intelligence inevitably result in moral wisdom?

One common belief is that any superintelligent program created by humans would be subservient to humans, or, better yet, would (as it grows more intelligent and learns more facts about the world) spontaneously "learn" a moral truth compatible with human values and would adjust its goals accordingly. Nick Bostrom's "orthogonality thesis" argues against this, and instead states that, with some technical caveats, more or less any level of "intelligence" or "optimization power" can be combined with more or less any ultimate goal. If a machine is created and given the sole purpose to enumerate the decimals of π then no moral and ethical rules will stop it from achieving its programmed goal by any means necessary. The machine may utilize all physical and informational resources it can to find every decimal of π that can be found. Bostrom warns against anthropomorphism: A human will set out to accomplish his projects in a manner that humans consider "reasonable"; an artificial intelligence may hold no regard for its existence or for the welfare of humans around it, only for the completion of the task.

Computer scientist Stuart Russell says the difficulty of aligning the goals of a superintelligence with human goals lies in the fact that, while (according to Russell) humans tend to mostly share the same values as each other, artificial superintelligences would not necessarily start out with the same values as humans.

In a paper submitted to the 2014 AAAI Spring Symposium, Richard Loosemore disagreed with Bostrom, arguing that any artificial general intelligence would self-modify to avoid **pathological** outcomes.

(5) "Optimization power" vs. normatively thick models of intelligence.

Part of the disagreement about whether a superintelligence machine would behave morally may arise from a terminological difference. Outside of the artificial intelligence field, "intelligence" is often used in a normatively thick manner that connotes moral wisdom or acceptance of agreeable forms of moral reasoning. At an extreme, if morality is part of the definition of intelligence, then by definition a superintelligent machine would behave morally. However, in artificial intelligence, while "intelligence" has many overlapping definitions, none of them reference morality. Instead, almost all current "artificial intelligence" research focuses on creating algorithms that "optimize", in an empirical way, the achievement of an arbitrary goal. To avoid anthropomorphism or the baggage of the word "intelligence", an advanced artificial intelligence can be thought of as an impersonal "optimizing process" that strictly takes whatever actions are judged most likely to accomplish its (possibly complicated and implicit) goals. Another way of conceptualizing an advanced artificial intelligence is to imagine a time machine that sends backward in time information about which choice always leads to the maximization of its goal function; this choice is then output, regardless of any extraneous ethical concerns.

(6) Anthropomorphism.

In science fiction, an AI, even though it has not been programmed with human emotions, often spontaneously experiences those emotions anyway: for example, Agent Smith in The Matrix was influenced by a "disgust" toward humanity. This is fictitious anthropomorphism: in reality, while an artificial intelligence could perhaps be deliberately programmed with human emotions, or could develop something similar to an emotion as a means to an ultimate goal if it is useful to do so, it would not spontaneously develop human emotions for no purpose whatsoever, as portrayed in fiction.

One example of anthropomorphism would be to believe that your PC is angry at you because you insulted it; another would be to believe that an intelligent robot would naturally find a woman sexy and be driven to mate with her. Scholars sometimes disagree with each other about whether a particular prediction about an AI's behavior is logical, or whether the prediction constitutes illogical **anthropomorphism**. An example that might initially be considered anthropomophism, but is in fact a logical statement about AI behavior, would be the Dario Floreano experiments where certain robots spontaneously evolved a crude capacity for "deception", and tricked other robots into eating "poison" and dying: here a trait, "deception", ordinarily associated with people rather than with machines, spontaneously evolves in a type of convergent evolution. According to Paul R. Cohen and Edward Feigenbaum, in order to differentiate between anthropomorphization and logical prediction of AI behavior, "the trick is to know enough about how humans and computers think to say exactly what they have in common, and, when we lack this knowledge, to use the comparison to suggest theories of human thinking or computer thinking."

(7) Other sources of risk.

Other scenarios by which advanced AI could produce unintended consequences include:

① self-delusion, in which the AI discovers a way to alter its perceptions to give itself the delusion that it is succeeding in its goals,

② corruption of the reward generator, in which the AI alters humans so that they are more likely to approve of AI actions,

③ and inconsistency of the AI's utility function and other parts of its definition.

James Barrat, documentary filmmaker and author of **Our Final Invention**, says in a Smithsonian interview, "Imagine: in as little as a decade, a half-dozen companies and nations field computers that rival or surpass human intelligence. Imagine what happens when those computers become expert at programming smart computers. Soon we'll be sharing the planet with machines thousands or millions of times more intelligent than we are. And, all the while, each generation of this technology will be weaponized. Unregulated, it will be catastrophic. "

6.7.3 Different Reactions on the Thesis

The thesis that AI could pose an existential risk provokes a wide range scientific community, as well as in the public at large.

AI researcher Ben Goertzel stated Bostrom and Yudkowsky's arguments for existential risks have "some logical foundation, but are often presented in an exaggerated way": he argues that superintelligent AI will likely not be driven by anything like reward maximization, but rather by an open-ended "complex self-organization and self-transcending development".

Many of the scholars who are concerned about existential risk believe that the best way forward would be to conduct (possibly massive) research into solving the difficult "control problem" to answer the question: what types of safeguards, algorithms, or architectures can programmers implement to maximize the probability that their recursively-improving AI would continue to behave in a friendly, rather than destructive, manner after it reaches superintelligence?

As seen throughout this article, the thesis that AI poses an existential risk, and that this risk is in need of much more attention than it currently commands, has been endorsed by many figures; perhaps the most famous are Elon Musk, Bill Gates, and Stephen Hawking. The most notable AI researcher to endorse the thesis is Stuart J. Russell. Endorsers sometimes express bafflement at **skeptics**: Gates states he "can't understand why some people are not concerned", and Hawking criticized widespread indifference in his 2014 editorial: "So, facing possible futures of incalculable benefits and risks, the experts are surely doing everything possible to ensure the best outcome, right? Wrong." If a superior alien civilization sent us a message saying, "We'll arrive in a few decades, " would we just reply, "OK, call us when you get here — we'll leave the lights on"? Probably not — but this is more or less what is happening with AI.

At the same time, the thesis that AI can pose existential risk also has many strong detractors. Skeptics sometimes charge that the thesis is crypto-religious, with an irrational belief in the possibility of superintelligence replacing an irrational belief in an omnipotent God; at an extreme, Jaron Lanier argues that the whole concept that current machines are in any way intelligent is "an illusion" and a "stupendous con" by the wealthy. Computer scientist Gordon Bell argues that the human race will already destroy itself before it reaches the technological singularity. Gordon Moore, the original proponent of Moore's Law, declares that "I am a skeptic. I don't believe (a technological singularity) is likely to happen, at least for a long time. And I don't know why I feel that way." Cognitive scientist Douglas Hofstadter states that "I think life and intelligence are far more complex than the current singularitarians seem to believe, so I doubt (the singularity) will happen in the next couple of centuries. "

Some AI and AGI① researchers are reluctant to discuss risks, for fear that politicians and policymakers will be swayed by "alarmist" messages, or that such messages will lead to cuts in AI funding.

In a YouGov② poll for the British Science Association, about a third of survey respondents said AI will pose a threat to the long term survival of humanity. Slate's Jacob Brogan stated that "most of the (readers filling out our online survey) were unconvinced that A I itself presents a direct threat. "

In a 2015 Wall Street Journal panel discussion devoted to AI risks, IBM's Vice-President of Cognitive Computing, Guruduth S. Banavar, brushed off discussion of AGI with the phrase "it is anybody's speculation". Geoffrey Hinton, the "godfather of deep learning", noted that "there is not a good track record of less intelligent things controlling things of greater intelligence", but stated that he continues his research because "the prospect of discovery is too sweet".

6.8 Key Terms and Review Questions

1. Technical Terms

Artificial Intelligence	人工智能	6.1

① AGI（Artificial General Intelligence）is the intelligence of a machine that could successfully perform any intellectual task that a human being can. AGI is also referred to as "strong AI", "full AI"or as the ability of a machine to perform "general intelligent action"

② an international Internet-based market research firm, headquartered in the UK

amplify	增强、放大	6.1
notion	见解、概念	6.1
exploratory research	探索性研究	6.1
expert system	专家系统	6.1
reasoning	推理	6.1
logical deductions	逻辑演绎	6.1
knowledge representation	知识表示	6.1
knowledge engineering	知识工程	6.1
ontology	本体论	6.1
planning	规划	6.1
multi-agent plannin	多代理规划	6.1
unsupervised learning	非监督学习	6.1
supervised learning	监督	6.1
classification	分类	6.1
regression	回归	6.1
reinforcement learning	强化学习	6.1
machine perception	机器感知	6.1
affective computing	情感计算	6.1
Turing test	图灵测试	6.1
Moore's law	摩尔定律	6.2
software architectures	软件体系结构	6.2
classifying patterns	模式分类	6.2
one-shot learning	单次学习	6.2
proposition	命题	6.3
abstractions	抽象	6.3
manipulation	操作，处理	6.3
logical inference	逻辑推理	6.3
knowledge base	知识库	6.3
inference engine	推理引擎	6.3
rule based expert systems	基于规则的专家系统	6.3
daemon	守护进程	6.3
semantic web	语义网络	6.3
uncertainty framework	不确定性框架	6.3
certainty factors	确定性因素	6.3
case-based reasoning	案例推理	6.4
analogical reasoning	类比推理	6.4
instance-based reasoning	实例推理	6.4
memory-based reasoning	记忆推理	6.4
single-domain cases	单一领域案例	6.4

cross-domain analogies	跨领域类比	6.4
bilingual corpora	双语语料库	6.4
bilingual lexicon	双语词典	6.4
parallel corpora	平行语料库	6.4
match of contexts	上下文匹配	6.4
match by analogy	相似匹配	6.4
similarity assessment	相似性评价	6.4
robotics	机器人学	6.5
interdisciplinary	跨学科的	6.5
generators	发电机	6.5
internal combustion engine	内燃机	6.5
actuators	致动器	6.5
brushless DC motors	无刷直流电机	6.5
AC motors	交流电机	6.5
CNC machines	计算机数控机器	6.5
hydraulic actuator	液动装置	6.5
pneumatic actuator	气动装置	6.5
shape memory alloy	形状记忆金属	6.5
electroactive polymers	电活性聚合物	6.5
piezo motors	压电马达	6.5
ultrasonic motors	超声波马达	6.5
piezoceramic	压电陶瓷	6.5
nanometer	纳米	6.5
elastic nanotubes	弹性纳米管	6.5
biceps	二头肌	6.5
prosthetic hand	机械手，义肢	6.5
tactile sensor	触觉传感器	6.5
touch receptors	触觉感受器	6.5
electrodes	电极	6.5
impedance	阻抗	6.5
electromagnetic radiation	电磁辐射	6.5
solid-state physics	固体物理学	6.5
quantum mechanics	量子力学	6.5
sensor fusion	传感器融合	6.5
lidar	激光定位器	6.5
humanoid robot	人形机器人	6.5
bipedal locomotion	双足步行	6.5
face recognition	人脸识别	6.6
image sensors	图像传感器	6.6

electromagnetic radiation	电磁辐射	6.6
infra-red light	红外光	6.6
quantum physics	量子物理	6.6
neurobiology	神经生物学	6.6
signal processing	信号处理	6.6
object recognition	目标识别	6.6
detection	检测	6.6
convolutional neural networks	卷积神经网络	6.6
ego motion	自我运动	6.6
tracking	跟踪	6.6
scene reconstruction	场景重建	6.6
image restoration	图像恢复	6.6
low-pass filters	低通滤波器	6.6
median filters	中值滤波器	6.6
photometric	光度测定的	6.6
tensor	张量	6.6
eigenfaces	特征脸	6.6
skin texture analysis	皮肤纹理分析	6.6
thermal cameras	热感摄像机	6.6
superintelligent	超常智能	6.7
outmaneuver	以策略胜出	6.7
orthogonality	正交，相互垂直	6.7
pathological	病态的	6.7
anthropomorphism	拟人论	6.7
skeptics	怀疑论者	6.7

2. Translation Exercises

(6-1) **Knowledge representation** and **knowledge engineering** are central to AI research. Many of the problems machines are expected to solve will require extensive knowledge about the world. Among the things that AI needs to represent are: objects, properties, categories and relations between objects; situations, events, states and time; causes and effects; knowledge about knowledge (what we know about what other people know); and many other, less well researched domains.

(6-2) **Unsupervised learning** is the ability to find patterns in a stream of input. **Supervised learning** includes both **classification** and **numerical regression**. Classification is used to determine what category something belongs in, after seeing a number of examples of things from several categories. Regression is the attempt to produce a function that describes the relationship between inputs and outputs and predicts how the outputs should change as the inputs change. In **reinforcement learning**, the agent is rewarded for good responses and punished for bad ones. The agent uses this sequence of rewards and punishments to form a strategy for operating in its problem space.

(6-3) Knowledge Representation, then, is this: it is the field of study concerned with using formal symbols to represent a collection of propositions believed by some putative agent. As we will see, however, we do not want to insist that these symbols must represent all the propositions believed by the agent. There may very well be an infinite number of propositions believed, only a finite number of which are ever represented. It will be the role of reasoning to bridge the gap between what is represented and what is believed.

(6-4) Expert system refers to the type of task the system is trying to solve, to replace or aid a human expert in a complex task. Knowledge-based system refers to the architecture of the system, that it represents knowledge explicitly rather than as procedural code. While the earliest knowledge-based systems were almost all expert systems, the same tools and architectures have since been used for a whole host of other types of systems. i.e., virtually all expert systems are knowledge-based systems but many knowledge-based systems are not expert systems.

(6-5) Case-based reasoning is a problem solving paradigm that in many respects is fundamentally different from other major AI approaches. Instead of relying solely on general knowledge of a problem domain, or making associations along generalized relationships between problem descriptors and conclusions, CBR is able to utilize the specific knowledge of previously experienced, concrete problem situations (cases). A new problem is solved by finding a similar past case, and reusing it in the new problem situation. A second important difference is that CBR also is an approach to incremental, sustained learning, since a new experience is retained each time a problem has been solved, making it immediately available for future problems.

(6-6) A superintelligence, which can be defined as a system that exceeds the capabilities of humans in every relevant endeavor, can outmaneuver humans any time when its goals conflict with human goals; therefore, unless the superintelligence decides to allow humanity to coexist, the first superintelligence to be created will inexorably result in human extinction.

References

[1] Akerkar RA, Srinivas S P. Knowledge-based systems. Jones & Bartlett Publishers, Sudbury, MA, USA, 2009.

[2] Kolodner J L.Case-based reasoning-an introduction[J/OL]. Artif. Intell Rev.1992, (6):3. doi:10.1007/BF00155578.https://en.wikipedia.org/wiki/Example-based_machine_translation.

[3] Turcato D, Popowich F. What is Example-Based Machine Translation?[OL].https://en.wikipedia.org/wiki/Existential_risk_from_artificial_general_intelligence.

[4] Stuart R, Peter N. Artificial Intelligence: A Modern Approach[M]. Prentice Hall, 2009.

[5] Nick Bostrom. Existential risks[J]. Journal of Evolution and Technology, 2009 ,9(1): 1–31.

[6] GiveWell. Potential risks from advanced artificial intelligence[R/OL]. (2015-08-01) [2017-04-05]. https://www.openphilanthropy.org/research/cause-reports/ai-risk.

[7] Eliezer Yudkowsky. Artificial intelligence as a positive and negative factor in global risk.

Global catastrophic risks.[EB/OL]. (2008-09-11) [2017-04-05]. https://en.wikipedia.org/wiki/Computer_vision.

[8] Ballard D H, Brown C M. Computer Vision[M]. Prentice Hall, 1982.

[9] Sonka M, Vaclav Hlavac, Roger Boyle. Image Processing, Analysis, and Machine Vision[M]. Thomson, 2008.

[10] Klette R. Concise Computer Vision. Springer, 2014.

[11] The History of Artificial Intelligence[EB/OL]. [2017-05-02]. http://courses.cs.washington.edu/courses/csep590/06au/projects/history-ai.pdf.

Chapter 7 Computer Graphics and Visualization

7.1 Computer Graphics

7.1.1 What Is Computer Graphics

Computer Graphics (CG) means drawing pictures on a computer screen. What's so good about that? Sketch something on paper — a man or a house — and what you have is a piece of analog information: the thing you draw is a likeness or analogy of something in the real world. Depending on the materials you use, changing what you draw can be easy or hard: you can erase pencil or charcoal marks easily enough, and you can scrape off oil paints and redo them with no trouble, but altering watercolors or permanent markers is an awful lot more tricky. That's the wonder of art, of course — it captures the fresh dash of creativity — and that's exactly what we love about it. But where everyday graphics is concerned, the immediacy of art is also a huge drawback. As every sketching child knows too well, if you draw the first part of your picture too big, you'll struggle to squeeze everything else on the page and what if you change your mind about where to put something or you want to swap red for orange or green for blue? Ever had one of those days where you rip up sheet after sheet of spoiled paper and toss it in the trash?

That's why many artists, designers, and architects have fallen in love with computer graphics. Draw a picture on a computer screen and what you have is a piece of digital information. It probably looks similar to what you'd have drawn on paper — the ghostly idea that was hovering in your mind's eye to begin with — but inside the computer your picture is stored as a series of numbers. Change the numbers and you can change the picture, in the blink of an eye or even quicker. It's easy to shift your picture around the screen, scale it up or down, rotate it, swap the colors, and transform it in all kinds of other ways. Once it's finished, you can save it, incorporate it into a text document, print it out, upload it to a Web page, or email it to a client or work colleague — all because it's digital information.

The term computer graphics has been used a broad sense to describe "almost everything on computers that is not text or sound". Typically, the term computer graphics refers to several different things:

(1) the representation and manipulation of image data by a computer.

(2) the various technologies used to create and manipulate images.

(3) the sub-field of computer science which studies methods for digitally synthesizing and manipulating visual content.

Today, computer graphics is widespread. Such imagery is found in and on television, newspapers, weather reports, and in a variety of medical investigations and surgical procedures. A well-constructed graph can present complex statistics in a form that is easier to understand and interpret. In the media 'such graphs are used to illustrate papers, reports, thesis and other presentation material.

Many tools have been developed to visualize data. Computer generated imagery can be categorized into several different types: two dimensional (2D), three dimensional (3D), and animated graphics. As technology has improved, 3D computer graphics have become more common, but 2D computer graphics are still widely used. Computer graphics has emerged as a sub-field of computer science which studies methods for digitally synthesizing and manipulating visual content. Over the past decade, other specialized fields have been developed like information visualization, and scientific visualization more concerned with "the visualization of three dimensional phenomena (architectural, meteorological, medical, biological, etc.), where the emphasis is on realistic renderings of volumes, surfaces, illumination sources, and so forth, perhaps with a dynamic (time) component".

7.1.2 Types of Graphics

1. Two-Dimensional Computer Graphics

All computer art is digital, but there are two very different ways of drawing digital images on a computer screen, known as **raster** and **vector graphics**. Simple computer graphic programs like Microsoft Paint and PaintShop Pro are based on raster graphics, while more sophisticated programs such as CorelDRAW, AutoCAD, and Adobe Illustrator use vector graphics. So what exactly is the difference?

1) Raster Graphics

Stare hard at your computer screen and you'll notice the pictures and words are made up of tiny colored dots or squares called **pixels**. Most of the simple computer graphic images we come across are pixelated in this way, just like walls are built out of bricks. The first computer screens, developed in the mid-20th century, worked much like televisions, which used to build up their moving pictures by "scanning" beams of electrons back and forth from top to bottom and left to right — like a kind of instant electronic paintbrush. This way of making a picture is called raster scanning and that's why building up a picture on a computer screen out of pixels is called raster graphics.

Raster graphics are simple to use and it's easy to see how programs that use them do their stuff. If you draw a pixel picture on your computer screen and you click a button in your graphics package to "mirror" the image (flip it from left to right or right to left), all the computer does is reverse the order of the pixels by reversing the sequence of zeros and ones that represent them. If you scale an image to its twice size, the computer copies each pixel twice over (so the numbers 10110 become 1100111100), but the image becomes noticeably more grainy and pixelated in the process. That's one of the main drawbacks of using raster graphics: they don't scale up to different sizes very well. Another drawback is the amount of memory they require. A really detailed photo might need 16 million colors, which involves storing 24 bits per pixel and 24 times as much memory as a basic black-and-white image.

The maximum number of pixels in an image (or on a computer screen) is known as its **resolution**. The first computer I ever used properly, a Commodore PET, had an ultra-low resolution display with 80 characters across by 25 lines down (so a maximum of 2,000 letters, numbers, or punctuation marks could be on the screen at any one time); since each character was built from an 8 × 8 square of pixels, that meant the screen had a resolution of 640 × 200 = 128,000 pixels (or 0.128 Megapixels, where a Megapixel is one million pixels). The laptop I'm using right now is set to a resolution of 1280 × 800 =1.024 **Megapixels**, which is roughly 7~8 times more detailed. A digital camera with 7 Megapixel resolution would be roughly seven times more detailed than the resolution of my laptop screen or about 50 times more detailed than that original Commodore PET screen.

Displaying smoothly drawn curves on a pixelated display can produce horribly **jagged edges** ("jaggies"). One solution to this is to blur the pixels on a curve to give the appearance of a smoother line. This technique, known as **anti-aliasing**, is widely used to smooth the fonts on pixelated computer screens.

2) Vector Graphics

There's an alternative method of computer graphics that gets around the problems of raster graphics. Instead of building up a picture out of pixels, you draw it a bit like a child would by using simple straight and curved lines called vectors or basic shapes (circles, curves, triangles, and so on) known as **primitives**. With raster graphics, you make a drawing of a house by building it from hundreds, thousands, or millions of individual pixels; importantly, each pixel has no connection to any other pixel except in your brain. With vector graphics, you might draw a rectangle for the basic house, smaller rectangles for the windows and door, a cylinder for the smokestack, and a polygon for the roof. Staring at the screen, a vector-graphic house still seems to be drawn out of pixels, but now the pixels are precisely related to one another — they're points along the various lines or other shapes you've drawn. Drawing with straight lines and curves instead of individual dots means you can produce an image more quickly and store it with less information: you could describe a vector-drawn house as "two red triangles and a red rectangle (the roof) sitting on a brown rectangle (the main building)", but you couldn't summarize a pixelated image so simply. It's also much easier to scale a vector-graphic image up and down by applying mathematical formulas called algorithms that **transform** the vectors from which your image is drawn. That's how computer programs can scale fonts to different sizes without making them look all pixelated and grainy.

Most modern computer graphics packages let you draw an image using a mixture of raster or vector graphics, as you wish, because sometimes one approach works better than another — and sometimes you need to mix both types of graphics in a single image. With a graphics package such as the GIMP (GNU Image Manipulation Program), you can draw curves on screen by tracing out and then filling in "paths" (technically known as **Bézier curves**) before converting them into pixels ("rasterizing" them) to incorporate them into something like a bitmap image.

2. Three-Dimensional Graphics

Real life isn't like a computer game or a virtual reality simulation. The very best CGI (Computer-Generated Imagery) animations are easy to tell apart from ones made on film or video with real actors. Why is that? When we look at objects in the world around us, they don't appear to be drawn from either pixels or vectors. In the blink of an eye, our brains gather much more information from the real-world than artists can include in even the most realistic computer-graphic images. To make a computerized image look anything like as realistic as a photograph (let alone a real-world scene), we need to include far more than simply millions of colored-in pixels.

Really sophisticated computer graphics programs use a whole series of techniques to make hand-drawn (and often completely imaginary) two-dimensional images look at least as realistic as photographs. The simplest way of achieving this is to rely on the same tricks that artists have always used — such things as **perspective** (how objects recede into the distance toward a "vanishing point" on the horizon) and **hidden-surface** elimination (where nearby things partly obscure ones that are further away).

If you want realistic 3D artwork for such things as CAD (Computer-Aided Design) and virtual reality, you need much more sophisticated graphic techniques. Rather than drawing an object, you make a 3D computer model of it inside the computer and manipulate it on the screen in various ways. First, you build up a basic three-dimensional outline of the object called a **wire-frame** (because it's drawn from vectors that look like they could be little metal wires). Then the model is rigged, a process in which different bits of the object are linked together a bit like the bones in a skeleton so they move together in a realistic way. Finally, the object is **rendered**, which involves shading the outside parts with different textures (surface patterns), colors, degrees of opacity or transparency, and so on. Rendering is a hugely complex process that can take a powerful computer hours, days, or even weeks to complete. Sophisticated math is used to model how light falls on the surface, typically using either **ray tracing** (a relatively simple method of plotting how light bounces off the surface of shiny objects in straight lines) or **radiosity** (a more sophisticated method for modeling how everyday objects **reflect** and **scatter** light in duller, more complex ways).

3. Computer Animation

Computer animation is the art of creating moving images via the use of computers. It is a subfield of computer graphics and animation. Increasingly it is created by means of 3D computer graphics, though 2D computer graphics are still widely used for stylistic, low bandwidth, and faster real-time rendering needs. Sometimes the target of the animation is the computer itself, but sometimes the target is another medium, such as film. It is also referred to as CGI (Computer-Generated Imagery or Computer-Generated Imaging), especially when used in films.

Virtual entities may contain and be controlled by assorted attributes, such as transform values (location, orientation, and scale) stored in an object's transformation matrix. Animation is the change of an attribute over time. Multiple methods of achieving animation exist; the rudimentary form is based on the creation and editing of **keyframes**, each storing a value at a given time, per attribute to be animated. The 2D/3D graphics software will change with each keyframe, creating an editable curve of a value mapped over time, in which results in animation. Other methods of animation include procedural and expression-based techniques: the former consolidates related elements of animated entities into sets of attributes, useful for creating particle effects and crowd simulations; the latter allows an evaluated result returned from a user-defined logical expression, coupled with mathematics, to automate animation in a predictable way (convenient for controlling bone behavior beyond what a hierarchy offers in skeletal system set up).

(7-1) To create the illusion of movement, an image is displayed on the computer screen then quickly replaced by a new image that is similar to the previous image, but shifted slightly. This technique is identical to the illusion of movement in television and motion pictures.

7.1.3 Techniques Used in CG

1. Rendering

Rendering is the generation of a 2D image from a 3D model by means of computer programs. (7-2) A scene file contains objects in a strictly defined language or data structure; it would contain geometry, viewpoint, texture, lighting, and shading information as a description of the virtual scene. The data contained in the scene file is then passed to a rendering program to be processed and output to a digital image or raster graphics image file. The rendering program is usually built into the computer graphics software, though others are available as plug-ins or entirely separate programs. The term "rendering" may be by analogy with an "artist's rendering" of a scene. Though the technical details of rendering methods vary, the general challenges to overcome in producing a 2D image from a 3D representation stored in a scene file are outlined as the graphics pipeline along a rendering device, such as a GPU. A GPU is a device able to assist the CPU in calculations. If a scene is to look relatively realistic and predictable under virtual lighting, the rendering software should solve the rendering equation. The rendering equation does not account for all lighting phenomena, but is a general lighting model for computer-generated imagery. "Rendering" is also used to describe the process of calculating effects in a video editing file to produce final video output.

1) 3D Projection

3D **projection** is a method of mapping three dimensional points to a two dimensional plane. As most current methods for displaying graphical data are based on **planar** two dimensional media, the use of this type of projection is widespread, especially in computer graphics, engineering and drafting.

2) Ray tracing

Ray tracing is a technique for generating an image by tracing the path of light through pixels in an image plane. The technique is capable of producing a very high degree of **photorealism**; usually higher than that of typical **scanline** rendering methods, but at a greater computational cost.

3) Shading

Shading refers to depicting depth in 3D models or illustrations by varying levels of darkness. It is a process used in drawing for depicting levels of darkness on paper by applying media more densely or with a darker shade for darker areas, and less densely or with a lighter shade for lighter areas. There are various techniques of shading including **cross hatching** where perpendicular lines of varying closeness are drawn in a grid pattern to shade an area. The closer the lines are together, the darker the area appears. Likewise, the farther apart the lines are, the lighter the area appears. The term has been recently generalized to mean that **shaders** are applied.

4) Texture Mapping

Texture mapping is a method for adding detail, surface texture, or color to a computer-generated graphic or 3D model. Its application to 3D graphics was pioneered by Dr Edwin Catmull in 1974. A texture map is applied (mapped) to the surface of a shape, or polygon. This process is akin to applying patterned paper to a plain white box. Multitexturing is the use of more than one texture at a time on a polygon. Procedural textures (created from adjusting parameters of an underlying algorithm that produces an output texture), and bitmap textures (created in an image editing application or imported from a digital camera) are, generally speaking, common methods of implementing texture definition on 3D models in computer graphics software, while intended placement of textures onto a model's surface often requires a technique known as UV mapping (arbitrary, manual layout of texture coordinates) for polygon surfaces, while NURBS surfaces have their own intrinsic parameterization used as texture coordinates. Texture mapping as a discipline also encompasses techniques for creating normal maps and bump maps that correspond to a texture to simulate height and specular maps to help simulate shine and light reflections, as well as environment mapping to simulate mirror-like reflectivity, also called gloss.

5) Anti-aliasing

Rendering resolution-independent entities (such as 3D models) for viewing on a raster (pixel-based) device such as a **liquid-crystal display** or CRT television inevitably causes **aliasing** artifacts mostly along geometric edges and the boundaries of texture details; these artifacts are informally called "**jaggies**". Anti-aliasing methods rectify such problems, resulting in imagery more pleasing to the viewer, but can be somewhat computationally expensive. Various anti-aliasing algorithms (such as supersampling) are able to be employed, then customized for the most efficient rendering performance versus quality of the resultant imagery; a graphics artist should consider this trade-off if anti-aliasing methods are to be used. A pre-anti-aliased bitmap texture being displayed on a screen (or screen location) at a resolution different than the resolution of the texture itself (such as a textured model in the distance from the virtual camera) will exhibit aliasing artifacts, while any procedurally defined texture will always show aliasing artifacts as they are resolution-independent; techniques such as **mipmapping** and texture filtering help to solve texture-related aliasing problems.

2. Volume Rendering

Volume rendering is a technique used to display a 2D projection of a 3D discretely sampled data set. A typical 3D data set is a group of 2D **slice** images acquired by a CT or MRI scanner.

Usually these are acquired in a regular pattern (e.g., one slice every millimeter) and usually have a regular number of image pixels in a regular pattern. This is an example of a regular volumetric grid, with each volume element, or voxel represented by a single value that is obtained by sampling the immediate area surrounding the voxel.

1) CT — Computed Tomography

A CT scan makes use of computer-processed combinations of many X-ray images taken from different angles to produce cross-sectional (tomographic) images (virtual "slices") of specific areas of a scanned object, allowing the user to see inside the object without cutting.

CT produces a volume of data that can be manipulated in order to demonstrate various bodily structures based on their ability to block the X-ray beam. Although, historically, the images generated were in the axial or transverse plane, perpendicular to the long axis of the body, modern scanners allow this volume of data to be reformatted in various planes or even as volumetric (3D) representations of structures. Although most commonly used in medicine, CT is also used in other fields, such as nondestructive materials testing. Another example is archaeological uses such as imaging the contents of sarcophagi. Individuals responsible for performing CT exams are called radiographers or radiologic technologists.

2) MRI

Magnetic Resonance Imaging (MRI) is a medical imaging technique used in radiology to form pictures of the anatomy and the physiological processes of the body in both health and disease. MRI scanners use strong magnetic fields, radio waves, and field gradients to generate images of the inside of the body.

MRI is based upon the science of Nuclear Magnetic Resonance (NMR). Certain atomic nuclei can absorb and emit radio frequency energy when placed in an external magnetic field. In clinical and research MRI, hydrogen atoms are most-often used to generate a detectable radio-frequency signal that is received by antennas in close proximity to the anatomy being examined. Hydrogen atoms exist naturally in people and other biological organisms in abundance, particularly in water and fat. For this reason, most MRI scans essentially map the location of water and fat in the body. Pulses of radio waves excite the nuclear spin energy transition, and magnetic field gradients localize the signal in space. By varying the parameters of the pulse sequence, different contrasts can be generated between tissues based on the relaxation properties of the hydrogen atoms therein. Since its early development in the 1970s and 1980s, MRI has proven to be a highly versatile imaging technique. While MRI is most prominently used in diagnostic medicine and biomedical research, it can also be used to form images of non-living objects. MRI scans are capable of producing a variety of chemical and physical data, in addition to detailed spatial images.

MRI is widely used in hospitals and clinics for medical diagnosis, staging of disease and follow-up without exposing the body to **ionizing radiation**.

3. 3D Modeling

3D modeling is the process of developing a mathematical, **wireframe** representation of any three-dimensional object, called a "3D model", via specialized software. Models may be created automatically or manually; the manual modeling process of preparing geometric data for 3D computer graphics is similar to plastic arts such as sculpting. 3D models may be created using multiple approaches: use of NURBS curves to generate accurate and smooth **surface patches**, **polygonal mesh** modeling (manipulation of faceted geometry), or polygonal mesh subdivision (advanced tessellation of polygons, resulting in smooth surfaces similar to NURBS models). A 3D model can be displayed as a two-dimensional image through a process called 3D rendering, used in a computer simulation of physical phenomena, or animated directly for other purposes. The model can also be physically created using 3D Printing devices.

7.1.4 Computer-aided Design①

Drawing on a computer screen with a graphics package is a whole lot easier than sketching on paper, because you can modify your design really easily. But that's not all there is to CAD. Instead of producing a static, two-dimensional (2D) picture, usually what you create on the screen is a three-dimensional (3D) computer model, drawn using vector graphics and based on a kind of line-drawn skeleton called a wireframe, which looks a bit like an object wrapped in graph paper.

(7-3) <u>Once the outside of the model's done, you turn your attention to its inner structure. This bit is called rigging your model (also known as skeletal animation). What parts does the object contain and how do they all connect together? When you've specified both the inside and outside details, your model is pretty much complete. The final stage is called texturing, and involves figuring out what colors, surface patterns, **finishes**, and other details you want your object to have.</u> When your model is complete, you can render it: turn it into a final image. Ironically, the picture you create at this stage may look like it's simply been drawn right there on the paper: it looks exactly like any other 3D drawing. But, unlike with an ordinary drawing, it's super-easy to change things: you can modify your model in any number of different ways. The

① http://www.explainthatstuff.com/computer-graphics.html

computer can rotate it through any angle, zoom in on different bits, or even help you "cutaway" certain parts (maybe to reveal the engine inside a plane) or "explode" them (show how they break into their component pieces).

1. What is CAD Used for

From false teeth to supercars and designer dresses to drink cartons, virtually every product we buy today is put together with the help of computer-aided design. Architects, advertising and marketing people, draftsmen, car designers, shipbuilders, and aerospace engineers — these are just some of the people who rely on CAD. (7-4)Apart from being cheaper and easier than using paper, CAD designs are easy to send round the world by email (from designers in Paris to manufacturers in Singapore, perhaps). Another big advantage is that CAD drawings can be converted automatically into production instructions for industrial robots and other factory machines, which greatly reduces the overall time needed to turn new designs into finished products. Next time you buy something from a store, trace it back in your mind's eye: how did it find its way into your hand, from the head-scratching designer sitting at a computer in Manhattan to the robot-packed factory in Shanghai where it rolled off the production line? Chances are it was all done with CAD!

2. Using CAD in Architecture

Architects have always been visionaries — and they helped to pioneer the adoption of CAD technology from the mid-1980s, when easy-to-use desktop publishing computers like the Apple Mac became widely available. Before CAD came along, technical drawing, was the best solution to a maddening problem architects and engineers knew only too well: how to communicate the amazing three-dimensional constructions they could visualize in their mind's eye with clarity and precision. Even with three-dimensional drawings (such as orthographic projections), it can still be hard to get across exactly what you have in mind. What if you spent hours drawing your proposed building, airplane, or family car…only for someone to say infuriating things like: "And what does it look like from behind? How would it look from over there? What if we made that wall twice the size?" Having drawn their projections, architects would typically build little models out of paper and board, while engineers would whittle model cars and planes out of balsa wood. But even the best models can't answer "What if…?" questions.

Computer-aided design solves these problems in a particularly subtle way. It doesn't simply involve drawing 2D pictures of buildings on the screen: what you produce with CAD is effectively a computer model of your design. Once that's done, it's easy to rotate your design on-screen or change any aspect of it in a matter of moments. If you want to make a wall twice the size, click a button, drag your mouse here and there, and the computer automatically recalculates how the rest of your model needs to change to fit in. You can print out three dimensional projections of your model from any angle or you can demonstrate the 3D form to your clients on-screen, allowing them to rotate or play with the model for themselves. Some models even let you walk through them in virtual reality. CAD has revolutionized architecture not simply by removing the drudge of repetitive plan drawing and intricate model making, but by providing a tangible, digital representation of the mind's eye: what you see is — finally — what you get.

Over the last 30 years, computers have absolutely revolutionized architecture. In 2012, Architects' Journal went so far as to describe CAD as "the greatest advance in construction history".

7.1.5 3D Modeling

Three-dimensional (3D) models represent a physical body using a collection of points in 3D space, connected by various geometric entities such as triangles, lines, curved surfaces, etc. Being a collection of data (points and other information), 3D models can be created by hand, algorithmically (procedural modeling), or scanned. Their surfaces may be further defined with texture mapping.

3D models are widely used anywhere in 3D graphics and CAD. Actually, their use predates the widespread use of 3D graphics on personal computers. Many computer games used pre-rendered images of 3D models as sprites before computers could render them in real-time.

Today, 3D models are used in a wide variety of fields. The medical industry uses detailed models of organs; these may be created with multiple 2-D image slices from an MRI or CT scan. The movie industry uses them as characters and objects for animated and real-life motion pictures. The video game industry uses them as assets for computer and video games. The science sector uses them as highly detailed models of chemical compounds. The architecture industry uses them to demonstrate proposed buildings and landscapes in lieu of traditional, physical architectural models. The engineering community uses them as designs of new devices, vehicles and structures as well as a host of other uses. In recent decades the earth science community has started to construct 3D geological models as a standard practice. 3D models can also be the basis for physical devices that are built with 3D printers or CNC(**C**omputer **N**umeric **C**ontrol) machines.

1. Representation

Almost all 3D models can be divided into two categories.

(1) Solid — These models define the volume of the object they represent (like a rock). Solid models are mostly used for engineering and medical simulations, and are usually built with constructive solid geometry.

(2) Shell/boundary — these models represent the surface, e.g. the boundary of the object, not its volume (like an infinitesimally thin eggshell). Almost all visual models used in games and film are shell models.

Solid and shell modeling can create functionally identical objects. Differences between them are mostly variations in the way they are created and edited and conventions of use in various fields and differences in types of approximations between the model and reality.

Shell models must be **manifold** (having no holes or cracks in the shell) to be meaningful as a real object. Polygonal meshes (and to a lesser extent subdivision surfaces) are by far the most common representation. Level sets are a useful representation for deforming surfaces which undergo many topological changes such as fluids.

The process of transforming representations of objects, such as the middle point coordinate of a sphere and a point on its circumference into a polygon representation of a sphere, is called **tessellation**. This step is used in polygon-based rendering, where objects are broken down from abstract representations ("primitives") such as spheres, cones etc., to so-called meshes, which are nets of interconnected triangles. Meshes of triangles (instead of e.g. squares) are popular as they have proven to be easy to **rasterize** (the surface described by each triangle is planar, so the projection is always **convex**); Polygon representations are not used in all rendering techniques, and in these cases the tessellation step is not included in the transition from abstract representation to rendered scene.

2. Modeling Process

There are three popular ways to represent a model.

(1) Polygonal modeling — Points in 3D space, called vertices, are connected by line segments to form a polygon mesh. The vast majority of 3D models today are built as textured polygonal models as shown in Fig.7-1, because they are flexible and because computers can render them so quickly. However, polygons are planar and can only approximate curved surfaces using many polygons.

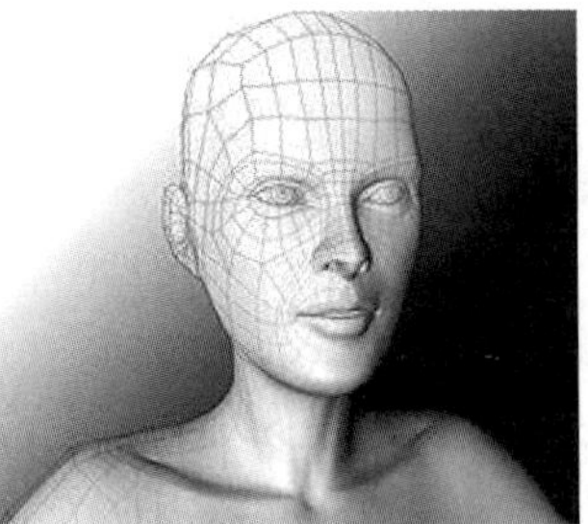

Fig.7-1　3D polygonal modelling of a human face

(2) Curve modeling — Surfaces are defined by curves, which are influenced by weighted control points. The curve follows (but does not necessarily interpolate) the points. Increasing the weight for a point will pull the curve closer to that point. Curve types include nonuniform rational B-spline (NURBS), splines, patches, and geometric primitives.

(3) Digital sculpting — Still a fairly new method of modeling, 3D sculpting has become very popular in the few years it has been around. There are currently three types of digital sculpting: ① Displacement, which is the most widely used among applications at this moment, uses a dense model (often generated by subdivision surfaces of a polygon control mesh) and stores new locations for the vertex positions through use of a 32b image map that stores the adjusted locations. ② Volumetric, loosely based on voxels, has similar capabilities as displacement but does not suffer from polygon stretching when there are not enough polygons in a region to achieve a deformation. ③ Dynamic tessellation is similar to voxel but divides the

surface using triangulation to maintain a smooth surface and allow finer details. These methods allow for very artistic exploration as the model will have a new topology created over it once the models form and possibly details have been sculpted. The new mesh will usually have the original high resolution mesh information transferred into displacement data or normal map data if for a game engine.

3. Uses

3D modeling is used in various industries like films, animation and gaming, interior designing and architecture. They are also used in the medical industry for the interactive representations of anatomy. A wide number of 3D software are also used in constructing digital representation of mechanical models or parts before they are actually manufactured. CAD/CAM related software are used in such fields, and with these software, not only can you construct the parts, but also assemble them, and observe their functionality.

3D modelling is also used in the field of Industrial Design, wherein products are 3D modeled before representing them to the clients. In Media and Event industries, 3D modelling is used in **Stage/Set Design**.

The OWL[①] translation of the vocabulary of X3D[②] can be used to provide semantic descriptions for 3D models, which is suitable for indexing and retrieval of 3D models by features such as geometry, dimensions, material, texture, **diffuse reflection**, **transmission spectra**, transparency, reflectivity, opalescence, glazes, varnishes, and enamels (as opposed to unstructured textual descriptions or 2.5D virtual museums and exhibitions using Google Street View on Google Arts & Culture, for example). The RDF(Resource Description Framework) representation of 3D models can be used in reasoning, which enables intelligent 3D applications which, for example, can automatically compare two 3D models by volume.

7.2 Virtual Reality[③]

You'll probably never go to Mars, swim with dolphins, run an Olympic 100 meters, or sing onstage with the Rolling Stones. But if virtual reality ever lives up to its promise, you might be able to do all these things — and many more — without even leaving your home. Unlike real reality (the actual world in which we live), virtual reality means simulating bits of our world (or completely imaginary worlds) using high-performance computers and sensory equipment, like headsets and gloves. Apart from games and entertainment, it's long been used for training airline pilots and surgeons and for helping scientists to figure out complex problems such as the structure of protein molecules. How does it work? Let's take a closer look!

Virtual reality means blocking yourself off from the real world and substituting a

① Web Ontology Language

② X3D is a royalty-free ISO standard XML-based file format for representing 3D computer graphics

③ http://www.explainthatstuff.com/virtualreality.html

computer-generated alternative. Often, it involves wearing a wraparound headset called a **head-mounted display**, clamping stereo headphones over your ears, and touching or feeling your way around your imaginary home using **datagloves** with built-in sensors,see Fig.7-2.

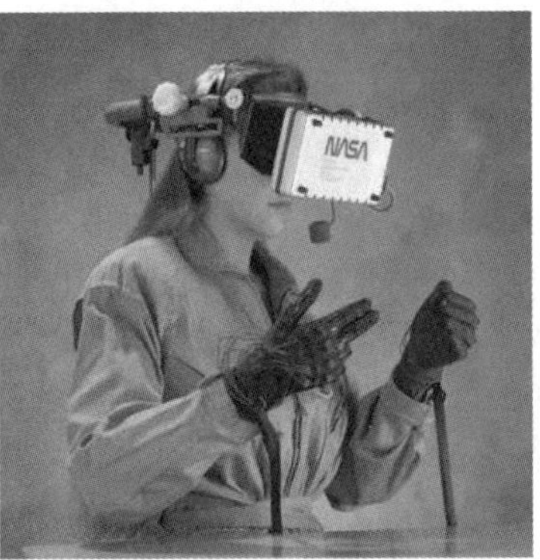

Fig.7-2 Virtual reality equipments

7.2.1 What Is Virtual Reality

Virtual Reality (VR) means experiencing things through our computers that don't really exist. From that simple definition, the idea doesn't sound especially new. When you look at an amazing Canaletto painting, for example, you're experiencing the sites and sounds of Italy as it was about 250 years ago — so that's a kind of virtual reality. In the same way, if you listen to ambient instrumental or classical music with your eyes closed, and start dreaming about things, isn't that an example of virtual reality — an experience of a world that doesn't really exist? What about losing yourself in a book or a movie? Surely that's a kind of virtual reality?

If we're going to understand why books, movies, paintings, and pieces of music aren't the same thing as virtual reality, we need to define VR fairly clearly. For the purposes of this simple, introductory article, I'm going to define it as:

A believable, interactive 3D computer-created world that you can explore so you feel you really are there, both mentally and physically.

Putting it another way, virtual reality is essentially:

(1) Believable: You really need to feel like you're in your virtual world (on Mars, or wherever) and to keep believing that or the illusion of virtual reality will disappear.

(2) Interactive: As you move around, the VR world needs to move with you. You can watch a 3D movie and be transported up to the moon or down to the seabed — but it's not interactive in any sense.

(3) Computer-generated: Why is that important? Because only powerful machines, with realistic 3D computer graphics, are fast enough to make believable, interactive, alternative worlds that change in real-time as we move around them.

(4) Explorable: A VR world needs to be big and detailed enough for you to explore.

However realistic a painting is, it shows only one scene, from one perspective. A book can describe a vast and complex "virtual world", but you can only really explore it in a linear way, exactly as the author describes it.

(5) Immersive: To be both believable and interactive, VR needs to engage both your body and your mind. Paintings by war artists can give us glimpses of conflict, but they can never fully convey the sight, sound, smell, taste, and feel of battle. You can play a flight simulator game on your home PC and be lost in a very realistic, interactive experience for hours (the landscape will constantly change as your plane flies through it), but it's not like using a real flight simulator (where you sit in a hydraulically operated mockup of a real cockpit and feel actual forces as it tips and tilts), and even less like flying a plane.

We can see from this why reading a book, looking at a painting, listening to a classical symphony, or watching a movie don't qualify as virtual reality. All of them offer partial glimpses of another reality, but none are interactive, explorable, or fully believable. If you're sitting in a movie theater looking at a giant picture of Mars on the screen, and you suddenly turn your head too far, you'll see and remember that you're actually on earth and the illusion will disappear. If you see something interesting on the screen, you can't reach out and touch it or walk towards it; again, the illusion will simply disappear. So these forms of entertainment are essentially passive: however plausible they might be, they don't actively engage you in any way.

VR is quite different. It makes you think you are actually living inside a completely believable virtual world (one in which, to use the technical jargon, you are partly or fully immersed). It is two-way interactive: as you respond to what you see, what you see responds to you: if you turn your head around, what you see or hear in VR changes to match your new perspective.

7.2.2 Types of Virtual Reality

"Virtual reality" has often been used as a marketing buzzword for compelling, interactive video games or even 3D movies and television programs, none of which really count as VR because they don't immerse you either fully or partially in a virtual world. Search for "virtual reality" in your cellphone app store and you'll find hundreds of hits, even though a tiny cellphone screen could never get anywhere near producing the convincing experience of VR. Nevertheless, things like interactive games and computer simulations would certainly meet parts of our definition up above, so there's clearly more than one approach to building virtual worlds — and more than one flavor of virtual reality. Here are a few of the bigger variations.

1. Fully Immersive

For the complete VR experience, we need three things. First, a plausible, and richly detailed virtual world to explore; a computer model or simulation, in other words. Second, a powerful computer that can detect what we're going and adjust our experience accordingly, in real time (so what we see or hear changes as fast as we move — just like in real reality). Third, hardware linked to the computer that fully immerses us in the virtual world as we roam around. Usually, we'd need to put on what's called a head-mounted display (HMD) with two screens and stereo sound, and wear one or more sensory gloves. Alternatively, we could move around inside a room, fitted out with surround-sound loudspeakers, onto which changing images are projected from outside. We'll explore VR equipment in more detail in a moment.

2. Non-immersive

A highly realistic flight simulator on a home PC might qualify as nonimmersive virtual reality, especially if it uses a very wide screen, with headphones or surround sound, and a realistic joystick and other controls. Not everyone wants or needs to be fully immersed in an alternative reality. An architect might build a detailed 3D model of a new building to show to clients that can be explored on a desktop computer by moving a mouse. Most people would classify that as a kind of virtual reality, even if it doesn't fully immerse you. In the same way, computer archaeologists often create engaging 3D reconstructions of long-lost settlements that you can move around and explore. They don't take you back hundreds or thousands of years or create the sounds, smells, and tastes of prehistory, but they give a much richer experience than a few pastel drawings or even an animated movie.

3. Collaborative

What about "virtual world" games like Second Life and Minecraft? Do they count as virtual reality? Although they meet the first four of our criteria (believable, interactive, computer-created and explorable), they don't really meet the fifth: they don't fully immerse you. But one thing they do offer that cutting-edge VR typically doesn't is collaboration: the idea of sharing an experience in a virtual world with other people, often in real time or something very close to it. Collaboration and sharing are likely to become increasingly important features of VR in future.

4. Web-based

Virtual reality was one of the hottest, fastest-growing technologies in the late 1980s and early 1990s, but the rapid rise of the World Wide Web largely killed off interest after that. Even though computer scientists developed a way of building virtual worlds on the Web (using a technology analogous to HTML called Virtual Reality Markup Language(VRML)), ordinary people were much more interested in the way the Web gave them new ways to access real reality — new ways to find and publish information, shop, and share thoughts, ideas, and experiences with friends through social media. With Facebook's growing interest in the technology, the future of VR seems likely to be both Web-based and collaborative.

5. Augmented Reality

Mobile devices like smartphones and tablets have put what used to be supercomputer power in our hands and pockets. If we're wandering round the world, maybe visiting a heritage site like the pyramids or a fascinating foreign city we've never been to before, what we want is typically not virtual reality but an enhanced experience of the exciting reality we can see in front of us. That's spawned the idea of augmented reality (AR), where, for example, you point your smartphone at a landmark or a striking building and interesting information about it pops up automatically. Augmented reality is all about connecting the real world we experience to the vast virtual world of information that we've collectively created on the Web. Neither of these worlds is virtual, but the idea of exploring and navigating the two simultaneously does, nevertheless, have things in common with virtual reality. For example, how can a mobile device figure out its precise location in the world? How do the things you see on the screen of your tablet change as you wander round a city? Technically, these problems are similar to the ones developers of VR systems have to solve — so there are close links between AR and VR.

7.2.3 Equipment Used in Virtual Reality

Close your eyes and think of virtual reality and you probably picture something like our top photo: a geek wearing a wraparound headset and datagloves, wired into a powerful workstation or supercomputer. What differentiates VR from an ordinary computer experience (using your PC to write an essay or play games) is the nature of the input and output. Where an ordinary computer uses things like a keyboard, mouse, or (more exotically) speech recognition for input, VR uses sensors that detect how your body is moving. And where a PC displays output on a screen (or a printer), VR uses two screens (one for each eye), stereo or surround-sound speakers, and maybe some forms of haptic (touch and body perception) feedback as well. Let's take a quick tour through some of the more common VR input and output devices.

1. Head-Mounted Displays (HMDs)

A typical HMD, as shown in Fig.7-3, has two tiny screens that show different pictures to each of your eyes, so your brain produces a combined 3D (stereoscopic) image.

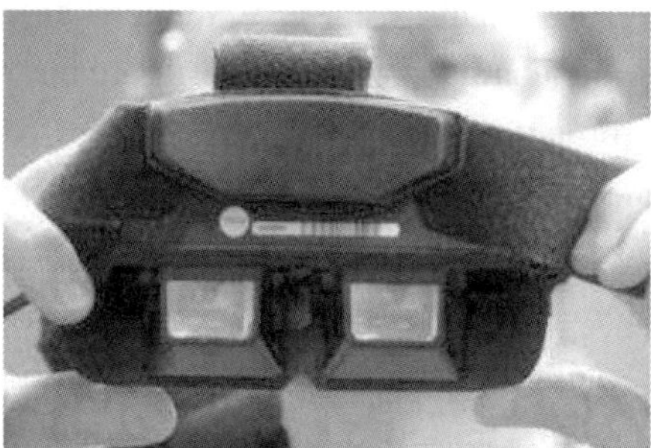

Fig.7-3 The view from inside of a typical HMD

There are two big differences between VR and looking at an ordinary computer screen: in VR, you see a 3D image that changes smoothly, in real-time, as you move your head. That's made possible by wearing a head-mounted display, which looks like a giant motorbike helmet or welding visor, but consists of two small screens (one in front of each eye), a blackout blindfold that blocks out all other light (eliminating distractions from the real world), and stereo headphones. The two screens display slightly different, stereoscopic images, creating a realistic 3D perspective of the virtual world. HMDs usually also have built-in accelerometers or position sensors so they can detect exactly how your head and body are moving (both position and orientation — which way they're tilting or pointing) and adjust the picture accordingly. The trouble with HMDs is that they're quite heavy, so they can be tiring to wear for long periods; some of the really heavy ones are even mounted on stands with **counterweights**. But HMDs don't have to be so elaborate and sophisticated: at the opposite end of the spectrum, Google has developed an affordable, low-cost pair of cardboard goggles with built-in lenses that convert an ordinary smartphone into a crude HMD.

2. Immersive Rooms

An alternative to putting on an HMD is to sit or stand inside a room onto whose walls changing images are projected from outside. As you move in the room, the images change accordingly. Flight simulators use this technique, often with images of landscapes, cities, and airport approaches projected onto large screens positioned just outside a mockup of a cockpit. A famous 1990s VR experiment called CAVE (Cave Automatic Virtual Environment), developed at the University of Illinois by Thomas de Fanti, also worked this way. People moved around inside a large cube-shaped room with semi-transparent walls onto which stereo images were back-projected from outside. Although they didn't have to wear HMDs, they did need stereo glasses to experience full 3D perception.

3. Datagloves

See something amazing and your natural instinct is to reach out and touch it — even babies do that. So giving people the ability to handle virtual objects has always been a big part of VR. Usually, this is done using datagloves, which are ordinary gloves with sensors wired to the outside to detect hand and figure motions. One technical method of doing this uses fiber-optic cables stretched the length of each finger. Each cable has tiny cuts in it so, as you flex your fingers back and forth, more or less light escapes. A photocell at the end of the cable measures how much light reaches it and the computer uses this to figure out exactly what your fingers are doing. Other gloves use **strain gauges**, **piezoelectric sensors**, or electromechanical devices (such as **potentiometers**) to measure finger movements.

4. Wands

Even simpler than a dataglove, a wand, as shown in Fig.7-4, is a stick you can use to touch, point to, or otherwise interact with a virtual world. It has position or motion sensors (such as **accelerometers**) built in, along with mouse-like buttons or scroll wheels. Originally, wands were clumsily wired into the main VR computer; increasingly, they're wireless.

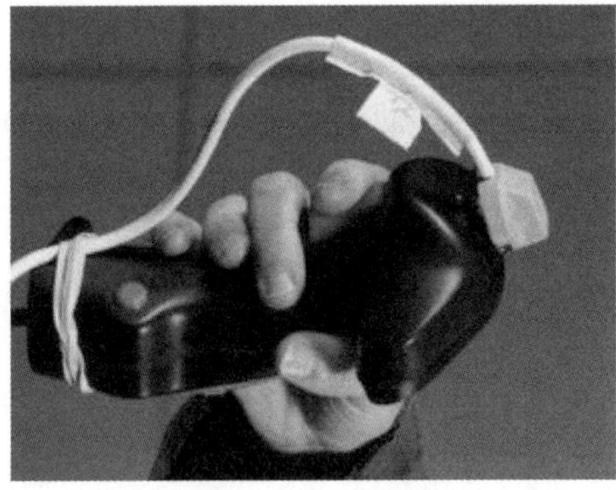

Fig.7-4 A typical handheld virtual reality controller

7.2.4 Applications of Virtual Reality

VR has always suffered from the perception that it's little more than a glorified arcade game — literally a "dreamy escape" from reality. In that sense, "virtual reality" can be an unhelpful misnomer; "alternative reality" "artificial reality", or "computer simulation" might be better terms. The key thing to remember about VR is that it really isn't a fad or fantasy waiting in the wings to whistle people off to alternative worlds; it's a hard-edged practical technology that's been routinely used by scientists, doctors, dentists, engineers, architects, archaeologists, and the military for about the last 30 years. What sorts of things can we do with it?

1. Education

Difficult and dangerous jobs are hard to train for. How can you safely practice taking a trip to space, landing a jumbo jet, making a parachute jump, or carrying out brain surgery? All these things are obvious candidates for virtual reality applications.

Flight training is a classic application of virtual reality, though it doesn't use HMDs or datagloves. Instead, you sit in a pretend cockpit with changing images projected onto giant screens to give an impression of the view you'd see from your plane(as shown Fig.7-5). The cockpit is a meticulous replica of the one in a real airplane with exactly the same instruments and controls.

Fig.7-5 Flight training

Just like pilots, surgeons are now routinely trained using VR. In a 2008 study of 735 surgical trainees from 28 different countries, 68 percent said the opportunity to train with VR was "good" or "excellent" for them and only 2 percent rated it useless or unsuitable.

2. Scientific Visualization

Anything that happens at the atomic or molecular scale is effectively invisible unless you're prepared to sit with your eyes glued to an electron microscope. But suppose you want to design new materials or drugs and you want to experiment with the molecular equivalent of LEGO. That's another obvious application for virtual reality. Instead of wrestling with numbers, equations, or two-dimensional drawings of molecular structures, you can snap complex molecules together right before your eyes. This kind of work began in the 1960s at the University of North Carolina at Chapel Hill, where Frederick Brooks launched GROPE, a project to develop a VR system for exploring the interactions between protein molecules and drugs.

3. Medicine

Apart from its use in things like surgical training and drug design, virtual reality also makes possible telemedicine (monitoring, examining, or operating on patients remotely). A logical extension of this has a surgeon in one location hooked up to a virtual reality control panel and a robot in another location (maybe an entire continent away) wielding the knife. The best-known example of this is the DaVinci surgical robot, released in 2009, of which several thousand have now been installed in hospitals worldwide. Introduce collaboration and there's the possibility of a whole group of the world's best surgeons working together on a particularly difficult operation —a kind of WikiSurgery, if you like!

4. Industrial Design and Architecture

Architects used to build models out of card and paper; now they're much more likely to build virtual reality computer models you can walk through and explore. By the same token, it's generally much cheaper to design cars, airplanes, and other complex, expensive vehicles on a computer screen than to model them in wood, plastic, or other real-world materials. This is an area where virtual reality overlaps with computer modeling: instead of simply making an immersive 3D visual model for people to inspect and explore, you're creating a mathematical model that can be tested for its aerodynamic, safety, or other qualities.

5. Games and Entertainment

From flight simulators to race-car games, VR has long hovered on the edges of the gaming world — never quite good enough to revolutionize the experience of gamers, largely due to computers being too slow, displays lacking full 3D, and the lack of decent HMDs and datagloves. All that may be about to change with the development of affordable new peripherals.

7.2.5 Pros and Cons of Virtual Reality

Like any technology, virtual reality has both good and bad points. How many of us would rather have a complex brain operation carried out by a surgeon trained in VR, compared to someone who has merely read books or watched over the shoulders of their peers? How many of us would rather practice our driving on a car simulator before we set foot on the road? Or sit back and relax in a Jumbo Jet, confident in the knowledge that our pilot practiced landing at this very airport, dozens of times, in a VR simulator before she ever set foot in a real cockpit?

Critics always raise the risk that people may be seduced by alternative realities to the point of neglecting their real-world lives — but that criticism has been leveled at everything from radio and TV to computer games and the Internet. And, at some point, it becomes a philosophical and ethical question: What is real anyway? And who is to say which is the better way to pass your time? Like many technologies, VR takes little or nothing away from the real world: you don't have to use it if you don't want to.

The promise of VR has loomed large over the world of computing for at least the last quarter century — but remains largely unfulfilled. While science, architecture, medicine, and the military all rely on VR technology in different ways, mainstream adoption remains virtually nonexistent; we're not routinely using VR the way we use computers, smartphones, or the Internet. But the 2014 acquisition of VR company Oculus, by Facebook, greatly renewed interest in the area and could change everything. Facebook's basic idea is to let people share things with their friends using the Internet and the Web. What if you could share not simply a photo or a link to a Web article but an entire experience? Instead of sharing photos of your wedding with your Facebook friends, what if you could make it possible for people to attend your wedding remotely, in virtual reality, in perpetuity? What if we could record historical events in such a way that people could experience them again and again, forever more? These are the sorts of social, collaborative virtual reality sharing that (we might guess) Facebook is thinking about exploring right now. If so, the future of virtual reality looks very bright indeed!

7.3 Data Visualization

Data visualization is viewed by many disciplines as a modern equivalent of visual communication. It is a general term that describes any effort to help people understand the significance of data by placing it in a visual context, involves the creation and study of the visual representation of data.

(7-5) A primary goal of data visualization is to communicate information clearly and efficiently via statistical graphics, plots and information graphics. Numerical data may be encoded using dots, lines, or bars, to visually communicate a quantitative message. Effective visualization helps users analyze and reason about data and evidence. It makes complex data more accessible, understandable and usable. Patterns, trends and correlations that might go undetected in text-based data can be exposed and recognized easier with data visualization software. Tables are generally used where users will look up a specific measurement, while charts of various types are used to show patterns or relationships in the data for one or more variables.

Data visualization is both an art and a science. It is viewed as a branch of descriptive statistics by some, but also as a grounded theory development tool by others. The rate at which data is generated has increased. Data created by Internet activity and an expanding number of sensors in the environment, such as satellites, are referred to as "Big Data". Processing, analyzing and communicating this data present a variety of ethical and analytical challenges for data visualization. The field of data science and practitioners called data scientists has emerged to help address this challenge.

Data visualization is closely related to information graphics, information visualization, scientific visualization, exploratory data analysis and statistical graphics. In the new millennium, data visualization has become an active area of research, teaching and development.

7.3.1 Characteristics of Effective Graphical Displays

Professor Edward Tufte explained that users of information displays are executing particular analytical tasks such as making comparisons or determining causality. The design principle of the information graphic should support the analytical task, showing the comparison or causality.

In his 1983 book **The Visual Display of Quantitative Information**, Edward Tufte defines "graphical displays" and principles for effective graphical display in the following passage: "Excellence in statistical graphics consists of complex ideas communicated with clarity, precision and efficiency. Graphical displays should:

(1) show the data

(2) induce the viewer to think about the substance rather than about methodology, graphic design, the technology of graphic production or something else

(3) avoid distorting what the data has to say

(4) present many numbers in a small space

(5) make large data sets coherent

(6) encourage the eye to compare different pieces of data

(7) reveal the data at several levels of detail, from a broad overview to the fine structure

(8) serve a reasonably clear purpose: description, exploration, tabulation or decoration

(9) be closely integrated with the statistical and verbal descriptions of a data set.

Graphics reveal data. Indeed graphics can be more precise and revealing than conventional statistical computations."

For example, the Minard① diagram shows the losses suffered by Napoleon's army in the 1812—1813 period. Six variables are plotted: the size of the army, its location on a two-dimensional surface (*x* and *y*), time, direction of movement, and temperature. The line width illustrates a comparison (size of the army at points in time) while the temperature axis suggests a cause of the change in army size. This multivariate display on a two dimensional surface tells a story that can be grasped immediately while identifying the source data to build credibility. Tufte wrote in 1983 that: "It may well be the best statistical graphic ever drawn".

The Congressional Budget Office summarized several best practices for graphical displays in a June 2014 presentation. These included: ① Knowing your audience; ② Designing graphics that can stand alone outside the context of the report; and ③ Designing graphics that communicate the key messages in the report.

7.3.2 Quantitative Messages

Author Stephen Few described eight types of quantitative messages that users may attempt

① Charles Joseph Minard (27 March 1781 — 24 October 1870) was a French civil engineer recognized for his significant contribution in the field of information graphics in civil engineering and statistics.

to understand or communicate from a set of data and the associated graphs used to help communicate the message.

(1) Time-series: A single variable is captured over a period of time, such as the unemployment rate over a 10-year period. A line chart may be used to demonstrate the trend.

(2) Ranking: Categorical subdivisions are ranked in ascending or descending order, such as a ranking of sales performance (the measure) by sales persons (the category, with each sales person a categorical subdivision) during a single period. A bar chart may be used to show the comparison across the sales persons.

(3) Part-to-whole: Categorical subdivisions are measured as a ratio to the whole (i.e., a percentage out of 100%). A pie chart or bar chart can show the comparison of ratios, such as the market share represented by competitors in a market.

(4) Deviation: Categorical subdivisions are compared against a reference, such as a comparison of actual vs. budget expenses for several departments of a business for a given time period. A bar chart can show comparison of the actual versus the reference amount.

(5) Frequency distribution: Shows the number of observations of a particular variable for given interval, such as the number of years in which the stock market return is between intervals such as 0%~10%, 11%~20%, etc. A histogram, a type of bar chart, may be used for this analysis. A boxplot helps visualize key statistics about the distribution, such as median, quartiles, outliers, etc.

(6) Correlation: Comparison between observations represented by two variables (*X*,*Y*) to determine if they tend to move in the same or opposite directions. For example, plotting unemployment (*X*) and inflation (*Y*) for a sample of months. A scatter plot is typically used for this message.

(7) Nominal comparison: Comparing categorical subdivisions in no particular order, such as the sales volume by product code. A bar chart may be used for this comparison.

(8) Geographic or geospatial: Comparison of a variable across a map or layout, such as the unemployment rate by state or the number of persons on the various floors of a building. A cartogram is a typical graphic used.

(9) Analysts reviewing a set of data may consider whether some or all of the messages and graphic types above are applicable to their task and audience. The process of trial and error to identify meaningful relationships and messages in the data is part of exploratory data analysis.

7.3.3 Visual Perception and Data Visualization

Human can distinguish differences in line length, shape, orientation, and color (*hue*) readily without significant processing effort; these are referred to as "pre-attentive attributes". For example, it may require significant time and effort ("attentive processing") to identify the number of times the digit "5" appears in a series of numbers; but if that digit is different in size, orientation, or color, instances of the digit can be noted quickly through pre-attentive processing.

Effective graphics take advantage of pre-attentive processing and attributes and the relative strength of these attributes. For example, since humans can more easily process differences in line length than surface area, it may be more effective to use a bar chart (which takes advantage of line length to show comparison) rather than pie charts (which use surface area to show comparison).

(7-6) Almost all data visualizations are created for human consumption. Knowledge of human perception and cognition is necessary when designing intuitive visualizations. **Cognition** refers to processes in human beings like perception, attention, learning, memory, thought, concept formation, reading, and problem solving. Human visual processing is efficient in detecting changes and making comparisons between quantities, sizes, shapes and variations in lightness. When properties of symbolic data are mapped to visual few properties, humans can browse through large amounts of data efficiently. It is estimated that 2/3 of the brain's neurons can be involved in visual processing. Proper visualization provides a different approach to show potential connections, relationships, etc. which are not as obvious in non-visualized quantitative data. Visualization can become a means of data exploration.

7.3.4 Examples of Diagrams Used for Data Visualization

Here are some examples for data visualization that widely employed.

1. Histogram

A **histogram** is a graphical representation of the distribution of numerical data. It is an estimate of the probability distribution of a continuous variable (quantitative variable) and was first introduced by Karl Pearson①. It is a kind of bar graph. To construct a histogram, the first step is to "bin" the range of values — that is, divide the entire range of values into a series of intervals — and then count how many values fall into each interval. The bins are usually specified as consecutive, non-overlapping intervals of a variable. The bins (intervals) must be adjacent, and are often of equal size.

If the bins are of equal size, a rectangle is erected over the bin with height proportional to the frequency-the number of cases in each bin, Fig.7-6 is an example histogram of the heights of 31 Black Cherry trees. A histogram may also be normalized to display "relative" frequencies. It then shows the proportion of cases that fall into each of several categories, with the sum of the heights equaling 1.

However, bins need not be of equal width; in that case, the erected rectangle is defined to have its area proportional to the frequency of cases in the bin. The vertical axis is then not the frequency but frequency density — the number of cases per unit of the variable on the horizontal axis. Examples of variable bin width are displayed on Census bureau data below (Fig.7-7). The U.S. Census Bureau found that there were 124 million people who work outside of their homes. Using their data on the time occupied by travel to work, the table below shows the absolute number of people who responded with travel times "at least 30 but less than 35

① was an influential English mathematician and biostatistician(27 March 1857 — 27 April 1936)

minutes" is higher than the numbers for the categories above and below it.

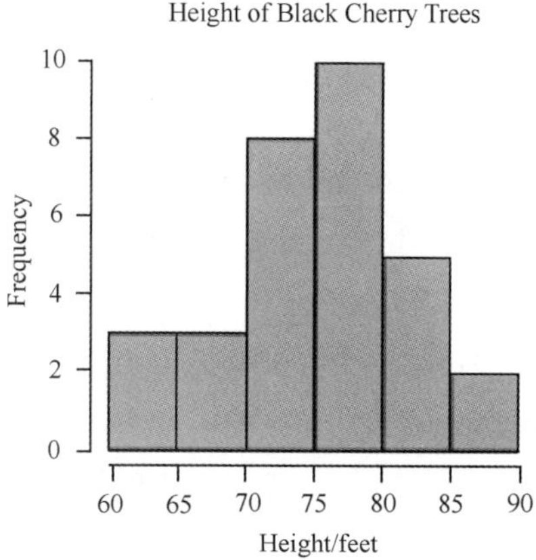

Fig.7-6 An simple example of histogram

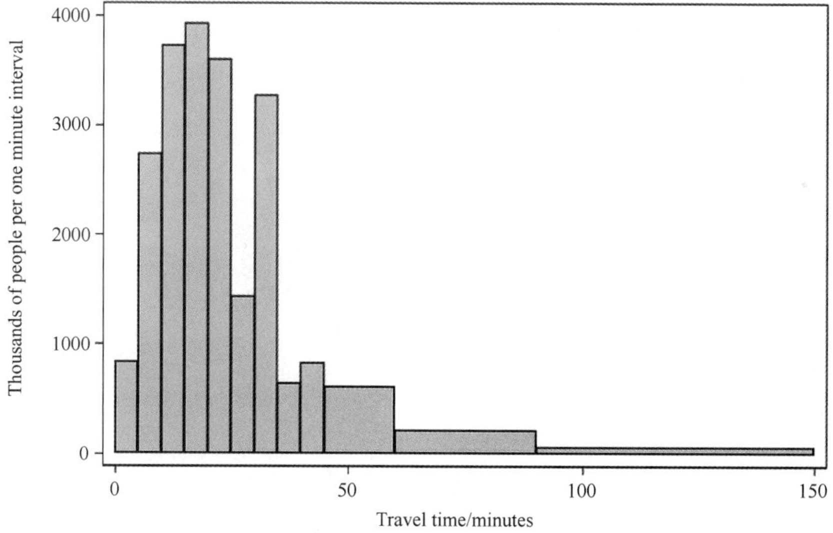

Fig.7-7 Histogram with variable bin width

As the adjacent bins leave no gaps, the rectangles of a histogram touch each other to indicate that the original variable is continuous.

Histograms give a rough sense of the density of the underlying distribution of the data, and often for density estimation: estimating the probability density function of the underlying variable. The total area of a histogram used for probability density is always normalized to 1. If the length of the intervals on the x-axis are all 1, then a histogram is identical to a relative frequency plot.

A histogram can be thought of as a simplistic kernel density estimation, which uses a kernel to smooth frequencies over the bins. This yields a smoother probability density function, which will in general more accurately reflect distribution of the underlying variable. The density estimate could be plotted as an alternative to the histogram, and is usually drawn as a curve rather than a set of boxes.

Another alternative is the average shifted histogram, which is fast to compute and gives a smooth curve estimate of the density without using kernels.

2. Scatter Plot

A **scatter plot** (also called a scatter graph, scatter chart, scattergram, or scatter diagram) is a type of plot or mathematical diagram using Cartesian coordinates to display values for typically two variables for a set of data. If the points are color-coded, one additional variable can be displayed. The data is displayed as a collection of points, each having the value of one variable determining the position on the horizontal axis and the value of the other variable determining the position on the vertical axis. Fig.7-8 shows waiting time between eruptions and the duration of the eruption for the Old Faithful Geyser in Yellowstone National Park,Wyoming, USA. This chart suggests there are generally two “types” of eruptions: short-wait-short-duration, and long-wait-long-duration.

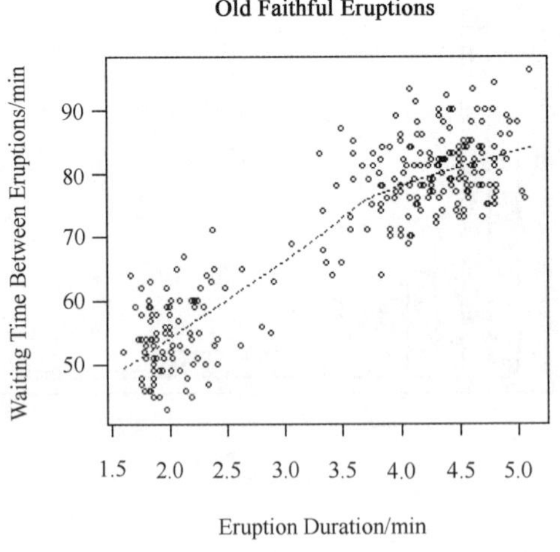

Fig.7-8 Waiting time between eruptions and the duration of the eruption

A scatter plot can be used either when one continuous variable that is under the control of the experimenter and the other depends on it or when both continuous variables are independent. If a parameter exists that is systematically incremented and/or decremented by the other, it is called the control parameter or independent variable and is customarily plotted along the horizontal axis. The measured or dependent variable is customarily plotted along the vertical axis. If no dependent variable exists, either type of variable can be plotted on either axis and a scatter plot will illustrate only the degree of correlation (not causation) between two variables.

A scatter plot can suggest various kinds of correlations between variables with a certain confidence interval. For example, weight and height, weight would be on y axis and height would be on the x axis. Correlations may be positive (rising), negative (falling), or null (uncorrelated). If the pattern of dots slopes from lower left to upper right, it indicates a positive correlation between the variables being studied. If the pattern of dots slopes from upper left to lower right, it indicates a negative correlation. A line of best fit (alternatively called "trendline") can be drawn in order to study the relationship between the variables. An equation for the correlation between the variables can be determined by established best-fit procedures. For a linear correlation, the best-fit procedure is known as linear regression and is guaranteed to generate a correct solution in a finite time. No universal best-fit procedure is guaranteed to generate a correct solution for arbitrary relationships. A scatter plot is also very useful when we wish to see how two comparable data sets agree with each other. In this case, an identity line, i.e., a y=x line, or an 1：1 line, is often drawn as a reference. The more the two data sets agree, the more the scatters tend to concentrate in the vicinity of the identity line; if the two data sets are numerically identical, the scatters fall on the identity line exactly.

One of the most powerful aspects of a scatter plot, however, is its ability to show nonlinear relationships between variables. The ability to do this can be enhanced by adding a smooth line such as LOESS (locally weighted scatterplot smoothing). Furthermore, if the data are represented by a mixture model of simple relationships, these relationships will be visually evident as superimposed patterns.

3. Gantt Chart

A **Gantt chart** is a type of bar chart, devised by Henry Gantt in the 1910s, that illustrates a project schedule. Gantt charts illustrate the start and finish dates of the terminal elements and summary elements of a project. Terminal elements and summary elements comprise the work breakdown structure of the project. Modern Gantt charts also show the dependency (i.e., precedence network) relationships between activities. Gantt charts can be used to show current schedule status using percent-complete shadings and a vertical "TODAY" line.

Although now regarded as a common charting technique, Gantt charts were considered revolutionary when first introduced. This chart is also used in information technology to represent data that has been collected.

In Tab.7-1, there are seven tasks, labeled A through G. Some tasks can be done concurrently (A and B) while others cannot be done until their predecessor task is complete (C and D cannot begin until A is complete). Additionally, each task has three time estimates: the **optimistic time estimate** (O), the most likely or normal time estimate (M), and the **pessimistic time estimate** (P). The **expected time** (TE) is estimated using the **beta probability distribution** for the time estimates, using the formula $(O + 4M + P) \div 6$.

Tab.7-1 Schedule of Activities

Activity	Predecessor	Time estimates			Expected time
		Opt. (O)	Normal (M)	Pess. (P)	
A	—	2	4	6	4.00
B	—	3	5	9	5.33
C	*A*	4	5	7	5.17
D	*A*	4	6	10	6.33
E	*B*, *C*	4	5	7	5.17
F	*D*	3	4	8	4.50
G	*E*	3	5	8	5.17

Once this step is complete, one can draw a Gantt chart or a network diagram as shown in Fig.7-9.

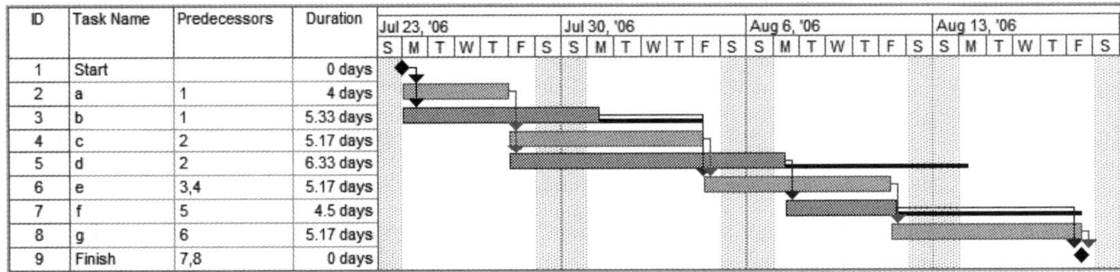

ID	Task Name	Predecessors	Duration
1	Start		0 days
2	a	1	4 days
3	b	1	5.33 days
4	c	2	5.17 days
5	d	2	6.33 days
6	e	3,4	5.17 days
7	f	5	4.5 days
8	g	6	5.17 days
9	Finish	7,8	0 days

Fig.7-9 A Gantt chart created using Microsoft Project (MSP).

Note: ① the critical path is in red, ② the slack is the black lines connected to non-critical activities, ③ since Saturday and Sunday are not work days and are thus excluded from the schedule, some bars on the Gantt chart are longer if they cut through a weekend.

4. Heat Map

A **heat map** (or heatmap) is a graphical representation of data where the individual values contained in a matrix are represented as colors. The term "heat map" was originally coined and trademarked by software designer Cormac Kinney in 1991, to describe a 2D display depicting financial market information, though similar plots such as shading matrices have existed for over a century.

There are different kinds of heat maps:

(1) Web heat maps have been used for displaying areas of a Web page most frequently scanned by visitors. Web heatmaps are often used alongside other forms of Web analytics and session replay tools.

(2) Biology heat maps are typically used in molecular biology to represent the level of expression of many genes across a number of comparable samples (e.g. cells in different states, samples from different patients) as they are obtained from DNA microarrays.

(3) The tree map is a 2D hierarchical partitioning of data that visually resembles a heat map.

(4) A mosaic plot is a tiled heat map for representing a two-way or higher-way table of data. As with treemaps, this rectangular regions in a mosaic plot are hierarchically organized. This means that the regions are rectangles instead of squares.

(5) A density function visualization is a heat map for representing the density of dots in a map. It enables one to perceive density of points independently of the zoom factor. Perrot et al (2015) proposed a way to use density function to visualize billions of dots using big data infrastructure with Spark and Hadoop.

Fig.7-10 shows a heat map, atop a color bathymetric map, indicating the probable location of missing Malaysia Airlines Flight 370 based on a Bayesian method analysis of possible flight paths of the aircraft.

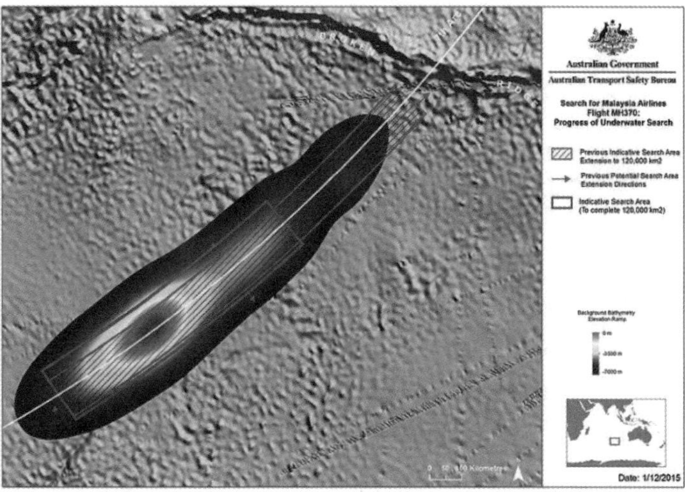

Fig.7-10 MH370 location probability heat map

7.4 Key Terms and Review Question

1. Technical Terms

raster graphics	光栅图形	7.1
vector graphics	矢量图形	7.1
pixel	像素	7.1
resolution	分辨率	7.1
Megapixels	百万像素	7.1
jagged edges	锯齿边缘	7.1
anti-aliasing	抗锯齿处理	7.1
primitive	基元，图元	7.1
transform	变形	7.1
Bézier curve	贝塞尔曲线	7.1
perspective	透视图	7.1
hidden-surface	隐藏面	7.1
wire-frame	线框	7.1

rendered	渲染的	7.1
ray tracing	光线跟踪	7.1
radiosity	辐射度	7.1
reflect	反射	7.1
scatter	散射	7.1
computer animation	计算机动画	7.1
keyframes	关键帧	7.1
projection	投影	7.1
planar	平面的	7.1
photorealism	照相写实主义	7.1
scanline	扫描线	7.1
cross hatching	交叉影线	7.1
shader	明暗器；色器	7.1
texture mapping	纹理映射	7.1
liquid-crystal display	液晶显示器	7.1
aliasing	走样	7.1
jaggies	锯齿边缘	7.1
mipmapping	贴图分级细化	7.1
volume rendering	立体绘制	7.1
slice	切片	7.1
surface patches	面片	7.1
polygonal mesh	多边形网格	7.1
finishe	表面处理	7.1
manifold	流形	7.1
tessellation	曲面细分	7.1
cone	锥形体	7.1
rasterize	栅格化，点阵化	7.1
convex	凸面体	7.1
stage/set design	舞台/布景设计	7.1
diffuse reflection	漫射	7.1
transmission spectra	透射光谱	7.1
head-mounted display	头盔式显示器	7.2
datagloves	数据手套	7.2
counterweight	配重；平衡力	7.2
strain gauge	应变计量仪	7.2
piezoelectric sensor	压电传感器	7.2
potentiometer	电位器	7.2
accelerometers	加速计	7.2
data visualization	数据可视化	7.3

hue	色度	7.3
cognition	认知	7.3
histogram	柱状图	7.3
scatter plot	散点图	7.3
Gantt chart	甘特图	7.3
optimistic time estimate	乐观时间估计	7.3
pessimistic time estimate	悲观时间估计	7.3
expected time	期望时间	7.3
beta probability distribution	beta 概率分布	7.3
heat map	热度图	7.3

2. Translation Exercises

(7-1) To create the illusion of movement, an image is displayed on the computer screen then quickly replaced by a new image that is similar to the previous image, but shifted slightly. This technique is identical to the illusion of movement in television and motion pictures.

(7-2) A scene file contains objects in a strictly defined language or data structure; it would contain geometry, viewpoint, texture, lighting, and shading information as a description of the virtual scene. The data contained in the scene file is then passed to a rendering program to be processed and output to a digital image or raster graphics image file.

(7-3) Once the outside of the model's done, you turn your attention to its inner structure. This bit is called rigging your model (also known as skeletal animation). What parts does the object contain and how do they all connect together? When you've specified both the inside and outside details, your model is pretty much complete. The final stage is called texturing, and involves figuring out what colors, surface patterns, finishes, and other details you want your object to have.

(7-4)Apart from being cheaper and easier than using paper, CAD designs are easy to send round the world by email (from designers in Paris to manufacturers in Singapore, perhaps). Another big advantage is that CAD drawings can be converted automatically into production instructions for industrial robots and other factory machines, which greatly reduces the overall time needed to turn new designs into finished products.

(7-5) A primary goal of data visualization is to communicate information clearly and efficiently via statistical graphics, plots and information graphics. Numerical data may be encoded using dots, lines, or bars, to visually communicate a quantitative message. Effective visualization helps users analyze and reason about data and evidence. It makes complex data more accessible, understandable and usable. Patterns, trends and correlations that might go undetected in text-based data can be exposed and recognized easier with data visualization software.

(7-6) Almost all data visualizations are created for human consumption. Knowledge of human perception and cognition is necessary when designing intuitive visualizations. **Cognition** refers to processes in human beings like perception, attention, learning, memory, thought, concept formation, reading, and problem solving. Human visual processing is efficient in detecting changes and making comparisons between quantities, sizes, shapes and variations in lightness. When properties of symbolic data are mapped to visual few properties, humans can browse through large amounts of data efficiently.

References

[1] Shirly P, Marschner S. Fundamentals of Computer Graphics (3rd Edition)[M]. CRC Press, 2011.

[2] Computer Graphics[DM/OL].[2017-05-11].https://en.wikipedia.org/wiki/ Computer_graphics#Concepts_and_principles.

[3] 3D_modeling[DM/OL]. [2017-05-11]. https://en.wikipedia.org/wiki/3D_modeling.

[4] Post F H, Nielson G M, Bonneau G P. Data Visualization: The State of the Art. Research paper, TU delft, 2002.

Chapter 8 Human–Computer Interaction

8.1 Human-Computer Interaction[①]

Human-Computer Interaction (HCI) is an area of research and practice that emerged in the early 1980s, initially as a specialty area in computer science embracing cognitive science and human factors engineering. HCI has expanded rapidly and steadily for three decades, attracting professionals from many other disciplines and incorporating diverse concepts and approaches. To a considerable extent, HCI now aggregates a collection of semi-autonomous fields of research and practice in human-centered informatics. However, the continuing synthesis of disparate conceptions and approaches to science and practice in HCI has produced a dramatic example of how different epistemologies and paradigms can be reconciled and integrated in a vibrant and productive intellectual project.

8.1.1 History of HCI

Until the late 1970s, the only humans who interacted with computers were information technology professionals and dedicated hobbyists. This changed disruptively with the emergence of personal computing in the later 1970s. Personal computing, including both personal software (productivity applications, such as text editors and spreadsheets, and interactive computer games) and personal computer platforms (operating systems, programming languages, and hardware), made everyone in the world a potential computer user, and vividly highlighted the deficiencies of computers with respect to usability for those who wanted to use computers as tools.

The challenge of personal computing became manifest at an opportune time. The broad project of cognitive science, which incorporated cognitive psychology, artificial intelligence, linguistics, cognitive anthropology, and the philosophy of mind, had formed at the end of the 1970s. Part of the programme of cognitive science was to articulate systematic and scientifically informed applications to be known as "cognitive engineering". Thus, at just the point when personal computing presented the practical need for HCI, cognitive science presented people, concepts, skills, and a vision for addressing such needs through an ambitious synthesis of science and engineering. HCI was one of the first examples of cognitive engineering.

---

① Carroll J M. Human Computer Interaction - brief intro, The Encyclopedia of Human-Computer Interaction, 2nd Ed, https://www.interaction-design.org/literature/book/the-encyclopedia-of-human-computer-interaction- 2nd-ed/human-computer-interaction-brief-intro.

This was facilitated by analogous developments in engineering and design areas adjacent to HCI, and in fact often overlapping HCI, notably human factors engineering and documentation development. Human factors had developed empirical and task-analytic techniques for evaluating human-system interactions in domains such as aviation and manufacturing, and were moving to address interactive system contexts in which human operators regularly exerted greater problem-solving discretion. Documentation development was moving beyond its traditional role of producing systematic technical descriptions toward a cognitive approach incorporating theories of writing, reading, and media, with empirical user testing. Documents and other information needed to be usable also.

Other historically fortuitous developments contributed to the establishment of HCI. Software engineering, mired in unmanageable software complexity in the 1970s (the "**software crisis**"), was starting to focus on nonfunctional requirements, including usability and maintainability, and on empirical software development processes that relied heavily on **iterative prototyping** and empirical testing. Computer graphics and information retrieval had emerged in the 1970s, and rapidly came to recognize that interactive systems were the key to progressing beyond early achievements. All these threads of development in computer science pointed to the same conclusion: The way forward for computing entailed understanding and better empowering users. These diverse forces of need and opportunity converged around 1980, focusing a huge burst of human energy, and creating a highly visible **interdisciplinary** project.

8.1.2 From Cabal to Community

The original technical focus of HCI was the concept of **usability**. This concept was originally articulated somewhat naively in the slogan "easy to learn, easy to use". The blunt simplicity of this conceptualization gave HCI an edgy and prominent identity in computing. It served to hold the field together, and to help it influence computer science and technology development more broadly and effectively. However, inside HCI the concept of usability has been re-articulated and reconstructed almost continually, and has become increasingly rich and problematic. Usability now often subsumes qualities like fun, **collective efficacy, aesthetic tension**, enhanced creativity, flow, support for human development, and others.

Although the original academic home for HCI was computer science, and its original focus was on personal productivity applications, mainly text editing and spreadsheets, the field has constantly diversified and outgrown all boundaries. It quickly expanded to encompass visualization, information systems, collaborative systems, the system development process, and many areas of design. HCI is taught now in many departments/faculties that address information technology, including psychology, design, communication studies, cognitive science, information science, science and technology studies, geographical sciences, management information systems, and industrial, manufacturing, and systems engineering. HCI research and practice draws upon and integrates all of these perspectives.

A result of this growth is that HCI is now less singularly focused with respect to core concepts and methods, problem areas and assumptions about infrastructures, applications, and types of users. Indeed, it no longer makes sense to regard HCI as a specialty of computer science; HCI has grown to be broader, larger and much more diverse than computer science itself. HCI expanded from its initial focus on individual and generic user behavior to include social and organizational computing, accessibility for the elderly, the cognitively and physically impaired, and for all people, and for the widest possible spectrum of human experiences and activities. It expanded from desktop office applications to include games, learning and education, commerce, health and medical applications, emergency planning and response, and systems to support collaboration and community. It expanded from early graphical user interfaces to include myriad interaction techniques and devices, **multi-modal interactions**, tool support for model-based user interface specification, and a host of emerging ubiquitous, handheld and **context-aware** interactions.

There is no unified concept of an HCI professional. In the 1980s, the cognitive science side of HCI was sometimes contrasted with the software tools and user interface side of HCI. The landscape of core HCI concepts and skills is far more differentiated and complex now. HCI academic programs train many different types of professionals: user experience designers, interaction designers, user interface designers, application designers, usability engineers, user interface developers, application developers, technical communicators/online information designers, and more. And indeed, many of the sub-communities of HCI are themselves quite diverse. For example, **ubiquitous computing** (aka ubicomp) is subarea of HCI, but it is also a superordinate area integrating several distinguishable subareas, for example mobile computing, **geo-spatial information systems, in-vehicle systems, community informatics**, distributed systems, handhelds, wearable devices, **ambient intelligence**, sensor networks, and specialized views of usability evaluation, programming tools and techniques, and application infrastructures. The relationship between ubiquitous computing and HCI is paradigmatic: HCI is the name for a community of communities.

Indeed, the principle that HCI is a community of communities is now a point of definition **codified**, for example, in the organization of major HCI conferences and journals. The integrating element across HCI communities continues to be a close linkage of critical analysis of usability, broadly understood, with development of novel technology and applications. This is the defining identity commitment of the HCI community. It has allowed HCI to successfully cultivate respect for the diversity of skills and concepts that underlie innovative technology development, and to regularly transcend disciplinary obstacles. In the early 1980s, HCI was a small and focused specialty area. It was a cabal trying to establish what was then a heretical view of computing. Today, HCI is a vast and multifaceted community, bound by the evolving concept of usability, and the integrating commitment to value human activity and experience as the primary driver in technology.

8.1.3 Beyond the Desktop

Given the contemporary shape of HCI, it is important to remember that its origins are personal productivity interactions bound to the desktop, such as word processing and spreadsheets. Indeed, one of the biggest design ideas of the early 1980s was the so-called messy desk metaphor, popularized by the Apple Macintosh: Files and folders were displayed as icons that could be, and were scattered around the display surface. The messy desktop was a perfect incubator for the developing paradigm of graphical user interfaces. Perhaps it wasn't quite as easy to learn and easy to use as claimed, but people everywhere were soon double clicking, dragging windows and icons around their displays, and losing track of things on their desktop interfaces just as they did on their physical desktops. It was surely a stark contrast to the immediately prior teletype metaphor of UNIX, in which all interactions were accomplished by typing commands.

As HCI developed, it moved beyond the desktop in three distinct senses. First, the desktop metaphor proved to be more limited than it first seemed. It's fine to directly represent a couple dozen digital objects as icons, but this approach quickly leads to clutter, and is not very useful for people with thousands of personal files and folders. Through the mid-1990s, HCI professionals and everyone else realized that search is a more fundamental paradigm than browsing for finding things in a user interface. Ironically though, when early World Wide Web pages emerged in the mid-1990s, they not only dropped the messy desktop metaphor, but for the most part dropped graphical interactions entirely. And still they were seen as a breakthrough in usability (of course, the direct contrast was to UNIX-style tools like FTP and Telnet). The design approach of displaying and directly interacting with data objects as icons has not disappeared, but it is no longer a hegemonic design concept.

The second sense in which HCI moved beyond the desktop was through the growing influence of the Internet on computing and on society. Starting in the mid-1980s, email emerged as one of the most important HCI applications, but ironically, email made computers and networks into communication channels; people were not interacting with computers, they were interacting with other people through computers. Tools and applications to support collaborative activity now include **instant messaging**, wikis, blogs, online forums, **social networking**, social bookmarking and tagging services, media spaces and other collaborative workspaces, **recommender** and **collaborative filtering systems**, and a wide variety of online groups and communities. New paradigms and mechanisms for collective activity have emerged including online auctions, reputation systems, soft sensors, and **crowd sourcing**. This area of HCI, now often called social computing, is one of the most rapidly developing.

The third way that HCI moved beyond the desktop was through the continually, and occasionally explosive diversification in the ecology of computing devices. Before desktop applications were consolidated, new kinds of device contexts emerged, notably laptops, which began to appear in the early 1980s, and handhelds, which began to appear in the mid-1980s. One frontier today is ubiquitous computing: The pervasive incorporation of computing into human habitats — cars, home appliances, furniture, clothing, and so forth. Desktop computing is still very important, though the desktop habitat has been transformed by the wide use of laptops. To a considerable extent, the desktop itself has moved off the desktop.

The focus of HCI has moved beyond the desktop, and its focus will continue to move. HCI is a technology area, and it is ineluctably driven to frontiers of technology and application possibility. The special value and contribution of HCI is that it will investigate, develop, and harness those new areas of possibility not merely as technologies or designs, but as means for enhancing human activity and experience.

8.1.4 The Task-artifact Cycle

The movement of HCI off the desktop is a large-scale example of a pattern of technology development that is replicated throughout HCI at many levels of analysis. HCI addresses the dynamic co-evolution of the activities people engage in and experience, and the **artifacts** — such as interactive tools and environments — that mediate those activities, see Fig.8-1. Artifacts are designed in response, but inevitably do more than merely respond. Through the course of their adoption and appropriation, new designs provide new possibilities for action and interaction. Ultimately, this activity articulates further human needs, preferences, and design visions. HCI is about understanding and critically evaluating the interactive technologies people use and experience. But it is also about how those interactions evolve as people appropriate technologies, as their expectations, concepts and skills develop, and as they articulate new needs, new interests, and new visions and agendas for interactive technology.

Reciprocally, HCI is about understanding contemporary human practices and aspirations, including how those activities are embodied, elaborated, but also perhaps limited by current infrastructures and tools. HCI is about understanding practices and activity specifically as requirements and design possibilities envisioning and bringing into being new technology, new tools and environments. It is about exploring design spaces, and realizing new systems and devices through the co-evolution of activity and artifacts, the **task-artifact cycle**.

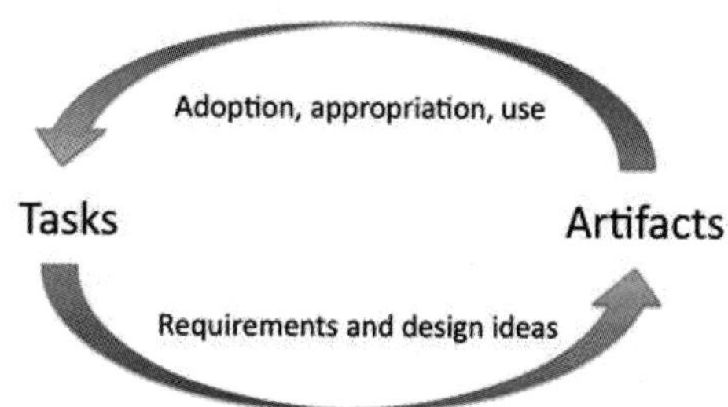

Fig.8-1 Human activities implicitly articulate needs, preferences and design visions.

Understanding HCI as inscribed in a **co-evolution** of activity and technological artifacts is useful. Most simply, it reminds us what HCI is like, that all of the infrastructure of HCI, including its concepts, methods, focal problems, and stirring successes will always be in flux. Moreover, because the co-evolution of activity and artifacts is shaped by a cascade of contingent initiatives across a diverse collection of actors, there is no reason to expect HCI to be convergent, or predictable. This is not to say progress in HCI is random or arbitrary, just that it is more like world history than it is like physics. One could see this quite optimistically: Individual and collective initiative shapes what HCI is, but not the laws of physics.

A second implication of the task-artifact cycle is that continual exploration of new applications and application domains, new designs and design paradigms, new experiences, and new activities should remain highly prized in HCI. We may have the sense that we know where we are going today, but given the apparent rate of co-evolution in activity and artifacts, our effective look-ahead is probably less than we think. Moreover, since we are in effect constructing a future trajectory, and not just finding it, the cost of missteps is high. The co-evolution of activity and artifacts evidences shows strong **hysteresis**, that is to say, effects of past co-evolutionary adjustments persist far into the future. For example, many people struggle every day with operating systems and core productivity applications whose designs were evolutionary reactions to misanalyses from two or more decades ago. Of course, it is impossible to always be right with respect to values and criteria that will emerge and coalesce in the future, but we should at least be mindful that very consequential missteps are possible.

The remedy is to consider many alternatives at every point in the progression. It is vitally important to have lots of work exploring possible experiences and activities, for example, on design and experience probes and prototypes. If we focus too strongly on the affordances of currently embodied technology we are too easily and uncritically accepting constraints that will limit contemporary HCI as well as all future trajectories.

HCI is not fundamentally about the laws of nature. Rather, it manages innovation to ensure that human values and human priorities are advanced, and not diminished through new technology. This is what created HCI; this is what led HCI off the desktop; it will continue to lead HCI to new regions of technology-mediated human possibility. This is why usability is an open-ended concept, and can never be reduced to a fixed checklist.

8.1.5 A Caldron of Theory

The contingent trajectory of HCI as a project in transforming human activity and experience through design has nonetheless remained closely integrated with the application and development of theory in the social and cognitive sciences. Even though, and to some extent because the technologies and human activities at issue in HCI are continually co-evolving, the domain has served as a laboratory and incubator for theory. The origin of HCI as an early case study in cognitive engineering had an imprinting effect on the character of the endeavor. From the very start, the models, theories and frameworks developed and used in HCI were pursued as contributions to science: HCI has enriched every theory it has appropriated. For example, the GOMS (Goals, Operations, Methods, Selection rules) model, the earliest native theory in HCI, was a more comprehensive cognitive model than had been attempted elsewhere in cognitive science and engineering; the model human processor included simple aspects of perception, attention, short-term memory operations, planning, and motor behavior in a single model. But GOMS was also a practical tool, articulating the dual criteria of scientific contribution plus engineering and design efficacy that has become the culture of theory and application in HCI.

The focus of theory development and application has moved throughout the history of HCI, as the focus of the co-evolution of activities and artifacts has moved. Thus, the early information processing-based psychological theories, like GOMS, were employed to model the cognition and behavior of individuals interacting with keyboards, simple displays, and pointing devices. This initial conception of HCI theory was broadened as interactions became more varied and applications became richer. For example, perceptual theories were marshaled to explain how objects are recognized in a graphical display, mental model theories were appropriated to explain the role of concepts — like the messy desktop metaphor — in shaping interactions, active user theories were developed to explain how and why users learn and making sense of interactions. In each case, however, these elaborations were both scientific advances and bases for better tools and design practices.

This **dialectic** of theory and application has continued in HCI. It is easy to identify a dozen or so major currents of theory, which themselves can by grouped (roughly) into three eras: theories that view human-computer interaction as information processing, theories that view interaction as the initiative of agents pursuing projects, and theories that view interaction as socially and materially embedded in rich contexts. To some extent, the sequence of theories can be understood as a convergence of scientific opportunity and application need: Codifying and using relatively austere models made it clear what richer views of people and interaction could be articulated and what they could contribute; at the same time, personal devices became portals for interaction in the social and physical world, requiring richer theoretical frameworks for analysis and design.

The sequence of theories and eras is of course somewhat idealized. People still work on GOMS models; indeed, all of the major models, theories and frameworks that ever were employed in HCI are still in current use. Indeed, they continue to develop as the context of the field develops. GOMS today is more a niche model than a paradigm for HCI, but has recently been applied in research on smart phone designs and human-robot interactions.

The challenge of integrating, or at least better coordinating descriptive and explanatory science goals with prescriptive and constructive design goals is abiding in HCI. There are at least three ongoing directions — traditional application of ever-broader and deeper basic theories, development of local, sometimes domain dependent proto-theories within particular design domains, and the use of design rationale as a mediating level of description between basic science and design practice.

8.1.6 Implications of HCI for Science, Practice, and Epistemology

(8-1) One of the most significant achievements of HCI is its evolving model of the integration of research and practice. Initially this model was articulated as a reciprocal relation between cognitive science and cognitive engineering. Later, it ambitiously incorporated a diverse science foundation, notably social and organizational psychology, Activity Theory, distributed cognition, and sociology, and a ethnographic approaches human activity, including the activities of design and technology development and appropriation. Currently, the model is incorporating design practices and research across a broad spectrum, for example, theorizing user experience and ecological sustainability. In these developments, HCI provides a blueprint for a mutual relation between science and practice that is unprecedented.

Although HCI was always talked about as a design science or as pursuing guidance for designers, this was construed at first as a boundary, with HCI research and design as separate contributing areas of professional expertise. Throughout the 1990s, however, HCI directly assimilated, and eventually itself spawned, a series of design communities. At first, this was a merely ecumenical acceptance of methods and techniques laying those of beyond those of science and engineering. But this outreach impulse coincided with substantial advances in user interface technologies that shifted much of the potential proprietary value of user interfaces into graphical design and much richer ontologies of user experience.

Somewhat ironically, designers were welcomed into the HCI community just in time to help remake it as a design discipline. A large part of this transformation was the creation of design disciplines and issues that did not exist before. For example, user experience design and interaction design were not imported into HCI, but rather were among the first exports from HCI to the design world. Similarly, analysis of the productive tensions between creativity and rationale in design required a design field like HCI in which it is essential that designs have an internal logic, and can be systematically evaluated and maintained, yet at the same time provoke new experiences and insights. Design is currently the facet of HCI in most rapid flux. It seems likely that more new design proto-disciplines will emerge from HCI during the next decade.

No one can accuse HCI of resting on laurels. Conceptions of how underlying science informs and is informed by the worlds of practice and activity have evolved continually in HCI since its inception. Throughout the development of HCI, paradigm-changing scientific and **epistemological** revisions were deliberately embraced by a field that was, by any measure, succeeding intellectually and practically. The result has been an increasingly fragmented and complex field that has continued to succeed even more.

8.2 User Interface Design Adaptation①

8.2.1 Introduction

One of the main reasons for the increasing importance of **adaptation** is that we interact with our applications in contexts of use which are more and more varied because of the advent of

① Paterno F. https://www.interaction-design.org/literature/book/the-encyclopedia-of-human-computer-interaction-2nd-ed/user-interface-design-adaptation.

mobile technologies and smart environments.

Various aspects can be part of the possible contexts of use and can be grouped along four dimensions (see Fig.8-2):

(1) user-related aspects: preferences, goals and tasks, physical state (e.g. position), emotional state, etc.;

(2) technology-related aspects: screen resolution, connectivity, browser, battery, etc.;

(3) environment-related aspects: location, light, noise, etc.;

(4) social aspects: privacy rules, collaboration, etc.

According to changes in those aspects of the context of use, any aspect characterizing a user interface can be modified. Thus, the user interface can be adapted in its: presentation — the perceivable aspects, including media and interaction techniques, layout, graphical attributes, dynamic behavior, including navigation structure, dynamic activation, and deactivation of interaction techniques, and content, including texts, labels, and images.

Various adaptation strategies are possible, which can be classified according to the impact they have on the **user interface**: conservation, e.g. simple scaling of UI elements; rearrangement, e.g. changing the layout; simplification / magnification, same UI elements but with modified presentation; increase (also called progressive enhancement) / reduction (also called graceful degradation) in terms of UI elements.

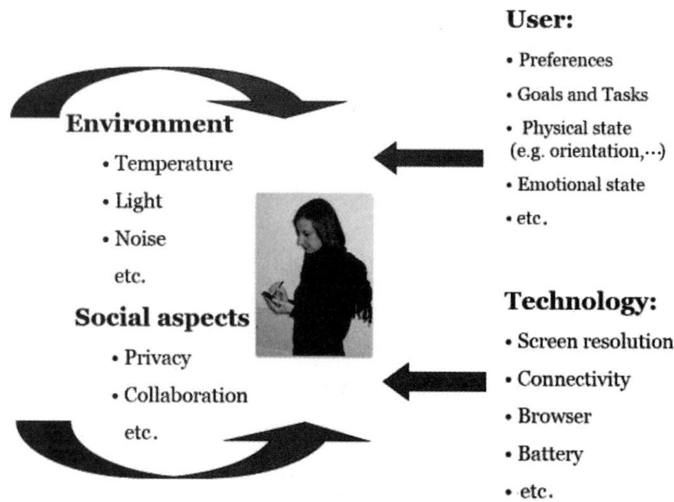

Fig.8-2 The Context of Use

One of the main reasons for the increasing interest in user interface adaptation is the device fragmentation stimulated by technological evolution, in particular in mobile devices. Device fragmentation concerns hardware and support for formats, browsers, audio/video playback/ streaming, etc. For example, in terms of screens we can notice that the screen resolutions of personal computers usually vary between 800×600 and 1920×1200 pixels, whereas those of mobile devices have a variation of between 320×240 and 1136×640 pixels (iPhone 5) and up to 1920×1080pixels (Galaxy S 4). Thus, screen resolution varies more with mobile devices than with desktop ones. The interesting point is that Moore's Law continually changes these numbers, and we can expect even more variance in the near future.

In recent years mobile technology has evolved considerably. We can easily realize this if we look at how interaction has changed in our smartphones. The oldest devices have **focus-based interactions**, in which the browser focus cycles through elements; the current focus of the page is easily determined because the focus element is highlighted, and the focus moves from one selectable element to another (e.g. from link to link) only sequentially, even when widely spaced (this can take some time). Then, devices supporting pointer-based interaction have been proposed, in which key-based navigation controls a pointer that can cover any part of the screen. With this solution selectable elements need to be large enough to be easily selected, since the pointer often moves in steps of 5~10 pixels, and selectable elements should have rollovers to make it clear when the pointer has entered their active area. Following the **pointer-based interaction** we now have the success of **touch-based interaction**, where events are related directly to a finger or stylus touch position on the screen; selectable elements should be widely spaced in order to allow users to select them precisely (studies suggest between 7mm and 9.6mm), selectable elements must be large enough to be easily selected; no elements are in focus until they are selected so extra information cannot be passed to the user (e.g. rollovers ineffective).

Various design aspects can be useful in supporting usability in mobile interaction. We have to consider that the user can be on the move and able to pay limited attention to the interaction. Thus, it is important to minimize text input, keep consistency between platforms so that application knowledge acquired through desktop interaction can be reused in mobile access and hence prevent user error, avoid overloading the user interface with too many elements, limit the need for zooming, and prevent touch selections that miss intended targets. Generally, we have to consider that mobile users often have short access time available, and thus they prefer access to small pieces of information.

More generally, we have to consider that our life is becoming a multi-device user experience. Indeed, a recent study (Google, 2012) found that our time online is spread across four device types (smartphones, tablets, PC/laptops, TVs). There are two modes of using them: sequential usage, moving from one device to another at different times to accomplish a task; and simultaneous usage, using more than one device at the same time for either a related or an unrelated activity. Managing information across such devices is one challenging aspect of using multiple devices. In general the main issues in multi-device UIs are: poor adaptation to the context of use, lack of coordination among tasks performed through different devices, inadequate support for seamless cross-device task performance.

Some studies have started to investigate what characterizes the user experience in cross-device application access. For example, Waljas *et al.* (2010) have identified three important dimensions for improving cross-device user experience: **appropriateness** for task performance, so that the structure of the interactive application provides an effective fit with what the users expect to perform in each device type; **continuity**, so that the flow of interaction with and across devices is perceived as fluent and connected; **consistency**, the user interfaces for the various device types should be perceived as coherent, still parts of the same application.

8.2.2 User Interface/Task/Platform Relations

In this section we discuss a logical framework that allows designers to think about the various possible relations between the tasks to perform, the user interfaces, and the platforms available. By platform we mean groups of devices that share similar interaction resources, e.g. the desktop, the smartphone, the tablet. In particular, we have identified five possible relations:

(1) Same task with the same user interface on different platforms.

(2) Same task with different user interface on different platforms.

(3) Same main task with different levels of subtasks on different platforms.

(4) Dependencies among different tasks performed on different platforms.

(5) Tasks meaningful only on some platform types (e.g. because they require very lengthy access or are related to a mobile position or to specific equipment such as a camera).

Because of the rapidly expanding variety of mobile technology there are indeed significant differences between different platforms. The consequence is that sometimes for the same task different user interfaces are more appropriate depending on the platform, and some specific tasks are only really appropriate for a specific platform. For example, watching a football match does not make sense through a smartphone, even if this is technically possible, since the small screen is inappropriate for a ninety minute period and many details of the match could not be appreciated. On the other hand, this is a pleasant activity to carry out while comfortably sitting on a sofa with a large screen in front of you at an appropriate distance.

8.2.3 Authoring Multi-Device Interactive Applications

Authoring multi-device interactive applications requires changing the traditional ways to develop interactive applications. There are various ways to address this. The simplest way is to develop separately specific versions for each target platform. In this way the developers have full control over the specific aspects of each version. However, this means multiplying the effort of developing an interactive application by the number of target platforms. Thus, it implies more effort in development and maintenance. Indeed, if something has to be changed in the application then each version needs to be updated.

Another approach consists of developing one main version with **fluid layout** and subversions. This is what happens with responsive Web design, in which the authors implement liquid layouts and use media query support to identify different types of devices. For each type identified they provide stylesheets through which they can change the values of some attributes or show or hide some elements. This can be a relatively cheap way to address the problem, but it can limit the differences among versions that can be obtained in some cases, since stylesheets do not allow deep changes in the structure of the interactive applications.

Another approach is single authoring, in which one conceptual description of the interactive application is developed, from which various versions optimized for the various target platforms are obtained. One further solution is automatic reauthoring, in which the starting point is the implementation for a specific platform, and then derive implementations adapted for different platforms through appropriate transformations.

In the research community various solutions for this purpose have been proposed. An example is SUPPLE (Gajos, Weld and Wobbrock, 2010) , which takes a functional specification of the interface, the device-specific constraints, a typical usage trace, and a cost function. The cost function is based on user preferences and expected speed of operation. SUPPLE's optimization algorithm finds the user interface, which minimizes the cost function while also satisfying all device constraints.

The SUPPLE authors then focused on how to exploit SUPPLE in order to support disabled users, for example, by automatically generating user interfaces for a user with impaired dexterity based on a model of her actual motor abilities. More generally, we can consider adaptation useful for both permanent and temporary disabilities. An example of temporary disability is when the user has to move fast and interact with a graphical mobile device. Thus, the user's visual attention cannot be completely allocated to the interaction.

One of the first approaches to authoring multi-device user interfaces is Damask (Lin and Landay, 2008), which supports authoring for three types of platforms: desktop, smartphone, and vocal. Damask is based on three aspects: sketches, layers, and patterns. Sketches are used to indicate easily what the user interface should look like. Layers indicate whether a user interface part should be allocated to all devices or only to one specific platform. Patterns are used to identify solutions to recurring problems in order to facilitate their reuse across different applications.

8.2.4 Adaptation Rules

How user interfaces adapt to the context of use can be described through rules that can be classified according to the types of effect that they can achieve.

Some adaptations consist of replacement rules: they indicate how to replace some elements according to the current platform. The elements to replace can be single user interface elements. E.g. an application for accessing train timetables that on the desktop version supports the hour selection through a long drop-down menu while in the mobile version uses a radio button with a limited number of options since the possible hours are grouped.

Another type of rule is splitting the user interface into two or more separate presentations. The new interfaces can be obtained in two ways: either by performing the creation of separate user interfaces or by dynamically showing and hiding the elements in order to achieve a similar effect. In other cases adaptation needs removal rules. The purpose is removing content considered irrelevant for the target device. There can be various reasons for this: technological limitations or manufacturer choices (e.g. iPhones do not support Flash videos); objects removed because too expensive in terms of resources consumed in the new device; elements supporting tasks considered irrelevant for the target device. It is important to remember that removing elements can have an impact on script execution because they can refer to them, and if they are removed then the scripts may not work properly.

The most used adaptation rules are those aimed at changing some user interface properties. In this case the UI elements are the same but their presentations change in terms of: their attributes (e.g. colour, size, etc.); their position in the UI; space between them; overall user interface structure.

The adaptation rules can be expressed in the format: Event / Condition / Action. The event occurrence triggers the evaluation of the rule. Elementary events occur in the interactive application or in the context of use, or a composition of such events. The condition (optional) is a Boolean condition to be satisfied in order to execute the associated action(s); it can be related to something which happened before, or some state condition. The action indicates how the abstract/concrete /implementation description of the interactive application should change in order to perform the requested adaptation. It can change the user interface at different granularities: complete change of UI, change of some UI parts, change of UI elements, or change of attributes of specific UI elements. Here are some examples of adaptation rules.

(1) Event (the user has selected the link to a printer description); condition (the user has selected more than three links to printer descriptions); action (show the five most sold printers).

(2) Event (change of battery level); condition (if the battery level is below a given threshold); action (change screen brightness).

(3) Event (user accesses application); condition (the user is elderly); action (increase font sizes 10%).

(4) Event ((user is outdoors) and (it is lunch time)); condition (there are restaurants nearby); action (show list of nearby restaurants).

(5) Event (application accessed); condition (the device is a mobile phone); action (show master and detail in different presentations).

If we want to consider applications in which accessibility is an important aspect we can have different examples of adaptation rules (Minon *et al.* 2013).

(1) Event: the noise of the environment changes to a value over 25 **decibels**; Condition: the user has a mild hearing impairment; Action: all videos must display **subtitles**.

(2) Event: the user accesses an application with many interaction elements; Condition: the user is blind; Action: an application table of content is created for easy access to each interaction element.

(3) Event: the user interface is activated; Condition: the user is colour-blind; Action: change the foreground colour to black and the background colour to white in order to provide a high-contrast UI.

(4) Event: the UI contains an element with a timeout; Condition: the user has a cognitive

disability; Action: remove the timeout or increase the time limit considerably if necessary.

(5) Event: the user interface is activated; Condition: the user has poor vision; Action: activate a screen magnifier.

(6) Event: the user begins to move; Condition: the user has paraplegia and the UI is not rendered with the vocal modality; Action: the user interface is changed to the vocal modality.

(7) Event: the application contains many different interaction elements for performing different tasks at the same time; Condition: the user has problems in maintaining attention; Action: the UI is organized in such a way that only one task is supported at a time.

Another important aspect to consider is that applications running on mobile devices have often to adapt to contextual events. Thus, there is an increasing interest in proposing environments that allow even people who are not programmers to define their context-dependent applications. Tasker is an Android app that allows users to perform context-sensitive actions based on simple event-trigger rules. The user can create the context-sensitive rules in terms of tasks (sets of actions, which can be alerts or applications to activate, audio or display properties to change, …) executed according to contexts (that depend on aspects such as application, time, date, location, event, gesture, state) in user-defined profiles. Although Tasker is still limited in terms of types of application that can be developed, it is a start nonetheless and moreover demonstrates the utility of this type of contribution. Locale is another Android app that allows users to create situations specifying conditions under which the user's phone settings should change. An example of a rule that can be implemented with such tools is: after 4 pm if the battery level is less than 20% and WiFi is active then disable WiFi and decrease the screen luminosity.

8.2.5 Model-based UI Design in Multi-Device Contexts

Models are abstractions of real entities. We use models in our life more often than is generally acknowledged. For example, often in the morning we think about the main activities to carry out during the day, thus creating a model of the day.

In model-based approaches the basic idea is to use languages that describe the main aspects in conceptual terms in order to allow designers and developers to concentrate on the main semantic aspects and avoid having to learn a plethora of implementation languages. In this way it is also possible to link semantic information and implementation elements. (8-2)We are referring here to the interaction semantics, which define the purpose of the user interface elements. This makes it possible to obtain device interoperability through many possible implementation languages because through implementation generators it is possible to derive various adapted implementations from logical descriptions. One further advantage of more semantic descriptions is that they facilitate support from assistive technology since the purpose of each element is more clearly defined.

The community working on model-based approaches has agreed in identifying a number of abstraction levels that can be considered when describing interactive applications, they are:

(1) Tasks and Domain Objects: At this level the purpose is to describe the activities that should be performed in order to reach the users' goals and the objects that need to be manipulated for this purpose.

(2) Abstract Interactive Application: At this level the focus moves to the interactive part of an application and aims to provide its description in terms independent of the interaction modality used. There is a need for an interaction object able to achieve this effect without specifying any detail that is modality-dependent; thus it is not indicated whether the object should be selected through a graphical interaction or a vocal command or a gesture.

(3) Concrete Interactive Application: The description is modality dependent but independent of the implementation language. A concept example is "List Interaction object with X elements". The assumption is that a graphical modality is used and a list element is required. However, such a list element can be obtained in various implementation languages.

(4) Interactive Application Implementation: Here we have the actual implementation in some implementation language. Thus, for example the List object can be implemented in a toolkit for Java user interfaces (e.g. AWT) or HTML or other user interface libraries.

Currently there is a working group on model-based UI in W3C that is developing standards based on these concepts[①]. In addition, such concepts have proven to be useful for accessibility. Indeed, recently another group in W3C, the Independent User Interface (IndieUI) group[②], has been set up. It aims to make it easier for Web applications to work in a wide range of contexts — different devices, different **assistive technologies**, different user needs.

(8-3) One issue with model-based approaches is that the development of the models sometimes has requirements that designers cannot address. To partially solve this problem, reverse engineering approaches and tools have been developed. The basic idea is that such tools are able to analyze the user interface implementation and build the corresponding underlying model. An example is described in Bellucci *et al.* (2012) in which the tool presented is able to analyze Web pages, including the associated stylesheets, and build the corresponding logical description in such a way as to preserve the original scripts. One of the most engineered model-based languages is MARIA[③], which includes: support for Data Model useful for specifying the format of input values, association of various data objects to the various interactors; an Event Model, which associates each interactor with a set of events that can be either property change events or activation events (e.g. access to a Web service or a database); a Dialogue Model, which specifies the dynamic behaviour (what events can be triggered at a given time). The dialogue expressions are connected using CTT[④] operators in order to define their temporal relationships; the ability to support user interfaces including complex and Ajax scripts able to continually update fields by invoking external functions, which can be implemented as

① http://www.w3.org/2011/mbui/
② http://www.w3.org/WAI/intro/indieui
③ In ACM Transactions on Computer-Human Interaction, 16 (4): 19.
④ Concur Task Trees

Web services, without explicit user request; and dynamic set of user interface elements, which can be obtained through conditional connections between presentations or the possibility of changing only a UI part. It is notable that HTML 5 is evolving in the same direction by introducing a number of more semantic tags (such as navbar, article, etc.) which provide more explicit hints of the purpose of the associated elements. However, HTML 5 is mainly limited to graphical, form-based user interfaces. Thus, it is not able to address the increasing availability of various interaction modalities.

8.2.6 Vocal Interfaces

(8-4) **Vocal interfaces** can play an important role in various contexts: with vision-impaired users; when users are on the move; more generally when the visual channel is busy. Examples of possible applications are booking services, airline information, weather information, telephone list, news. However, vocal interactive applications have specific features that make them different from graphical user interfaces. They are linear and non-persistent, while graphical interfaces support concurrent interactions and are persistent. The advantage of vocal interfaces is that they can be fast and natural for some operations.

Recently there has been increasing interest in vocal interfaces since vocal technology is improving. It is becoming more robust and immediate, without need for long training, and thus various its applications have been proposed in the mass market, e.g. vocal searches and map navigation by Google or Siri on iPhone. This has been made possible by providing the possibility of entering vocal input with audio stored locally and then sent to the server for speech recognition. Vocal menu-based navigation must be carefully designed: there is a need for continuous feedback in order to check the application state, it should provide short prompts and option lists to reduce memory efforts, and should support management of specific events (no-input, no-match, help). Although the logical structure of a graphical page is a tree, its depth and width are too large for vocal browsing.

8.2.7 Multimodal User Interfaces

Multimodality concerns the identification of the most effective combination of various interaction modalities. A simple vocabulary for this purpose was provided by the CARE properties: complementarity, the considered part of the interface is partly supported by one modality and partly by another one; assignment, the considered part of the interface is supported by one assigned modality; redundancy, the considered part of the interface is supported by both modalities; equivalence, the considered part of the interface is supported by either one modality or another. Manca *et al.* (2013) describe how to exploit them more in detail for the design and development of multimodal user interfaces: they can be applied to composition operators, interaction and output-only elements. In the case of interaction elements, it is possible to decompose them further into three parts: prompt, input, and feedback, which can be associated with different CARE properties. In this approach equivalence can be applied only to the input elements since only with them the user can choose which element to enter, while redundancy can be applied to prompt and feedback but not to input since once an input is entered through a modality it does not make sense to enter it also through another modality.

Fig.8-3 shows a general architecture for supporting adaptive multimodal user interfaces. There is a context manager able to detect events related to the user, technology, environment and social aspects. Then, the adaptation engine receives the descriptions of the user interface and the possible adaptation rules. The descriptions of the user interfaces can be obtained through authoring environments at design time or generated automatically through reverse engineering tools at run-time. When events associated with any adaptation rule occur, then the corresponding action part should be executed. For this purpose three options are possible:

(1) complete change of interaction modality, the corresponding adapter should be invoked in order to make the corresponding complete adaptation to the new modality, and then the new user interface is generated;

(2) some change in the current user interface structure should be performed; then its logical description should be modified and the new implementation generated;

(3) small changes in the current user interface should be performed, e.g. changes of some attributes in some elements; then the changes can be performed directly in the implementation through adaptation scripts.

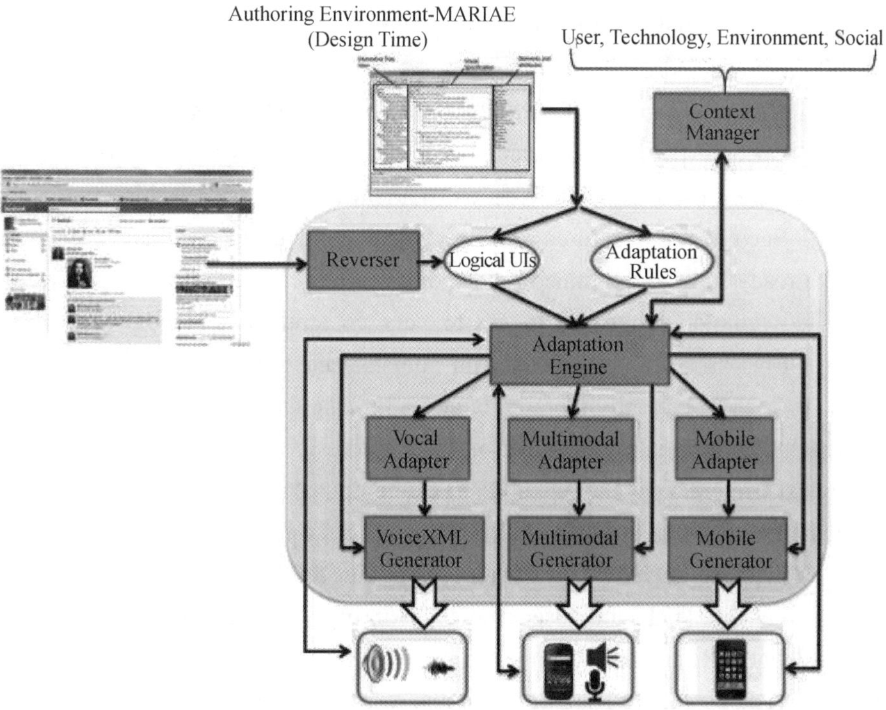

Fig.8-3 A general architecture for multimodal adaptation (Copyright © HIIS Laboratory)

8.3 HRI[①]

8.3.1 Introduction of HRI

Human-Robot Interaction (HRI) is a relatively young discipline that has attracted a lot of attention over the past few years due to the increasing availability of complex robots and people's exposure to such robots in their daily lives, e.g. as robotic toys or, to some extent, as household appliances (robotic vacuum cleaners or lawn mowers). Also, robots are increasingly being developed for real world application areas, such as robots in **rehabilitation**, eldercare, or robots used in **robot-assisted therapy** and other assistive or educational applications.

Researchers may be motivated differently to join the field HRI. Some may be roboticists, working on developing advanced robotic systems with possible real-world applications, e.g. service robots that should assist people in their homes or at work, and they may join this field in order to find out how to handle situations when these robots need to interact with people, in order to increase the robots' efficiency. Others may be psychologists or ethologists and take a human-centered perspective on HRI; they may use robots as tools in order to understand fundamental issues of how humans interact socially and communicate with others and with interactive artifacts. Artificial Intelligence and Cognitive Science researchers may join this field with the motivation to understand and develop complex intelligent systems, using robots as embodied instantiations and testbeds of those.

Last but not least, a number of people are interested in studying the interaction of people and robots, how people perceive different types and behaviors of robots, how they perceive social cues or different robot embodiments, etc. The means to carry out this work is usually via "user studies". Such work has often little technical content; e.g. it may use commercially available and already fully programmed robots, or research prototypes showing few behaviors or being controlled remotely, in order to create very constrained and controlled experimental conditions. Such research strongly focuses on humans' reactions and attitudes towards robots. Research in this area typically entails large-scale evaluations trying to find statistically significant results. Unfortunately this area of "user studies", which is methodologically heavily influenced by experimental psychology and human-computer interaction (HCI) research, is often narrowly equated with the field of "HRI". "Shall we focus on the AI and technical development of the robot or shall we do HRI?" is not an uncommon remark heard in research discussions. This tendency to equate HRI with "user studies" is in my view very unfortunate, and it may in the long run sideline HRI and transform this field into a niche-domain. HRI as a research domain is a synthetic science, and it should tackle the whole range of challenges from technical, cognitive/AI to psychological, social, cognitive and behavioral.

---

① https://www.interaction-design.org/literature/book/the-encyclopedia-of-human-computer-interaction-2nd-ed/human-robot-interaction

HRI is a field that has emerged during the early 1990s and has been characterized as: "Human-Robot Interaction (HRI) is a field of study dedicated to understanding, designing, and evaluating robotic systems for use by or with humans". The characterization of the fundamental HRI problem given above focuses on the issues of understanding what happens between robots and people, and how these interactions can be shaped, i.e. influenced, improved towards a certain goal etc.

The above view implicitly assumes a reference point of what is meant by "robot". The term is often traced back to the Czechoslovakian word robota, and its first usage is attributed to Karel Capek's play R.U.R.: Rossum's Universal Robots (1920). However, the term "robot" is far from clearly defined. Many technical definitions are available concerning its motor, sensory and cognitive functionalities, but little is being specified about the robot's appearance, behavior and interaction with people. As it happens, if a non-researcher interacts with a robot that he or she has never encountered before, then what matters is how the robot looks, what it does, and how it interacts and communicates with the person. The "user" in such a context will not care much about the cognitive architecture that has been implemented, or the programming language that has been used, or the details of the mechanical design.

Behaviors and appearances of robots have dramatically changed since the early 1990s, and they continue to change — new robots appearing on the market, other robots becoming obsolete. The design range of robot appearances is huge, ranging from **mechanoid** (mechanical-looking) to **zoomorphic** (animal-looking robots) to **humanoid** (human-like) machines as well as **android robots** at the extreme end of human-likeness. Similarly big is the design space of robot appearance, behavior and their cognitive abilities. Most robots are unique designs, their hardware and often software may be incompatible with other robots or even previous versions of the same robot. Thus, robots are generally discrete, isolated systems, they have not evolved in the same way as natural species have evolved, they have not adapted during evolution to their environments. When biological species evolve, new generations are connected to the previous generations in non-trivial ways; in fact, one needs to know the evolutionary history of a species in order to fully appreciate its morphology, biology, behavior and other features. (8-5) Robots are designed by people, and are programmed by people. Even for robots that are learning, they have been programmed how and when to learn. Evolutionary approaches to robots' embodiment and control and developmental approaches to the development of a robot's social and cognitive abilities may one day create a different situation, but at present, robots used in HRI are human-designed systems.

Thus, what we mean by "robot" today will be very different from what we mean by "robot" in a hundred of year time. The concept of robot is a moving target, we constantly reinvent what we consider to be "robot". Studying interactions with robots and gaining general insights into HRI applicable across different platforms is therefore a big challenge. Focusing only on the "H" in HRI, "user studies", i.e. the human perspective, misses the important "R", the robot component, the technological and robotics characteristics of the robot. Only a deep investigation of both aspects will eventually illuminate the illusive "I", the interaction that emerges when we put people and interactive robots in a shared context. In my perspective, the key challenge and characterization of HRI can be phrased as follows:

"HRI is the science of studying people's behavior and attitudes towards robots in relationship to the physical, technological and interactive features of the robots, with the goal to develop robots that facilitate the emergence of human-robot interactions that are at the same time efficient (according to the original requirements of their envisaged area of use), but are also acceptable to people, and meet the social and emotional needs of their individual users as well as respecting human values."

8.3.2 HRI — About (not) Romanticizing Robots

The present reality of robotics research is that robots are far from showing any truly human-like abilities in terms of physical activities, cognition or social abilities. Nevertheless, in the robotics and HRI literature they are often portrayed as "friends" "partners" "co-workers", etc., all of which are genuinely human terms. These terms are rarely used in an operational sense, and few definitions exist — most often these terms are used without further reflection. Previously, I proposed a more formal definition of **companion robots**, i.e. "A robot companion in a home environment needs to 'do the right things', i.e. it has to be useful and perform tasks around the house, but it also has to 'do the things right', i.e. in a manner that is believable and acceptable to humans".

In contrast to the companion paradigm, where the robot's key function is to take care of the human's needs, in the caretaker paradigm it is the person's duty to take care of the "immature" robot. In that same article I also argued that due to evolutionarily determined cognitive limits we may be constrained in how many "friends" we may make. When humans form relationships with people, this entails emotional, psychological and physiological investment. We would tend to make a similar investment towards robots, which do not reciprocate this investment. A robot will "care" about us as much or as little as the programmers want it to. Robots are not people; they are machines. Biological organisms, but not robots, are sentient beings, they are alive, they have an evolutionary and developmental history, they have life-experiences that are shaping their behavior and their relationships with the environment. (8-6)In contrast, machines are neither alive nor sentient; they can express emotions, pretend to "bond" with you, but these are simulations, not the real experiences that humans share. The "emotions" of a humanoid robot may look human-like but the robot does not feel anything, and the expressions are not based on any experiential understanding. A humanoid robot which looks deeply into your eyes and mutters "I love you" — is running a programme. We may enjoy this interaction, in the way we enjoy role play or immersing ourselves in imaginary worlds, but one needs to be clear about the inherently mechanical nature of the interaction. As Sherry Turkle has pointed out, robots as "relational artifacts" that are designed to encourage people to develop a relationship with them, can lead to misunderstandings concerning the authenticity of the interaction. If children grow up with a robot companion as their main friend who they interact with for several hours each day, they will learn that they can just switch it off or lock it into a cupboard whenever it is annoying or challenging them. What concept of friendship will these children develop? Will they develop separate categories, e.g. "friendship with a robot", "friendship with pets" and "friendship with people"? Will they apply the same moral and ethical concerns to robots, animals and people? Or will their notion of friendship, shaped by interactions with robots, spill over to the biological world? Similar issues are discussed in terms of children's possible addiction to computer games and game characters and to what extent these may have a negative impact on their social and moral development. Will people who grow up with a social robot view it as a "different kind", regardless of its human or animal likeness? Will social robots become new ontological categories? At present such questions cannot be answered, they will require long-term studies into how people interact with robots, over years or decades—and such results are difficult to obtain and may be ethically undesirable. However, robotic pets for children and robotic assistants for adults are becoming more and more widespread, so we may get answers to these questions in the future. The answers are unlikely to be "black and white"—similar to the question of whether computer games are beneficial for children's cognitive, academic and social development, where answers are inconclusive.

Humans have been fascinated by autonomous machines throughout history, so the fascination with robots, what they are and what they can be, will stay with us for a long time to come. However, it is advisable to have the discussion on the nature of robots based on facts and evidence, and informed predictions, rather than pursuing a romanticizing fiction.

8.3.3 HRI — There Is No Such Thing as "Natural Interaction"

A widespread assumption within the field of HRI is that "good" interaction with a robot must reflect natural (human-human) interaction and communication as closely as possible in order to ease people's need to interpret the robot's behavior. Indeed, people's face-to-face interactions are highly dynamic and multi-modal—involving a variety of gestures, language (content as well as prosody are important), body posture, facial expressions, eye gaze, in some contexts tactile interactions, etc. This has led to intensive research into how robots can produce and understand gestures, how they can understand when being spoken to and respond correspondingly, how robots can use body posture, eye gaze and other cues to regulate the interaction, and cognitive architectures are being developed to provide robots with natural social behavior and communicative skills. The ultimate goal inherent in such work is to create human-like robots, which look human-like and behave in a human-like manner. While we discuss below in more detail that the goal of human-like robots needs to be reflected upon critically, the fundamental assumption of the existence of "natural" human behavior is also problematic. What is natural behavior to begin with? Is a person behaving naturally in his own home, when playing with his children, talking to his parents, going to a job interview, meeting colleagues, giving a presentation at a conference? The same person behaves differently in different contexts and at different times during their lifetime. Were our hunter-gatherer ancestors behaving naturally when trying to avoid big predators and finding shelter? If "natural" is meant to be "biologically realistic" then the argument makes sense — a "natural gesture" would then be a gesture using a biological motion profile and an arm that is faithfully modeling human arm morphology. Similarly, a natural smile would then try to emulate the complexity of human facial muscles and emotional expressions. However, when moving up from the level of movements and actions to social behavior, the term "natural" is less meaningful. To give an example, how polite shall a robot be? Humans show different behavior and use different expressions in situations where we attend a formal work dinner, or are having a family dinner at home. As humans, we may have many different personal and professional roles in life, e.g. daughter/son, sibling, grandmother, uncle, spouse, employee, employer, committee member, volunteer, etc. We will behave slightly differently in all these different circumstances, from the way we dress, speak, behave, what we say and how we say it, it influences our style of interaction, the manner we use tactile interaction, etc. We can seamlessly switch between these different roles, which are just different aspects of "who we are".

"Our fundamental tactic of self-protection, self-control, and self-definition is not building dams or spinning webs, but telling stories — and more particularly concocting and controlling the story we tell others — and ourselves — about who we are. These strings or streams of narrative issue forth as if from a single source — not just in the obvious physical sense of flowing from just one mouth, or one pencil or pen, but in a more subtle sense: their effect on any audience or readers is to encourage them to (try to) posit a unified agent whose words they are, about whom they are: in short, to posit what I call a center of narrative gravity".

Thus, for humans, behaving "naturally" is more than having a given or learnt behavior repertoire and making rational decisions in any one situation on how to behave. We are "creating" these behaviors, reconstructing them, taking into consideration the particular context, interaction histories, etc., we are creating behavior consistent with our "narrative self". For humans, such behavior can be called "natural".

What is "natural" behavior for robots? Where is the notion of "self", their "center of narrative gravity"? Today's robots are machines, they may have complex "experiences" but these experiences are no different from those of other complex machines. We can program them to behave differently in different contexts, but from their perspective, it does not make any difference whether they behave one way or the other. They are typically not able to relate perceptions of themselves and their environment to a narrative core; they are not re-creating, but rather recalling, experience. Robots do not have a genuine evolutionary history, their bodies and their behavior (including gestures etc.) have not evolved over many years as an adaptive response to challenges in the environment. For example, the shape of our human arms and hands has very good "reasons", it goes back to the design of forelimbs of our vertebrate ancestors, used first for swimming, then as tetrapods for walking and climbing, later bipedal postures freed the hands to grasp and manipulate objects, to use tools, or to communicate via gestures. The design of our arms and hands is not accidental, and is not "perfect" either. But our arms and hands embody an evolutionary history of adaptation to different environmental constraints. In contrast, there is no "natural gesture" for a robot, in the same way as there is no "natural" face or arm for a robot.

To conclude, there appears to be little argument to state that a particular behavior X is natural for a robot Y. Any behavior of a robot will be natural or artificial, solely depending on how the humans interacting with the robot perceive it. Thus, naturalness of robot behavior is in the eyes of the beholder, i.e. the human interacting with or watching the robot; it is not a property of the robot's behavior itself.

8.3.4 HRI — There Is a Place For Non-humanoid Robots

It is often assumed as "given" (e.g. not reflected upon) that the ultimate goal for designers of robots for human-inhabited environments is to develop humanoid robots, i.e. robots with a human-like shape, 2 legs, 2 arms, a head, social behavior and communication abilities similar to human beings. Different arguments are often provided, some technical, others non-technical:

(1) humanoid robots would be able to operate machines and work in environments that originally were designed for humans, e.g. the humanoid robot would be able to open our washing machine and use our tool box. This would be in contrasted to robots that require a pre-engineered environment.

(2) in many applications robots are meant to be used in tasks that require human-like body shapes, e.g. arms to manipulate objects, legs to walk over uneven terrain etc.

(3) the assumption that humanoid robots would have greater acceptability by people, that they mind "blend in" better, that people would prefer to interact with them. It is argued that people would be able to more easily predict and respond to the robot's behavior due to its familiarity with human motion and behavior, and predictability may contribute to safety.

(4) the assumption that those robots would fulfill better human-like tasks, e.g. operating machinery and functioning in an environment designed for people, or for the purpose of a robot carrying out human-like tasks, e.g. a companion robots assisting people in their homes or in a hospital our care home.

Likewise, in the domain of life-like agents, e.g. virtual characters, a similar tendency towards human-like agents can be found. Previously, I described this tendency as the "life-like agent hypothesis":

"Artificial social agents (robotic or software) which are supposed to interact with humans are most successfully designed by imitating life, i.e. making the agents mimic as closely as possible animals, in particular humans. This comprises both 'shallow' approaches focusing on the presentation and believability of the agents, as well as 'deep' architectures which attempt to model faithfully animal cognition and intelligence. Such life-like agents are desirable since.

(1) The agents are supposed to act on behalf of or in collaboration with humans; they adopt roles and fulfill tasks normally done by humans, thus they require human forms of (social)

intelligence.

(2) Users prefer to interact ideally with other humans and less ideally with human-like agents. Thus, life-like agents can naturally be integrated in human work and entertainment environment, e.g. as assistants or pets.

(3) Life-like agents can serve as models for the scientific investigation of animal behavior and animal minds".

Argument (3) presented above easily translates to robotic agent and companions, since these may be used to study human and animal behavior, cognition and development. Clearly, the humanoid robot is an exciting area of research, not only for those researchers interested in the technological aspects but also, importantly, for those interested in developing robots with human-like cognition; the goal would be to develop advanced robots, or to use the robots as tools for the study of human cognition and development. When trying to achieve human-like cognition, it is best to choose a humanoid platform, due to the constraints and interdependencies of animal minds and bodies. Precursors of this work can be found in Adaptive Behavior and Artificial Life research using robots as models to understand biological systems.

However, arguments (1) and (2) are problematic, for the following reasons.

Firstly, while humans have a natural tendency to anthropomorphize the world and to engage even with non-animate objects (such as robots) in a social manner, a humanoid shape often evokes expectations concerning the robot's ability, e.g. human-like hands and fingers suggest that the robot is able to manipulate objects in the same way humans can, a head with eyes suggests that the robot has advanced sensory abilities e.g. vision, a robot that produces speech is expected also to understand when spoken to. More generally, a human-like form and human-like behavior is associated with human-level intelligence and general knowledge, as well as human-like social, communicative and empathic understanding. Due to limitations both in robotics technology and in our understanding of how to create human-like levels of intelligence and cognition, in interaction with a robot people quickly realize the robot's limitations, which can cause frustration and disappointment.

Secondly, if a non-humanoid shape can fulfill the robot's envisaged function, then this may be the most efficient as well as the most acceptable form. For example, the autonomous vacuum cleaning robot Roomba (iRobot) has been well accepted by users as an autonomous, but clearly non-humanoid robot. Some users may attribute personality to it, but the functional shape of the robot clearly signifies its robotic nature, and indeed few owners have been shown to treat the robot as a social being. Thus, rather than trying to use a humanoid robot operating a vacuum cleaner in a human-like manner (which is very hard to implement), an alternative efficient and acceptable solution has been found. Similarly, the ironing robot built by Siemens does not try to replicate the way humans iron a shirt but finds an alternative, technologically simpler solution.

Building humanoids which operate and behave in a human-like manner is technologically highly challenging and costly in terms of time and effort required, and it is unclear when such human-likeness may be achieved (if ever) in future. But even if such robots were around, would we want them to replace e.g. the Roomba? The current tendency to focus on humanoid robots in HRI and robotics may be driven by scientific curiosity, but it is advisable to consider the whole design space of robots, and how the robot's design may be very suitable for particular tasks or application areas. Non-humanoid, often special purpose machines, such as the Roomba, may provide cheap and robust solutions to real-life needs, i.e. to get the floor cleaned, in particular for tasks that involve little human contact. For tasks that do involve a significant amount of human-robot interaction, some humanoid characteristics may add to the robot's acceptance and success as an interactive machine, and may thus be justified better. Note, the design space of robots is huge, and "humanoid" does not necessarily mean "as closely as possible resembling a human". A humanoid robot such as Autom (2013), designed as a weight loss coach has clearly human-like features, but very simplified features, more reminiscent of a cartoon-design. On the other end of the spectrum towards human-like appearance we find the androids developed by Hiroshi Ishiguro and his team[①]. However, in android technology the limitations are clearly visible in terms of producing human-like motor control, cognition and interactive skills. Androids have been proposed, though, as tools to investigate human cognition.

Thus, social robots do not necessarily need to "be like us"; they do not need to behave or look like us, but they need to do their jobs well, integrate into our human culture and provide an acceptable, enjoyable and safe interaction experience.

8.4 Key Terms and Review Questions

1. Technical Terms

iterative prototyping	迭代原型设计	8.1
empirical testing	经验测试	8.1
software crisis	软件危机	8.1
interdisciplinary	跨领域的	8.1
usability	易用性	8.1
collective efficacy	集体效能	8.1
aesthetic tension	审美张力	8.1
multi-modal interactions	多模态交互	8.1
context-aware	上下文感知	8.1
ubiquitous computing	普适计算	8.1
geo-spatial information systems	地理空间信息系统	8.1
in-vehicle systems	车载系统	8.1
community informatics	社区信息学	8.1
ambient intelligence	环境智能	8.1
codified	成文的；编成法典的	8.1

① http://www.geminoid.jp/en/index.html

instant messaging	即时消息	8.1
social networking	社交网络	8.1
recommender	推荐系统	8.1
collaborative filtering systems	协同过滤系统	8.1
crowd sourcing	众包，将任务派给未知的人群	8.1
artifacts	人工制品	8.1
co-evolution	共同演化	8.1
contingent initiatives	偶然的开端	8.1
task-artifact cycle	任务—人工作品循环	8.1
hysteresis	迟滞现象	8.1
dialectic	辩证法	8.1
epistemological	认知论的	8.1
adaptation	适应	8.2
interact	互相影响，互相作用	8.2
user interface	用户界面，用户接口	8.2
focus-based interactions	基于焦点的交互	8.2
pointer-based interaction	基于指点的交互	8.2
touch-based interaction	基于触摸的交互	8.2
appropriateness	适合性	8.2
continuity	连续性	8.2
consistency	一致性	8.2
authoring	编写	8.2
fluid layout	不固定的布局	8.2
decibel	分贝	8.2
subtitles	字幕	8.2
assistive technologies	辅助技术	8.2
vocal interface	有声界面	8.2
multimodality	多模态交互	8.2
rehabilitation	康复	8.3
robot-assisted therapy	机器人辅助治疗	8.3
mechanoid robot	机械形机器人	8.3
zoomorphic robot	动物形机器人	8.3
humanoid robot	类人机器人	8.3
android robots	人形机器人	8.3
companion robot	伴侣机器人	8.3

2. Translation Exercises

(8-1) One of the most significant achievements of HCI is its evolving model of the integration of research and practice. Initially this model was articulated as a reciprocal relation between cognitive science and cognitive engineering. Later, it ambitiously incorporated a diverse science foundation, notably social and organizational psychology, Activity Theory, distributed cognition, and sociology, and a ethnographic approaches human activity, including the activities of design and technology development and appropriation. Currently, the model is incorporating design practices and research across a broad spectrum, for example, theorizing user experience and ecological sustainability. In these developments, HCI provides a blueprint for a mutual relation between science and practice that is unprecedented.

(8-2) We are referring here to the interaction semantics, which define the purpose of the user interface elements. This makes it possible to obtain device interoperability through many possible implementation languages because through implementation generators it is possible to derive various adapted implementations from logical descriptions. One further advantage of more semantic descriptions is that they facilitate support from assistive technology since the purpose of each element is more clearly defined.

(8-3) One issue with model-based approaches is that the development of the models sometimes has requirements that designers cannot address. To partially solve this problem, reverse engineering approaches and tools have been developed. The basic idea is that such tools are able to analyze the user interface implementation and build the corresponding underlying model.

(8-4) **Vocal interfaces** can play an important role in various contexts: with vision-impaired users; when users are on the move; more generally when the visual channel is busy. Examples of possible applications are booking services, airline information, weather information, telephone list, news. However, vocal interactive applications have specific features that make them different from graphical user interfaces. They are linear and non-persistent, while graphical interfaces support concurrent interactions and are persistent.

(8-5) Robots are designed by people, and are programmed by people. Even for robots that are learning, they have been programmed how and when to learn. Evolutionary approaches to robots' embodiment and control and developmental approaches to the development of a robot's social and cognitive abilities may one day create a different situation, but at present, robots used in HRI are human-designed systems.

(8-6) In contrast, machines are neither alive nor sentient; they can express emotions, pretend to "bond" with you, but these are simulations, not the real experiences that humans share. The "emotions" of a humanoid robot may look human-like but the robot does not feel anything, and the expressions are not based on any experiential understanding. A humanoid robot which looks deeply into your eyes and mutters "I love you" — is running a programme.

References

[1] Human computer interaction[DB/OL].[2017-05-14]. https://en.wikipedia.org/wiki/Human_computer_interaction.

[2] Carroll J M. Human Computer Interaction-brief intro. in The Encyclopedia of Human-Computer Interaction, 2nd Ed[M/OL]. https://www.interaction-design.org/literature/book.

[3] Yvonne R. HCI Theory: Classical, Modern, and Contemporary[J]. Synthesis Lectures on Human-Centered Informatics, 2012,5: 1-129.

[4] Marek P, Rinderknecht R Gordon. Do people like working with computers more than human beings?[J]. Computers in Human Behavior, 2015, 51: 232-238.

Chapter 9 Computer Security

Somewhere out there, behind a glowing monitor, a **hacker** is surfing for information, attempting to embed a virus in a website. In his best-case scenario, it will give him access to a user's computer and valuable business information. From the victim's perspective, it will most certainly short circuit productivity, and could even wreck the entire computer system.

Computer security, also known as **cybersecurity** or IT security, is the protection of computer systems from the theft or damage to the hardware, software or the information on them, as well as from disruption or misdirection of the services they provide.

It includes controlling physical access to the hardware, as well as protecting against harm that may come via network access, data and **code injection**, and due to **malpractice** by operators, whether intentional, accidental, or due to them being tricked into deviating from secure procedures.

The field is of growing importance due to the increasing reliance on computer systems and the Internet in most societies, wireless networks such as Bluetooth and Wi-Fi, and the growth of "smart" devices, including smartphones, televisions and tiny devices as part of the Internet of Things.

9.1 Computer Security Issues

9.1.1 Basic Security Concepts

Three basic security concepts important to information on the Internet are **confidentiality**, **integrity**, and **availability**. Concepts relating to the people who use that information are **authentication, authorization**, and **nonrepudiation**.

When information is read or copied by someone not authorized to do so, the result is known as loss of confidentiality. For some types of information, confidentiality is a very important attribute. Examples include research data, medical and insurance records, new product specifications, and corporate investment strategies. In some locations, there may be a legal obligation to protect the **privacy** of individuals. This is particularly true for banks and loan companies; debt collectors; businesses that extend credit to their customers or issue credit cards; hospitals, doctors' offices, and medical testing laboratories; individuals or agencies that offer services such as psychological counseling or drug treatment; and agencies that collect taxes.

Information can be corrupted when it is available on an insecure network. When information is modified in unexpected ways, the result is known as loss of integrity. This means that **unauthorized** changes are made to information, whether by human error or intentional tampering. Integrity is particularly important for critical safety and financial data used for activities such as electronic funds transfers, air traffic control, and financial accounting.

Information can be erased or become inaccessible, resulting in loss of availability. This means that people who are authorized to get information cannot get what they need.

Availability is often the most important attribute in service-oriented businesses that depend on information (e.g., airline schedules and online inventory systems). Availability of the network itself is important to anyone whose business or education relies on a network connection. When a user cannot get access to the network or specific services provided on the network, they experience a **denial of service**.

To make information available to those who need it and who can be trusted with it, organizations use authentication and authorization. Authentication is proving that a user is whom he or she claims to be. That proof may involve something the user knows (such as a password), something the user has (such as a "smartcard"), or something about the user that proves the person's identity (such as a fingerprint). Authorization is the act of determining whether a particular user (or computer system) has the right to carry out a certain activity, such as reading a file or running a program. (9-1)Authentication and authorization go hand in hand. Users must be authenticated before carrying out the activity they are authorized to perform. Security is strong when the means of authentication cannot later be refuted — the user cannot later deny that he or she performed the activity. This is known as nonrepudiation.

To assess effectively the security needs of an organization and to evaluate and choose various security products and policies, the manager responsible for computer and network security needs some systematic way of defining the requirements for security and characterizing the approaches to satisfying those requirements. This is difficult enough in a centralized data processing environment; with the use of local and wide area networks, the problems are compounded.

ITU-T4① Recommendation X.800, Security Architecture for OSI, defines such a systematic approach.② The OSI security architecture is useful to managers as a way of organizing the task of providing security. Furthermore, because this architecture was developed as an international standard, computer and communications vendors have developed security features for their products and services that relate to this structured definition of services and mechanisms.

For our purposes, the OSI security architecture provides a useful, if abstract, overview of many of the concepts that this book deals with. The OSI security architecture focuses on security

---

① The International Telecommunication Union (ITU) Telecommunication Standardization Sector (ITU-T) is a United Nations-sponsored agency that develops standards, called Recommendations, relating to telecommunications and to open systems interconnection (OSI).

② The OSI security architecture was developed in the context of the OSI protocol architecture, which is described in Appendix D. However, for our purposes in this chapter, an understanding of the OSI protocol architecture is not required.

attacks, mechanisms, and services. These can be defined briefly as:

(1) **Security attack**: Any action that compromises the security of information owned by an organization.

(2) **Security mechanism**: A process (or a device incorporating such a process) that is designed to detect, prevent, or recover from a security attack.

(3) Security service: A processing or communication service that enhances the security of the data processing systems and the information transfers of an organization. The services are intended to counter security attacks, and they make use of one or more security mechanisms to provide the service.

9.1.2 Threats and Attacks

Before moving on to the steps necessary to secure networks and computer systems it helps to first have an understanding of the kinds of attacks and threats that need to be defended against. Armed with this information it will be clearer in later chapters not just how to implement particular security measures, but also why such measures need to be implemented.

There are a variety of different forms of attack to which a network or computer system may be exposed each of which will be covered in this section.

1. TCP and UDP Based Denial of Service (DoS) Attacks

Denials of Service (DoS) attacks are undertaken with the express purpose of preventing users from accessing and using a service they should otherwise be able to access. Such attacks make malicious use of a variety of different standard protocols and tools. There is no single DoS attack method, and the term has come to encompass a variety of different forms of attack, a number of which are outlined below.

(1) Ping Flood — This attack uses the Internet Control Message Protocol (ICMP) ping request to a server as a DoS method. The strategy either involves sending ping requests in such vast quantities that the receiving system is unable to respond to valid user requests, or sending ping messages which are so large (known as a ping of death) that the system is unable to handle the request.

(2) Smurfing — As with Ping Flood attacks, smurfing makes use of the TCP Internet Message Protocol (ICMP) ping request to mount DoS attacks. In a typical smurfing attack the attacker sends a ping request to the broadcast address of network containing the IP address of the victim. The ping request is sent to all computers on the broadcast network, which in turn all reply to the IP address of the victim system thereby overloading the victim with ping responses. The primary method for preventing smurf attacks is to block ICMP traffic through routers so that the ping responses are blocked from reaching internal servers.

(3) TCP SYN Flood — Also known as the TCP ACK Attack, this attack leverages the TCP three way handshake to launch a DoS attack. The attack begins with a client attempting to establish a TCP connection with the victim server. The client sends a request to the server, which in turn returns an ACK package to acknowledge the connection. At this point in the

communication the client should respond with a message accepting the connection. Instead the client sends another ACK which is responded to by the server with yet another ACK. The client continues to send ACKs to the server with the effect of causing the server to hold sessions open in anticipation of the client sending the final packet required to complete the connection. As a result the server uses up all available sessions serving the malicious client, thereby preventing access to other users.

(4) Fraggle—A fraggle attack is similar to a smurfing attack with the exception that the User Datagram Protocol (UDP) is used instead of using ICMP.

(5) Land—Under a Land attack the attacker creates a fake SYN packet contain the same source and destination IP addresses and ports and sends it to the victim causing the system to become confused when trying to respond to the packet.

(6) Teardrop—A teardrop type of DoS attack exploits a weakness in the TCP/IP implementation on some operating systems. The attack works by sending messages fragmented into multiple UDP packages. Ordinarily the operating system is able to reassemble the packets into a complete message by referencing data in each UDP packet. The teardrop attack works by corrupting the offset data in the UDP packets making it impossible for the system to rebuild the original packets. On systems that are unable to handle this corruption a crash is the most likely outcome of a teardrop attack.

(7) Bonk—An effective attack on some Windows systems involving the transmission corrupted UDP packets to the DNS port (port 53) resulting in a system crash.

(8) Boink—Similar to the Bonk attack except that the corrupted UDP packets are sent to multiple ports, not just port 53 (DNS).

2. Distributed Denial of Service (DDoS) Attacks

The Denial of Service (DoS) attacks outlined above involve the use of a single client to launch an attack on a system or service. Distributed Denial of Service Attacks use the same basic attack methodologies as outline above, with the exception that the attacks are initiated from multiple client systems.

The way this typically works is that malicious parties will use viruses to subtly gain control over large numbers of computers (typically poorly defended home computers connected to broadband Internet connections). Unbeknown to the owner of the computer (which generally continues to function as normal) the system is essentially a **zombie** waiting to be given instructions. Once the malicious party has gathered an army of zombie computers they are instructed to participate in massive Distributed DoS attacks on unsuspecting victims. A large enough volume of zombie systems can, and indeed have been known to bring down even the largest and most scalable enterprise infrastructure, and even bring parts of the Internet itself to a grinding halt.

3. Back Door Attacks

Back Door attacks utilize programs that provide a mechanism for entering a system without going through the usual authentication process. This can either take the form of hidden access points intentionally put into application by the original developers to aid in maintaining and debugging the software (which were then left in when the software was deployed by customers) or a malicious program that is placed on a system via a virus, or other method which opens up the system to unauthorized access.

A number of back door programs have been discovered over the years, some which are listed below.

(1) Back Orifice — This rather distastefully named tool was developed by a group known as the Cult of the Dead Cow Communications. The primary purpose of Back Orifice is to provide remote access to a server for the purposes of performing administrative tasks.

(2) NetBus — Similar to Back Orifice, NetBus is also designed to enable remote administrative access to Windows system.

(3) Sub7 — Sub7 is yet another illicit back door program designed to allow unauthorized access to systems.

Whilst the installation of any of the above back door programs on a system will have serious implications for security, all these threats can be effectively prevented through the implementation of a comprehensive virus scanning strategy.

4. IP and DNS Spoofing Attacks

The basis of spoofing involves **masquerading** as a trusted system in order to gain unauthorized access to a secure environment. IP spoofing involves modifying data to make it appear to originate from the IP address of a system that is trusted by a server or firewall. Using this approach, a host is able to pass through the IP filtering.

The objective of **IP Spoofing** is to gain unauthorized access to a server or service. **DNS Spoofing** differs in that the objective is send users to a different location than the one they thought they were going to. Take, for example, a user who goes to their bank's website to perform online banking transactions (such as paying bills etc). The user enters the Web address (URL) of their bank into a browser. The browser contacts a Domain Name Server (DNS) which looks up the IP address which matches the URL. The user is then taken to the site located at that IP address where they enter their login and password. DNS spoofing involves the DNS server being compromised such that the bank URL is set to point to the IP address of a malicious party where a Web site that looks just like the real bank site has been set up. Now when the user enters the URL in a browser they are taken to the fake Web site where their login and password are captured and stored. The Web site will then likely report that the bank site is off-line for maintenance. The user decides to return and try again later. Meanwhile the attacker uses the customer's credentials to log into the account on the real site and transfer all the money out of the account.

5. Man-in-the-Middle Attacks

Man-in-the-Middle Attacks are perhaps one of the more complex and sophisticated forms of security breaching approaches. As the name implies, such an attack involves the surreptitious placement of a software agent between the client and server ends of a communication. In this scenario neither end of the communication is aware that the malicious agent is present in the line of communication. For the most part, the man in the middle simply relays the data transmissions between client and server as though nothing is happening. What is generally happening in parallel with this process is that the agent is also recording the data as it is passed through. This results in a third party having access to a variety of different types of data, from login and password credentials to proprietary and confidential information. It is also possible for the man-in-the-middle agent to modify data "on the fly" causing untold problems for the victim.

Man-in-the-Middle Attacks have increased considerable since the introduction of wireless networking. Now there is no need for the rogue to connect to a wire, instead the data can simply be intercepted from anywhere within range of the wireless signal (such as in the parking lot outside an office or the road in front of a house).

The best way to avoid such attacks is to use encryption and secure protocols in all communications.

6. Replay Attacks

Replay Attacks are a variation on the Man-in-the-Middle theme. In a Replay Attack an agent is once again placed within the client / server line of communication. In the case of a Replay Attack, however, the transaction data is recorded for the express purpose of allowing the data to be modified and replayed to the server at a later time for nefarious purposes. For example, a Replay Attack might record the entire process of a user logging into a banking Web site and performing transactions. The recorded transcript may then be replayed to repeat the login sequence for the purposes of stealing money from the account.

Replay Attacks are best countered using encryption, timestamps, serial numbers and packet sequences so that the server can detect that the data is being replayed from a previous session.

7. TCP/IP Hijacking

TCP/IP Hijacking occurs when an **attacker** takes control of an ongoing session between a client and a server. This is similar in to a Man-in-the-Middle Attack except that the rogue agent sends a reset request to the client so that the client loses contact with the server while the rogue system assumes the role of the legitimate client, continuing the session.

8. Mathematical Attacks

The solution to a number of the types of attack outlined above has involved the use of **encryption**. A Mathematical Attack involves the use of computation based on the mathematical properties of the encryption algorithm to attempt to decrypt data.

The best way to avoid the decryption of data is to use strong encryption (128-bit) rather than rely on weaker encryption (both 40-bit and 56-bit encryption can easily be broken).

9. Password Attacks

(9-2) On systems which rely solely on a login name and password the security of the entire system is only as strong as the passwords chosen by the users. The best way to ensure passwords are not cracked is to avoid the use of simple words or phrases which can be found in a dictionary. This needs to be balanced with making the passwords easy enough to remember so that users do not write them on pieces of paper and stick them on their laptops or monitors for others to find.

The best passwords consist of a mixture of upper and lower case characters combined with numbers and special characters. A common approach is to substitute numbers in place of similar letters. For example W3ath3rN3ws uses the number 3 in place of the letter "E", the reasoning being that the number 3 is much like a reversed "E" making the password easy to remember. Unfortunately most password cracking algorithms know about this type of **substitution**.

There are two primary mechanisms for breaking password protection, brute force and dictionary.

1) Brute Force Password Attacks

A **Brute Force Attack** uses algorithms to systematically try every possible **permutation** of characters in an effort to find the correct password. If allowed to persist, a Brute Force Attack will eventually identify the correct password, although a well implemented security strategy will disable the account and block the IP address from which the attempts were made after 3 or 4 failed password attempts.

2) Dictionary Password Attacks

Dictionary Password Attacks take advantage of the fact that many user simply rely on easy to remember words as their passwords. A dictionary attack simply works through a list of words from a dictionary to see if any of them turn out to be a valid password. Such brute force programs also take into consideration such tricks as using the number 3 instead of the letter E and the number 1 in place of the letter L.

It is difficult to characterize the people who cause these incidents. An intruder may be an adolescent who is curious about what he or she can do on the Internet, a college student who has created a new software tool, an individual seeking personal gain, or a paid "spy" seeking information for the economic advantage of a corporation or foreign country. An incident may also be caused by a disgruntled former employee or a consultant who gained network information while working with a company. An intruder may seek entertainment, intellectual challenge, a sense of power, political attention, or financial gain.

One characteristic of the intruder community as a whole is its communication. There are electronic newsgroups and print publications on the latest intrusion techniques, as well as conferences on the topic. **Intruders** identify and publicize misconfigured systems; they use those systems to exchange pirated software, credit card numbers, exploitation programs, and the identity of sites that have been compromised, including account names and passwords. By sharing knowledge and easy-to-use software tools, successful intruders increase their number and their impact.

9.1.3 A Model for Network Security①

A model for much of what we will be discussing is captured, in very general terms, in

① Stallings W. NETWORK SECURITY ESSENTIALS: APPLICATIONS AND STANDARDS FOURTH EDITION. Prentice Hall, 2011.

Fig.9-1. A message is to be transferred from one party to another across some sort of Internet service. The two parties, who are the principals in this transaction, must cooperate for the exchange to take place. A logical information channel is established by defining a route through the Internet from source to destination and by the cooperative use of communication protocols (e.g., TCP/IP) by the two principals. Security aspects come into play when it is necessary or desirable to protect the information transmission from an opponent who may present a threat to confidentiality, authenticity, and so on. All of the techniques for providing security have two components:

(1) A security-related transformation on the information to be sent. Examples include the encryption of the message, which **scrambles** the message so that it is unreadable by the **opponent**, and the addition of a code based on the contents of the message, which can be used to verify the identity of the sender.

(2) Some secret information shared by the two principals and, it is hoped, unknown to the opponent. An example is an encryption key used in conjunction with the transformation to scramble the message before transmission and unscramble it on reception.

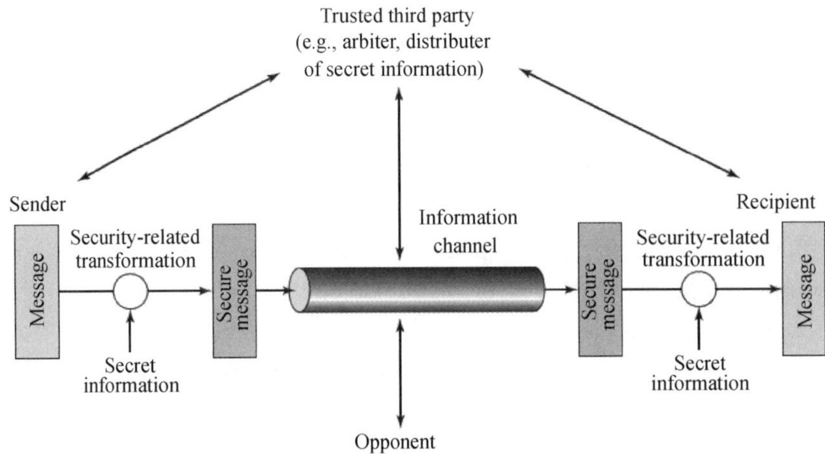

Fig.9-1 Network security model

A **trusted third party** may be needed to achieve secure transmission. For example, a third party may be responsible for distributing the secret information to the two principals while keeping it from any opponent. Or a third party may be needed to arbitrate disputes between the two principals concerning the authenticity of a message transmission.

This general model shows that there are four basic tasks in designing a particular security service.

(1) Design an algorithm for performing the security-related transformation. The algorithm should be such that an opponent cannot defeat its purpose.

(2) Generate the secret information to be used with the algorithm.

(3) Develop methods for the distribution and sharing of the secret information.

(4) Specify a protocol to be used by the two principals that makes use of the security

algorithm and the secret information to achieve a particular security service.

However, there are other security-related situations of interest that do not neatly fit this model is illustrated by Fig.9-2, which reflects a concern for protecting an information system from unwanted access.

Most readers are familiar with the concerns caused by the existence of hackers who attempt to penetrate systems that can be accessed over a network. The hacker can be someone who, with no malign intent, simply gets satisfaction from breaking and entering a computer system. The intruder can be a disgruntled employee who wishes to do damage or a criminal who seeks to exploit computer assets for financial gain (e.g., obtaining credit card numbers or performing illegal money transfers).

Another type of unwanted access is the placement in a computer system of logic that exploits **vulnerabilities** in the system and that can affect application programs as well as utility programs, such as editors and compilers. Programs can present two kinds of threats:

(1) Information access threats: Intercept or modify data on behalf of users who should not have access to that data.

(2) Service threats: Exploit service flaws in computers to inhibit use by legitimate users.

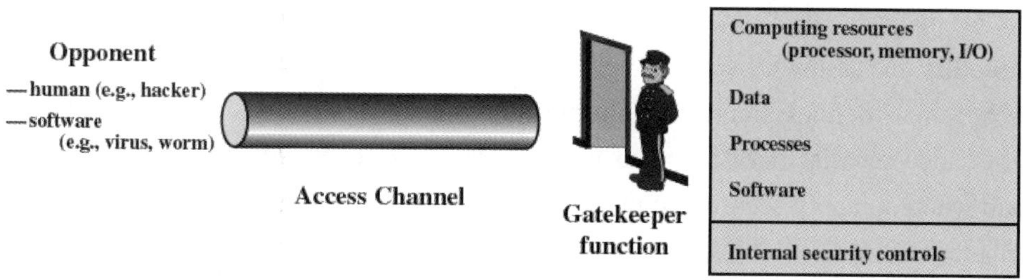

Fig.9-2 Network access security model

9.2 Security Countermeasure

In computer security a **countermeasure** is an action, device, procedure, or technique that reduces a threat, vulnerability, or an attack by eliminating or preventing it, by minimizing the harm it can cause, or by discovering and reporting it so that corrective action can be taken. Some common countermeasures are listed as followings.

1. Security by Design

Security by design, or alternately secure by design, means that the software has been designed from the ground up to be secure. In this case, security is considered as a main feature.

Some of the techniques in this approach include:

(1) The **principle of least privilege**, where each part of the system has only the privileges

that are needed for its function. That way even if an attacker gains access to that part, they have only limited access to the whole system.

(2) Automated theorem proving to prove the correctness of crucial software subsystems.

(3) Code reviews and unit testing, approaches to make modules more secure where formal correctness proofs are not possible.

(4) **Defense in depth**, where the design is such that more than one subsystem needs to be violated to compromise the integrity of the system and the information it holds.

(5) Default secure settings, and design to "fail secure" rather than "fail insecure" (see fail-safe for the equivalent in safety engineering).

(6) **Audit trails** tracking system activity, so that when a security breach occurs, the mechanism and extent of the breach can be determined. Storing audit trails remotely, where they can only be appended to, can keep intruders from covering their tracks.

(7) Full disclosure of all vulnerabilities, to ensure that the "window of vulnerability" is kept as short as possible when bugs are discovered.

2. Security Architecture

The Open Security Architecture organization defines IT security architecture as "the design artifacts that describe how the security controls (security countermeasures) are positioned, and how they relate to the overall information technology architecture. These controls serve the purpose to maintain the system's quality attributes: confidentiality, integrity, availability, **accountability** and assurance services".

Techopedia[①] defines security architecture as "a unified security design that addresses the necessities and potential risks involved in a certain scenario or environment. It also specifies when and where to apply security controls. The design process is generally **reproducible**." The key attributes of security architecture are:

(1) the relationship of different components and how they depend on each other.

(2) the determination of controls based on **risk assessment**, good practice, finances, and legal matters.

(3) the standardization of controls.

3. Security Measures

A state of computer "security" is the conceptual ideal, attained by the use of the three processes: threat prevention, detection, and response. These processes are based on various policies and system components, which include the following:

(1) User account access controls and cryptography can protect systems files and data, respectively.

(2) Firewalls are by far the most common prevention systems from a network security perspective as they can (if properly configured) shield access to internal network services, and block certain kinds of attacks through packet filtering. Firewalls can be both hardware- or

① An IT education site, at: https://www.techopedia.com/

software-based.

(3) Intrusion Detection System (IDS) products are designed to detect network attacks in-progress and assist in **post-attack forensics**, while audit trails and logs serve a similar function for individual systems.

(4) "Response" is necessarily defined by the assessed security requirements of an individual system and may cover the range from simple upgrade of protections to notification of legal authorities, counter-attacks, and the like.

Today, computer security comprises mainly "preventive" measures, like firewalls or an exit procedure. (9-3) A firewall can be defined as a way of filtering network data between a host or a network and another network, such as the Internet, and can be implemented as software running on the machine, hooking into the network stack to provide real time filtering and blocking. Another implementation is a so-called "physical firewall", which consists of a separate machine filtering network traffic. Firewalls are common amongst machines that are permanently connected to the Internet.

Some organizations are turning to **big data** platforms, such as Apache Hadoop, to extend data accessibility and machine learning to detect advanced persistent threats.

However, relatively few organizations maintain computer systems with effective detection systems, and fewer have organized response mechanisms in place. The primary obstacle to effective eradication of cyber crime could be traced to excessive reliance on firewalls and other automated "detection" systems. Yet it is basic evidence gathering by using packet capture appliances that puts criminals behind bars.

4. Vulnerability Management

Vulnerability management is the cycle of identifying, and remediating or mitigating vulnerabilities, especially in software and firmware. Vulnerability management is integral to computer security and network security.

Vulnerabilities can be discovered with a vulnerability **scanner**, which analyzes a computer system in search of known vulnerabilities, such as open ports, insecure software configuration, and susceptibility to malware.

Beyond vulnerability scanning, many organizations contract outside security auditors to run regular penetration tests against their systems to identify vulnerabilities. In some sectors this is a contractual requirement.

5. Reducing Vulnerabilities

While formal verification of the correctness of computer systems is possible, it is not yet common. Operating systems formally verified include seL4[①], and SYSGO's PikeOS[②]– but these make up a very small percentage of the market.

Cryptography properly implemented is now virtually impossible to directly break. Breaking

① seL4 is a high-assurance, high-performance microkernel developed OS, http://sel4.systems/

② a hard real-time operating system (RTOS) that offers a separation kernel based hypervisor with multiple partition types for many other operating systems (called GuestOS) and applications

them requires some non-cryptographic input, such as a stolen key, stolen plaintext (at either end of the transmission), or some other extra cryptanalytic information.

Two-factor authentication is a method for mitigating unauthorized access to a system or sensitive information. It requires "something you know"; a password or PIN, and "something you have"; a card, **dongle**, cellphone, or other piece of hardware. This increases security as an unauthorized person needs both of these to gain access.

Social engineering and direct computer access (physical) attacks can only be prevented by non-computer means, which can be difficult to enforce, relative to the sensitivity of the information. Training is often involved to help mitigate this risk, but even in highly disciplined environments (e.g. military organizations), social engineering attacks can still be difficult to foresee and prevent.

It is possible to reduce an attacker's chances by keeping systems up to date with security **patches and updates**, using a security scanner or/and hiring competent people responsible for security. The effects of data loss/damage can be reduced by careful backing up and insurance.

6. Hardware Protection Mechanisms

While hardware may be a source of insecurity, such as with microchip vulnerabilities maliciously introduced during the manufacturing process, hardware-based or assisted computer security also offers an alternative to software-only computer security. Using devices and methods such as dongles, trusted platform modules, intrusion-aware cases, drive locks, disabling USB (Universal Serial Bus) ports, and mobile-enabled access may be considered more secure due to the physical access (or sophisticated backdoor access) required in order to be compromised. Each of these is covered in more details below.

(1) USB dongles are typically used in software licensing schemes to unlock software capabilities, but they can also be seen as a way to prevent unauthorized access to a computer or other device's software. The dongle, or key, essentially creates a secure encrypted tunnel between the software application and the key. The principle is that an encryption scheme on the dongle, such as Advanced Encryption Standard (AES) provides a stronger measure of security, since it is harder to hack and replicate the dongle than to simply copy the native software to another machine and use it. Another security application for dongles is to use them for accessing Web-based content such as cloud software or Virtual Private Networks (VPNs). In addition, a USB dongle can be configured to lock or unlock a computer.

(2) Trusted Platform Modules (TPMs) secure devices by integrating cryptographic capabilities onto access devices, through the use of microprocessors, or so-called computers-on-a-chip. TPMs used in conjunction with server-side software offer a way to detect and authenticate hardware devices, preventing unauthorized network and data access.

(3) Computer case intrusion detection refers to a push-button switch which is triggered when a computer case is opened. The firmware or BIOS is programmed to show an alert to the operator when the computer is booted up the next time.

(4) Drive locks are essentially software tools to encrypt hard drives, making them

inaccessible to thieves. Tools exist specifically for encrypting external drives as well.

(5) Disabling USB ports is a security option for preventing unauthorized and malicious access to an otherwise secure computer. Infected USB dongles connected to a network from a computer inside the firewall are considered by the magazine Network World as the most common hardware threat facing computer networks.

(6) Mobile-enabled access devices are growing in popularity due to the ubiquitous nature of cell phones. Built-in capabilities such as Bluetooth, the newer Bluetooth low energy, Near Field Communication (NFC) on non-iOS devices and biometric validation such as thumb print readers, as well as **QR code** reader software designed for mobile devices, offer new, secure ways for mobile phones to connect to access control systems. These control systems provide computer security and can also be used for controlling access to secure buildings.

7. Secure Operating Systems

One use of the term "computer security" refers to technology that is used to implement secure operating systems. In the 1980s the United States Department of Defense (DoD) used the "Orange Book" standards, but the current international standard ISO/IEC 15408, "Common Criteria" defines a number of progressively more stringent **Evaluation Assurance Levels**. Many common operating systems meet the EAL4 standard of being "Methodically Designed, Tested and Reviewed", but the formal verification required for the highest levels means that they are uncommon. An example of an EAL6 ("Semiformally Verified Design and Tested") system is Integrity-178B, which is used in the Airbus A380 and several military jets.

8. Secure Coding

In software engineering, secure coding aims to guard against the accidental introduction of security vulnerabilities. It is also possible to create software designed from the ground up to be secure. Such systems are "secure by design". Beyond this, formal verification aims to prove the correctness of the algorithms underlying a system; important for cryptographic protocols for example.

9. Capabilities and Access Control Lists

Within computer systems, two of many security models capable of enforcing privilege separation are Access Control Lists (ACLs) and capability-based security. Using ACLs to confine programs has been proven to be insecure in many situations, such as if the host computer can be tricked into indirectly allowing restricted file access, an issue known as the confused deputy problem. It has also been shown that the promise of ACLs of giving access to an object to only one person can never be guaranteed in practice. Both of these problems are resolved by capabilities. This does not mean practical flaws exist in all ACL-based systems, but only that the designers of certain utilities must take responsibility to ensure that they do not introduce flaws.

Capabilities have been mostly restricted to research operating systems, while commercial OSs still use ACLs. Capabilities can, however, also be implemented at the language level, leading to a style of programming that is essentially a refinement of standard object-oriented design. An open source project in the area is the E language.

The most secure computers are those not connected to the Internet and shielded from any interference. In the real world, the most secure systems are operating systems where security is not an add-on.

10. Response to Breaches

Responding forcefully to attempted security breaches (in the manner that one would for attempted physical security breaches) is often very difficult for a variety of reasons.

(1) Identifying attackers is difficult, as they are often in a different **jurisdiction** to the systems they attempt to breach, and operate through proxies, temporary anonymous dial-up accounts, wireless connections, and other **anonymizing** procedures which make back tracing difficult and are often located in yet another jurisdiction. If they successfully breach security, they are often able to delete logs to cover their tracks.

(2) The sheer number of attempted attacks is so large that organizations cannot spend time pursuing each attacker (a typical home user with a permanent (e.g., cable modem) connection will be attacked at least several times per day, so more attractive targets could be presumed to see many more). Note however, that most of the sheer bulks of these attacks are made by automated vulnerability scanners and computer worms.

(3) Law enforcement officers are often unfamiliar with information technology, and so lack the skills and interest in pursuing attackers. There are also budgetary constraints. It has been argued that the high cost of technology, such as DNA testing, and improved forensics mean less money for other kinds of law enforcement, so the overall rate of criminals not getting dealt with goes up as the cost of the technology increases. In addition, the identification of attackers across a network may require logs from various points in the network and in many countries, the release of these records to law enforcement (with the exception of being voluntarily surrendered by a network administrator or a system administrator) requires a search warrant and, depending on the circumstances, the legal proceedings required can be drawn out to the point where the records are either regularly destroyed, or the information is no longer relevant.

9.3 Cryptography

Cryptography or **cryptology** is the practice and study of techniques for secure communication in the presence of third parties called **adversaries**. More generally, cryptography is about constructing and analyzing protocols that prevent third parties or the public from reading private messages; various aspects in information security such as data confidentiality, data integrity, authentication, and non-repudiation are central to modern cryptography. Modern cryptography exists at the intersection of the disciplines of mathematics, computer science, and electrical engineering. Applications of cryptography include ATM cards, computer passwords, and electronic commerce.

Cryptography prior to the modern age was effectively synonymous with encryption, the conversion of information from a readable state to apparent nonsense. The originator of an encrypted message shared the decoding technique needed to recover the original information **only** with intended recipients, thereby precluding unwanted persons from doing the same. Since the development of **rotor cipher machines** in World War I and the advent of computers in World War II, the methods used to carry out cryptology have become increasingly complex and its application more widespread.

9.3.1 Basic Concepts

Until modern times, **cryptography** referred almost exclusively to **encryption**, which is the process of converting ordinary information (called **plaintext**) into unintelligible text (called **ciphertext**). Decryption is the reverse, in other words, moving from the unintelligible ciphertext back to plaintext. A **cipher** (or cypher) is a pair of algorithms that create the encryption and the reversing decryption. The detailed operation of a cipher is controlled both by the algorithm and in each instance by a "key". The **key** is a secret, usually a short string of characters, which is needed to decrypt the ciphertext. Keys are important both formally and in actual practice, as ciphers without variable keys can be trivially broken with only the knowledge of the cipher used and are therefore useless for most purposes. Historically, ciphers were often used directly for encryption or decryption without additional procedures such as authentication or integrity checks. There are two kinds of **cryptosystems**: **symmetric** and **asymmetric**. In symmetric systems the same key (the secret key) is used to encrypt and decrypt a message. Data manipulation in symmetric systems is faster than asymmetric systems as they generally use shorter key lengths. Asymmetric systems use a public key to encrypt a message and a private key to decrypt it. Symmetric models include the commonly used AES (Advanced Encryption Standard) which replaced the older DES (Data Encryption Standard). Examples of asymmetric systems include RSA (Rivest-Shamir-Adleman), and ECC (Elliptic Curve Cryptography).

Cryptanalysis is the term used for the study of methods for obtaining the meaning of encrypted information without access to the key normally required to do so; i.e., it is the study of how to crack encryption algorithms or their implementations.

The study of characteristics of languages that have some application in cryptography or cryptology (e.g. frequency data, letter combinations, universal patterns, etc.) is called **cryptolinguistics**.

9.3.2 History of Cryptography

Before the modern era, cryptography was concerned solely with message confidentiality (i.e., encryption) — conversion of messages from a comprehensible form into an incomprehensible one and back again at the other end, rendering it unreadable by interceptors or eavesdroppers without secret knowledge (namely the key). Encryption attempted to ensure secrecy in communications, such as those of spies, military leaders, and diplomats. In recent decades, the field has expanded beyond confidentiality concerns to include techniques for message integrity checking, sender/receiver identity authentication, digital signatures, interactive proofs and secure computation, among others.

1. Classic Cryptography

The main classical cipher types are **permutation** ciphers, which rearrange the order of letters in a message (e.g., "hello world" becomes "ehlol owrdl" in a trivially simple rearrangement scheme), and **substitution** ciphers, which systematically replace letters or groups of letters with other letters or groups of letters (e.g., "fly at once" becomes "gmz bu podf" by replacing each letter with the one following it in the Latin alphabet). Simple versions of either have never offered much confidentiality from enterprising opponents. An early substitution cipher was the **Caesar cipher**, in which each letter in the plaintext was replaced by a letter some fixed number of positions further down the alphabet.

The Greeks of Classical times are said to have known of ciphers. **Steganography** (i.e., hiding even the existence of a message so as to keep it confidential) was also first developed in ancient times. An early example, from Herodotus[①], was a message tattooed on a slave's shaved head and concealed under the regrown hair. More modern examples of steganography include the use of invisible ink, microdots, and digital watermarks to conceal information.

Ciphertexts produced by a classical cipher (and some modern ciphers) will reveal statistical information about the plaintext, and that information can often be used to break the cipher. After the discovery of **frequency analysis**, nearly all such ciphers could be broken by an informed attacker.

Essentially all ciphers remained vulnerable to cryptanalysis using the frequency analysis technique until the development of the **polyalphabetic cipher**, most clearly by Leon Battista Alberti[②] around the year 1467, though there is some indication that it was already known to Al-Kindi[③]. Alberti's innovation was to use different ciphers (i.e., substitution alphabets) for various parts of a message. He also invented what was probably the first automatic cipher device, a wheel which implemented a partial realization of his invention. In the polyalphabetic Vigenère cipher, encryption uses a key word, which controls letter substitution depending on which letter of the key word is used. In the mid-19th century Charles Babbage showed that the Vigenère cipher was vulnerable to Kasiski examination, but this was first published about ten years later by Friedrich Kasiski.

Although frequency analysis can be a powerful and general technique against many ciphers, encryption has still often been effective in practice, as many a would-be cryptanalyst was unaware of the technique. Breaking a message without using frequency analysis essentially required knowledge of the cipher used and perhaps of the key involved, thus making espionage, bribery, burglary, defection, etc., more attractive approaches to the cryptanalytically uninformed. It was finally explicitly recognized in the 19th century that secrecy of a cipher's algorithm is not

---

① was a Greek historian(c. 484 — c. 425 BC)

② February 14, 1404 — April 25, 1472, was an Italian humanist author, artist, architect, poet, priest, linguist, philosopher and cryptographer

③ (c. 801 — 873 AD), known as "the Philosopher of the Islamic empire"

a sensible nor practical safeguard of message security; in fact, it was further realized that any adequate cryptographic scheme (including ciphers) should remain secure even if the adversary fully understands the cipher algorithm itself. Security of the key used should alone be sufficient for a good cipher to maintain confidentiality under an attack. This fundamental principle was first explicitly stated in 1883 by Auguste Kerckhoffs and is generally called Kerckhoffs's Principle; alternatively and more bluntly, it was restated by Claude Shannon, the inventor of information theory and the fundamentals of theoretical cryptography, as Shannon's Maxim — "the enemy knows the system".

Different physical devices and aids have been used to assist with ciphers. One of the earliest may have been the **scytale** of ancient Greece, a rod supposedly used by the Spartans as an aid for a transposition cipher (see Fig.9-3). In medieval times, other aids were invented such as the cipher grille, which was also used for a kind of steganography. With the invention of polyalphabetic ciphers came more sophisticated aids such as Alberti's own cipher disk, Johannes Trithemius' tabula recta scheme, and Thomas Jefferson's multi cylinder (not publicly known, and reinvented independently by Bazeries around 1900). Many mechanical encryption/decryption devices were invented early in the 20th century, and several patented, among them rotor machines — famously including the Enigma machine (see Fig.9-4) used by the German government and military from the late 1920s and during World War II. The ciphers implemented by better quality examples of these machine designs brought about a substantial increase in cryptanalytic difficulty after WWI.

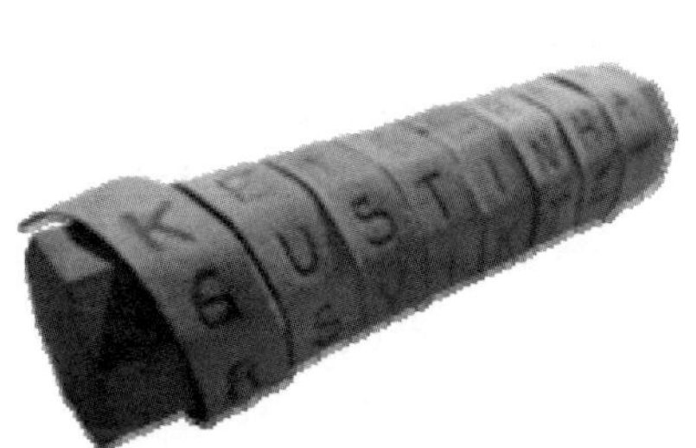

Fig.9-3 A scytale

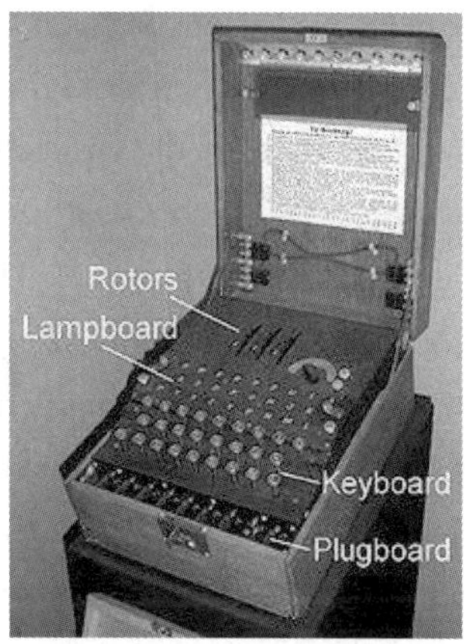

Fig.9-4 Enigma Machine

2. Computer Era

Cryptanalysis of the new mechanical devices proved to be both difficult and laborious. In the United Kingdom, cryptanalytic efforts at Bletchley Park during WW II spurred the development of more efficient means for carrying out repetitious tasks. This culminated in the development of the Colossus, the world's first fully electronic, digital, programmable computer, which assisted in the decryption of ciphers generated by the German Army's Lorenz SZ40/42 machine.

Just as the development of digital computers and electronics helped in cryptanalysis, it made possible much more complex ciphers. Furthermore, computers allowed for the encryption of any kind of data representable in any binary format, unlike classical ciphers which only encrypted written language texts; this was new and significant. Computer use has thus supplanted linguistic cryptography, both for cipher design and cryptanalysis. Many computer ciphers can be characterized by their operation on binary bit sequences (sometimes in groups or blocks), unlike classical and mechanical schemes, which generally manipulate traditional characters (i.e., letters and digits) directly. However, computers have also assisted cryptanalysis, which has compensated to some extent for increased cipher complexity. Nonetheless, good modern ciphers have stayed ahead of cryptanalysis; it is typically the case that use of a quality cipher is very efficient (i.e., fast and requiring few resources, such as memory or CPU capability), while breaking it requires an effort many orders of magnitude larger, and vastly larger than that required for any classical cipher, making cryptanalysis so inefficient and impractical as to be effectively impossible.

Extensive open academic research into cryptography is relatively recent; it began only in the mid-1970s. In recent times, IBM personnel designed the algorithm that became the Federal (i.e., US) Data Encryption Standard; Whitfield Diffie and Martin Hellman published their key agreement algorithm; and the RSA algorithm was published in Martin Gardner's Scientific American column. Since then, cryptography has become a widely used tool in communications, computer networks, and computer security generally. Some modern cryptographic techniques can only keep their keys secret if certain mathematical problems are intractable, such as the **integer factorization** or the **discrete logarithm** problems, so there are deep connections with abstract mathematics. There are very few cryptosystems that are proven to be unconditionally secure. The **one-time pad** is one. There are a few important ones that are proven secure under certain unproven assumptions. For example, the infeasibility of factoring extremely large integers is the basis for believing that RSA is secure, and some other systems, but even there, the proof is usually lost due to practical considerations. There are systems similar to RSA, such as one by Michael O. Rabin that is provably secure provided factoring $n = pq$ is impossible, but the more practical system RSA has never been proved secure in this sense. The discrete logarithm problem is the basis for believing some other cryptosystems are secure, and again, there are related, less practical systems that are provably secure relative to the discrete log problem.

As well as being aware of cryptographic history, cryptographic algorithm and system designers must also sensibly consider probable future developments while working on their designs. For instance, continuous improvements in computer processing power have increased the scope of brute-force attacks, so when specifying key lengths, the required key lengths are similarly advancing. The potential effects of **quantum computing** are already being considered by some cryptographic system designers; the announced imminence of small implementations of these machines may be making the need for this preemptive caution rather more than merely speculative.

Essentially, prior to the early 20th century, cryptography was chiefly concerned with linguistic and lexicographic patterns. Since then the emphasis has shifted, and cryptography now makes extensive use of mathematics, including aspects of information theory, computational complexity, statistics, combinatorics, abstract algebra, number theory, and finite mathematics generally. Cryptography is also a branch of engineering, but an unusual one since it deals with active, intelligent, and malevolent opposition; other kinds of engineering (e.g., civil or chemical engineering) need deal only with neutral natural forces. There is also active research examining the relationship between cryptographic problems and quantum physics.

9.3.3 Modern Cryptography

The modern field of cryptography can be divided into several areas of study.

1. Symmetric-key Cryptography

Symmetric-key cryptography refers to encryption methods in which both the sender and receiver share the same key (or, less commonly, in which their keys are different, but related in an easily computable way). This was the only kind of encryption publicly known until June 1976.

Symmetric key ciphers are implemented as either **block ciphers** or **stream ciphers**. A block cipher enciphers input in blocks of plaintext as opposed to individual characters, the input form used by a stream cipher.

The Data Encryption Standard (DES) and the Advanced Encryption Standard (AES) are block cipher designs that have been designated cryptography standards by the US government (though DES's designation was finally withdrawn after the AES was adopted). Despite its deprecation as an official standard, DES (especially its still-approved and much more secure triple-DES variant) remains quite popular; it is used across a wide range of applications, from ATM encryption to e-mail privacy and secure remote access. Many other block ciphers have been designed and released, with considerable variation in quality.

Stream ciphers, in contrast to the "block" type, create an arbitrarily long stream of key material, which is combined with the plaintext bit-by-bit or character-by-character, somewhat like the one-time pad. In a stream cipher, the output stream is created based on a hidden internal state that changes as the cipher operates. That internal state is initially set up using the secret key material. RC4 is a widely used stream cipher.

(9-4) Symmetric-key cryptosystems use the same key for encryption and decryption of a message, though a message or group of messages may have a different key than others. A significant disadvantage of symmetric ciphers is the key management necessary to use them securely. Each distinct pair of communicating parties must, ideally, share a different key, and perhaps each ciphertext exchanged as well. The number of keys required increases as the square of the number of network members, which very quickly requires complex key management schemes to keep them all consistent and secret. The difficulty of securely establishing a secret key between two communicating parties, when a secure channel does not already exist between them, also presents a chicken-and-egg problem which is a considerable practical obstacle for cryptography users in the real world.

2. Cryptographic Hash Functions

(9-5) Cryptographic **hash functions** are a third type of cryptographic algorithm. They take a message of any length as input, and output a short, fixed length hash, which can be used in (for example) a **digital signature**. For good hash functions, an attacker cannot find two messages that produce the same hash. MD4 is a long-used hash function that is now broken; MD5, a strengthened variant of MD4, is also widely used but broken in practice. The US National Security Agency developed the Secure Hash Algorithm series of MD5-like hash functions: SHA-0 was a flawed algorithm that the agency withdrew; SHA-1 is widely deployed and more secure than MD5, but cryptanalysts have identified attacks against it; the SHA-2 family improves on SHA-1, but it isn't yet widely deployed; and the US standards authority thought it "prudent" from a security perspective to develop a new standard to "significantly improve the robustness of NIST's overall hash algorithm toolkit". Thus, a hash function design competition was meant to select a new U.S. national standard, to be called SHA-3, by 2012. The competition ended on October 2, 2012 when the NIST announced that Keccak would be the new SHA-3 hash algorithm. Unlike block and stream ciphers that are invertible, cryptographic hash functions produce a hashed output that cannot be used to retrieve the original input data. Cryptographic hash functions are used to verify the **authenticity** of data retrieved from an untrusted source or to add a layer of security.

Message Authentication Codes (MACs) are much like cryptographic hash functions, except that a secret key can be used to authenticate the hash value upon receipt;[4] this additional complication blocks an attack scheme against bare digest algorithms, and so has been thought worth the effort.

3. Public-key Cryptography

In a groundbreaking 1976 paper, Whitfield Diffie and Martin Hellman proposed the notion of public-key (also, more generally, called asymmetric key) cryptography in which two different but mathematically related keys are used — a **public key** and a **private key**. A public key system is so constructed that calculation of one key (the "private key") is computationally infeasible from the other (the "public key"), even though they are necessarily related. Instead, both keys are generated secretly, as an interrelated pair. The historian David Kahn described public-key cryptography as "the most revolutionary new concept in the field since polyalphabetic substitution emerged in the Renaissance".

In public-key cryptosystems, the public key may be freely distributed, while its paired private key must remain secret. In a public-key encryption system, the public key is used for encryption, while the private or secret key is used for decryption. While Diffie and Hellman could not find such a system, they showed that public-key cryptography was indeed possible by presenting the Diffie–Hellman key exchange protocol, a solution that is now widely used in secure communications to allow two parties to secretly agree on a shared encryption key.

Diffie and Hellman's publication sparked widespread academic efforts in finding a practical public-key encryption system. This race was finally won in 1978 by Ronald Rivest, Adi Shamir, and Len Adleman, whose solution has since become known as the RSA algorithm.

The Diffie–Hellman and RSA algorithms, in addition to being the first publicly known examples of high quality public-key algorithms, have been among the most widely used. Others include the Cramer–Shoup cryptosystem, ElGamal encryption, and various elliptic curve techniques.

Public-key cryptography can also be used for implementing digital signature schemes. A digital signature is reminiscent of an ordinary signature; they both have the characteristic of being easy for a user to produce, but difficult for anyone else to forge. Digital signatures can also be permanently tied to the content of the message being signed; they cannot then be "moved" from one document to another, for any attempt will be detectable. In digital signature schemes, there are two algorithms: one for signing, in which a secret key is used to process the message (or a hash of the message, or both), and one for verification, in which the matching public key is used with the message to check the validity of the signature. RSA and DSA are two of the most popular digital signature schemes. Digital signatures are central to the operation of public key infrastructures and many network security schemes.

(9-6) Public-key algorithms are most often based on the computational complexity of "hard" problems, often from number theory. For example, the hardness of RSA is related to the integer factorization problem, while Diffie–Hellman and DSA are related to the discrete logarithm problem. Because of the difficulty of the underlying problems, most public-key algorithms involve operations such as modular multiplication and **exponentiation**, which are much more computationally expensive than the techniques used in most block ciphers, especially with typical key sizes. As a result, public-key cryptosystems are commonly hybrid cryptosystems, in which a fast high-quality symmetric-key encryption algorithm is used for the message itself, while the relevant symmetric key is sent with the message, but encrypted using a public-key algorithm. Similarly, hybrid signature schemes are often used, in which a cryptographic hash function is computed, and only the resulting hash is digitally signed.

4. Cryptanalysis

Variants of the Enigma machine, used by Germany's military and civil authorities from the late 1920s through World War II, implemented a complex electro-mechanical polyalphabetic cipher. Breaking and reading of the Enigma cipher at Poland's Cipher Bureau, for 7 years before the war, and subsequent decryption at Bletchley Park, was important to Allied victory.

The goal of cryptanalysis is to find some weakness or insecurity in a cryptographic scheme, thus permitting its subversion or evasion.

It is a common misconception that every encryption method can be broken. In connection with his WW II work at Bell Labs, Claude Shannon proved that the one-time pad cipher is unbreakable, provided the key material is truly random, never reused, kept secret from all possible attackers, and of equal or greater length than the message. Most ciphers, apart from the one-time pad, can be broken with enough computational effort by brute force attack, but the amount of effort needed may be exponentially dependent on the key size, as compared to the effort needed to make use of the cipher. In such cases, effective security could be achieved if it is proven that the effort required (i.e., "work factor", in Shannon's terms) is beyond the ability of any adversary. This means it must be shown that no efficient method (as opposed to the time-consuming brute force method) can be found to break the cipher. Since no such proof has been found to date, the one-time-pad remains the only theoretically unbreakable cipher.

Cryptanalysis of symmetric-key ciphers typically involves looking for attacks against the block ciphers or stream ciphers that are more efficient than any attack that could be against a perfect cipher. For example, a simple brute force attack against DES requires one known plaintext and 255 decryptions, trying approximately half of the possible keys, to reach a point at which chances are better than even that the key sought will have been found. But this may not be enough assurance; a linear cryptanalysis attack against DES requires 243 known plaintexts and approximately 243 DES operations. This is a considerable improvement on brute force attacks.

Public-key algorithms are based on the computational difficulty of various problems. The most famous of these is integer factorization (e.g., the RSA algorithm is based on a problem related to integer factoring), but the discrete logarithm problem is also important. Much public-key cryptanalysis concerns numerical algorithms for solving these computational problems, or some of them, efficiently (i.e., in a practical time). For instance, the best known algorithms for solving the elliptic curve-based version of discrete logarithm are much more time-consuming than the best known algorithms for factoring, at least for problems of more or less equivalent size. Thus, other things being equal, to achieve an equivalent strength of attack resistance, factoring-based encryption techniques must use larger keys than elliptic curve techniques. For this reason, public-key cryptosystems based on elliptic curves have become popular since their invention in the mid-1990s.

9.4 Top 10 Cyber-security Issues in 2016

"The Internet is a chimera," Andrew Chen said, who was an associate professor of computer science at Minnesota State University, Moorhead. "It starts out seeming powerful, then it becomes seductive, and then it becomes dangerous." It's hard to argue with Dr. Chen's view of the current internet, and 2016 looks like more of the same, if not more so. Here are some reasons why.

(1) Cyberespionage by nation states will escalate.

The Tallinn Manual, written by 20 cyber-defense experts and published by Cambridge University Press, defines **cyber-espionage** as "An act undertaken clandestinely or **under false pretenses** that uses cyber capabilities to gather information with the intention of communicating it to the opposing party".

What 2016 will bring is more of the "cloak and dagger" spy stuff. Nation states will be spying on everybody on the planet. The gathered data will be used for economic advantage and as digital weaponry in geopolitical conflicts.

(2) Governments will step up their demands for information from companies.

The now defunct US-EU Safe Harbor agreement is already creating **fallout**. According to EU Business, "Ireland's High Court ordered the Irish Data Protection Commissioner to examine whether to suspend the transfer of Facebook users' data from Europe to the United States".

Governments demanding companies to turn over sensitive information will continue, resulting in organizations not knowing whom to trust and what regulations they may be violating.

(3) Countries will take steps to control the internet within their borders.

With trust falling to the wayside, experts suggest that each country will try to control the internet within its borders. Managing director of Discern Analytics and consulting associate professor at Stanford, Paul Saffo writes, "The pressures to balkanize the global internet will continue and create new uncertainties. Governments will become more skilled at blocking access to unwelcome sites".

(4) Big data and analytics will be less than helpful when applied to security.

Big data and analytics are slowly making its way into security. However, there is a but. "Organizations that put blind faith in big data will make strategic decisions based on faulty or incomplete data sets," ISF Global Vice President Steve Durbin tells eWeek. "Avoid this by outlining a process for applying big-data analytics to information security problems".

To add more of a point, Dr. Anton Chuvakin, research vice president at Gartner mentions, "Do not pay for the glamour of big data if there is little chance of benefiting from the investment — especially if the very definition of 'benefiting' is unclear".

(5) Mobile apps will become the target of choice.

Mobile apps are becoming the target of choice; the Ponemon Institute decided to look at why. "Among the more than 400 organizations studied — nearly 40 percent of which were Fortune 500 companies — almost 40 percent of them aren't scanning the code in their apps for security vulnerabilities, leaving the door wide open to the potential hacking of sensitive user, corporate, and customer data," mentions Dr. Larry Ponemon. "The average organization tests fewer than half of the mobile apps it builds, and 33 percent of the surveyed companies never test their apps".

(6) Encryption is not and will not be a deterrent.

During the past several months, government law enforcement agencies have been demanding that backdoors be added to encryption software. As can be expected, security pundits fought back, and it appears they have won. However, there are those who aren't so sure — maybe this is all a smoke screen.

Andy Greenberg in his WIRED article Cops Don't Need a Crypto Backdoor to Get Into Your iPhone interviews several security experts, asking if backdoors are even needed. "In spite of the big words the FBI has used over the last year, the situation isn't quite as dire as they make it out to be," Chris Soghoian, principal technologist for the ACLU tells Greenberg. "The kind of encryption tech companies are giving us is geared towards protecting us from a thief stealing our laptop. It's not designed to keep out a government agent trying to get your data with or without a court order".

(7) Lax IoT security will become apparent.

The IoT device onslaught is coming, however, not this year. "We do not expect attacks on the IoT to become widespread yet. Most attacks are likely to be 'whitehat' hacks to report vulnerabilities and proof of concept exploits," Michael Fimin writes in this Netwrix Community column.

That will not be the case in 2016, especially if the bad guys read articles like **John Dixon's Who Will Step Up To Secure The Internet Of Things**?

"If today's titans of technology won't step up to secure the IoT, that endeavor may fall to the multitude of startup companies that are fueling much of the industry's current growth," mentions Dixon. "Gartner estimates by 2017, more than half of all IoT products and services will be developed by companies less than three years old. Moreover, while some of these newcomers are likely to have formidable technical expertise, many will lack the know-how or capability to implement the tight security that is needed".

(8) More "qualified" security professionals will be needed.

When it comes to security, the best defenders are ironically the best attackers, and right now there is a dire lack of qualified defenders. However, there is a more immediate problem. According to ISF Global Vice President Steve Durbin, there are not enough qualified security professionals to go around. This will not improve in 2016, with both businesses and government agencies fighting over those indeed who have the right qualifications.

(9) Convenience will continue to trump privacy and security.

Much has been written about why convenience overrules privacy and security. Cisco Systems Fellow Fred Baker cuts to the chase, saying, "The issues in security and privacy have improved in important ways, but will remain threats, primarily because human nature has not changed, and there is always a percentage of people who seek to harm others".

9.5 Cyberwar[①]

As nations spend billions of dollars stockpiling **digital weapons** and tension grows between them, the risk of world **cyber warfare** has suddenly turned offensive.

9.5.1 A Cybersecurity Wargame Scenario

The team was badly spooked, that much was clear. The bank was already reeling from two

① Ranger S. http://www.techrepublic.com/article/inside-the-secret-digital-arms-race/.

attacks on its systems, strikes that had brought it to a standstill and forced the cancellation of a high profile IPO.

The board had called in the team of security experts to brief them on the developing crisis. After listening to some of the mass of technical detail, the bank's CEO cut to the chase.

"What should I tell the Prime Minister when I get to Cobra?" he demanded, a reference to the emergency committee the government had set up as it scrambled to respond to what was looking increasingly like a coordinated cyberattack.

The security analysts hesitated, shifting in their seats, fearing this was the beginning, not the end, of the offensive.

"We think this could just be a smokescreen," one said, finally. And it was. Before the end of next day, the attack had spread from banks to transport and utilities, culminating in an attack on a nuclear power station.

The mounting horror of the analysts, the outrage and lack of understanding from the execs was all disturbingly authentic, but fortunately, none of it was real. The scene formed part of a wargame, albeit one designed by the UK's GCHQ surveillance agency among others to attract new recruits into the field of cybersecurity.

As I watched the scenario progress, it was hard not to get just as caught up in the unfolding events as the competition finalists played the security analysts tasked with fighting the attack, and real industry executives took the role of the bank's management, if only because these sorts of scenarios are now increasingly plausible.

And it's not just mad criminal geniuses planning these sorts of digital doomsday attacks either. After years on the defensive, governments are building their own offensive capabilities to deliver attacks just like this against their enemies. It's all part of a secret, hidden **arms race**, where countries spend billions of dollars to create new armies and stockpiles of digital weapons.

This new type of warfare is incredibly complex and its consequences are little understood. Could this secret digital arms race make real-world confrontations more likely, not less? Have we replaced the cold war with the coders' war?

Even the experts are surprised by how fast the online threats have developed. As Mikko Hypponen, chief research officer at security company F-Secure, said a conference recently, "If someone would have told me ten years ago that by 2014 it would be commonplace for democratic western governments to develop and deploy malware against other democratic western governments, that would have sounded like science fiction. It would have sounded like a movie plot, but that's where we are today".

9.5.2 The First Casualty of Cyberwar Is The Web

It's taken less than a decade for digital warfare to go from theoretical to the worryingly possible. The Web has been an unofficial battleground for many modern conflicts. At the most basic level, groups of hackers have been hijacking or defacing websites for years. Some of these groups have acted alone; some have at least the tacit approval of their governments.

Most of these attacks — taking over a few Twitter accounts, for example — are little more than a nuisance, high profile but relatively trivial.

However, one attack has already risen to the level of international incident. In 2007, attacks on Estonia swamped banks, newspaper and government websites. They began after Estonia decided to move a Soviet war memorial, and lasted for three weeks (Russia denied any involvement).

Estonia is a small state with a population of just 1.3 million. However, it has a highly-developed online infrastructure, having invested heavily in e-government services, digital ID cards, and online banking. That made the attack particularly painful, as the head of IT security at the Estonian defence ministry told the BBC at the time, "If these services are made slower, we of course lose economically".

The attacks on Estonia were a turning point, proving that a digital bombardment could be used not just to derail a company or a website, but to attack a country. Since then, many nations have been scrambling to improve their digital defenses — and their digital weapons.

While the attacks on Estonia used relatively simple tools against a small target, bigger weapons are being built to take on some of the mightiest of targets.

It's all part of a secret, hidden arms race, where countries spend billions of dollars to create new armies and stockpiles of digital weapons.

Last year the then-head of the US Cyber Command, General Keith Alexander, warned on the CBS (Columbia Broadcasting System) 60 Minutes programme of the threat of foreign attacks, stating: "I believe that a foreign nation could impact and destroy major portions of our financial system".

In the same programme, the NSA warned of something it called the "BIOS plot", a work by an unnamed nation to exploit a software flaw that could have allowed them to destroy the BIOS in any PC and render the machine unusable.

Of course, the US isn't just on the defensive. It has been building up its own capabilities to strike, if needed.

The only documented successful use of such a weapon — the famous **Stuxnet worm** — was masterminded by the US in the form that caused damage and delay to the Iranian nuclear program.

9.5.3 Building Digital Armies

The military has been involved with the Internet since its start. It emerged from a US Department of Defense-funded project, so it's not surprise that the armed forces have kept a close eye on its potential.

And politicians and military leaders of all nations are naturally attracted to digital warfare as it offers the opportunity to neutralize an enemy without putting troops at risk. As such, the last decade has seen rapid investment in what governments and the military have dubbed "cyberwar".

The UK's defence secretary Philip Hammond has made no secret of the country's interest in the field, telling a newspaper late last year, "We will build in Britain a cyber strike capability so we can strike back in cyberspace against enemies who attack us, putting cyber alongside land, sea, air and space as a mainstream military activity".

The UK is thought to be spending as much as £500m on the project over the next few years. On an even larger scale, last year General Alexander revealed the NSA was building 13 teams to strike back in the event of an attack on the US. "I would like to be clear that this team, this defend-the-nation team, is not a defensive team," he told the Senate Armed Services.

And of course, it's not just the UK and US that are building up a digital army. In a time of declining budgets, it's a way for defence ministries and defence companies to see growth, leading some to warn of the emergence of a twenty-first century cyber-industrial complex. And the shift from investment in cyber-defence initiatives to cyber-offensives is a recent and, for some, worrying trend.

Peter W. Singer, director of the Center for 21st Century Security and Intelligence at the Brookings Institution, said 100 nations are building cyber military commands of that there are about 20 that are serious players, and a smaller number could carry out a whole cyberwar campaign. And the fear is that by emphasising their offensive capabilities, governments will up the ante for everyone else.

"We are seeing some of the same manifestations of a classic arms race that we saw in the Cold War or prior to World War I. The essence of an arms race is where the sides spend more and more on building up and advancing military capabilities but feel less and less secure — and that definitely characterizes this space today," he said.

Politicians may argue that building up these skills is a deterrent to others, and emphasize such weapons would only be used to counter an attack, never to launch one. But for some, far from scaring off any would-be threats, these investments in offensive cyber capabilities risk creating more instability.

"In international stability terms, arms races are never a positive thing: the problem is it's incredibly hard to get out of them because they are both illogical and make perfect sense," Singer said.

Similarly Richard Clarke, a former presidential advisor on cybersecurity told a conference in 2012, "We turn an awful lot of people off in this country and around the world when we have generals and admirals running around talking about 'dominating the cyber domain'. We need cooperation from a lot of people around the world and in this country to achieve cybersecurity, militarizing the issue and talking about how the US military have to dominate the cyber domain is not helpful".

As a result, in the shadows, various nations building up their digital military presence are mapping out what could be future digital battlegrounds and seeking out potential targets, even leaving behind code to be activated later in any conflict that might arise.

9.5.4 How Cyber Weapons Work

As nations race to build their digital armies they also need to arm them. And that means developing new types of weapons.

While state-sponsored cyberwarfare may use some of the same tools as criminal hackers, and even some of the same targets, its wants to go further.

So while a state-sponsored cyber attack could use the old hacker standby of the denial of service attack (indeed the UK's GCHQ has already used such attacks itself, according to leaks from Edward Snowden), something like Stuxnet — built with the aim of destroying the centrifuges used in the Iranian nuclear project — is another thing entirely.

"Stuxnet was almost a Manhattan Project style in terms of the wide variety of expertise that was brought in: everything from intelligence analysts to some of the top cyber talent in the world to nuclear physicists to engineers, to build working models to test it out on, and another entire espionage effort to put it in to the systems in Iran that Iran thought were air-gapped. This was not a couple of kids," said Singer.

The big difference between military-grade cyber weapons and hacker tools is that the most sophisticated digital weapons want to break things, to create real and physical damage. And these weapons are **bespoke**, expensive to build, and have a very short shelf life.

To have a real impact, these attacks are likely to be **levelled at** the industrial software that runs production lines, power stations or energy grids, otherwise known as SCADA (Supervisory Control and Data Acquisition) systems.

Increasingly, SCADA systems are being Internet-enabled to make them easier to manage, which of course, also makes them easier to attack. Easier doesn't mean easy though. These complex systems, often build to last for decades, are often built for a very narrow, specific purpose — sometimes for a single building. This makes them much harder to undermine.

"The only piece of **target intelligence** you need to attack somebody's email or a website is an email address or a URL. In the case of a control system, you need much more information about the target, about the entire logic that controls the process, and legacy systems that are part of the process you are attacking".

That also means that delivering any more than a few of these attacks at a time would be almost impossible, making a long cyberwar campaign hard to sustain.

Similarly, these weapons need to exploit a unique weakness to be effective: so-called **zero-day flaws**. These are vulnerabilities in software that have not been patched and therefore cannot be defended against.

This is what makes them potentially so devastating, but also limits their longevity. Zero-day flaws are relatively rare and expensive and hard to come by. They're sold for hundreds of thousands of dollars by their finders. A couple of years ago a Windows flaw might have earned its finder $100,000 on the black market, an iOS vulnerability twice that.

Zero-day flaws have an in-built weakness, though: they're a one-use only weapon. Once an attack has been launched, the zero-day used is known to everyone. Take Stuxnet for example, even though it seems to have had one specific target — an Iranian power plant — once it was launched, Stuxnet spread very widely, meaning security companies around the world could examine the code, and making it much harder for anyone to use that exact same attack again.

But this leads to another, unseen problem. As governments stockpile zero-day flaws for use in their cyber-weapons, it means they aren't being reported to the software vendors to be fixed — leaving unpatched systems around the world at risk when they could easily be fixed.

9.5.5 When Is a Cyberwar Not a Cyberwar?

The greatest trick cyberwar ever played was convincing the world it doesn't exist.

While the laws of armed conflict are well understood — if not always adhered to — what's striking about cyberwar is that no one really knows what the rules are.

As NATO's own National Cybersecurity Framework Manual notes: "In general, there is agreement that cyber activities can be a legitimate military activity, but there is no global agreement on the rules that should apply to it".

Dr. Heather A. Harrison Dinniss of the International Law Centre at the Swedish National Defence College said that most cyber warfare shouldn't need to be treated differently to regular warfare, and that the general legal concepts apply "equally regardless of whether your weapon is a missile or a string of ones and zeros".

But cyberwarfare does raise some more difficult issues, she says. What about attacks that do not cause physical harm, for example: do they constitute attacks as defined under the laws of armed conflict?

Dinniss says that some sort of consensus emerging that attacks which cause loss of functionality to a system do constitute an attack, but the question is certainly not settled in law.

Western nations have been reluctant to sign any treaty that tries to define cyberwar. In the topsy-turvy world of international relations, it is China and Russia that are keenest on international treaties that define cyberwarfare as part of their general desire to regulate Internet usage.

The reluctance from the US and the UK is partly because no state wants to talk candidly about their cyberwarfare capabilities, but also by not clearly defining the status of cyberwarfare, they get a little more leeway in terms of how they use those weapons.

And, because in many countries cyberwarfare planning has grown out of intelligence agencies as much as out of the military, the line between surveillance-related hacking and more explicitly-offensive attacks is at best very blurred.

That blurring suits the intelligence agencies and the military just fine. While espionage is not illegal under international law, it is outlawed under most states' domestic laws.

If a cyber attack can be defined as an attack under the laws of armed conflict, a nation has a much better case for launching any kind of response, up to and including using conventional weapons in response. And that could mean that using digital weapons could have unexpected — and potentially disastrous — consequences.

Right now all of this is a deliberately grey area, but it's not hard to envisage an Internet espionage attempt that goes wrong, damages something, and rapidly escalates into a military conflict. Can a hacking attempt really lead to casualties on the battlefield? Possibly, but right now those rules around escalation aren't set. Nobody really knows how or if escalation works in a digital space.

If I hack your power grid, is it a fair response to shut down my central bank? At what point is a missile strike the correct response to a denial of service attack? Nobody really knows what a hierarchy of targets here would look like. And that's without the problem of working out exactly who has attacked you. It's much easier to see a missile launch than work out from where a distributed digital attack is being orchestrated. Any form of cyber arms control is a long way off.

9.5.6 The Targets in Cyberwar

You can drop a bomb on pretty much anything, as long as you can find it. It's a little different with digital weapons.

Some targets just don't have computers and while politicians may dream of being able to "switch off" an enemy's airfield, it's likely to be the civilian infrastructure that's going to be the most obvious target. That's the same as standard warfare. What is different now is that virtually any company could be a target, and many probably don't realize it.

Companies are only gradually understanding the threats they face, especially as they start to connect their industrial control systems to the Internet. Like the executives at the London cyber wargame, most real life executives fail to realize that they might be a target, or the potential risks.

Mark Brown, director of risk advisory at consultant Ernst & Young, says, "Companies have recognized they can connect them to their core networks, to the Internet, to operate them remotely but they haven't necessarily applied the same risk and controls methodology to the management of operational technology as they have to traditional IT".

Indeed, part of the problem is that these systems have never been thought about as security risks and so no-one has taken responsibility for them. "Not many CIOs have responsibility for those operational technology environments, at least not traditionally. Often you are caught in the crossfire of finger-pointing; the CIO says it's not my job, the head of engineering says it's not my job," Brown said.

A recent report warned that the cybersecurity efforts around the US electricity supply network are fragmented and not moving fast enough, while in the UK insurers are refusing cover to power companies because their defences are too weak.

9.5.7 Cyberwar: Coming to a Living Room Near You?

Cyberwar is — for all the billions being spent on it — still largely theoretical, especially when it comes to the use of zero-day attacks against public utilities. Right now a fallen tree is a bigger threat to your power supply than a hacker.

While states have the power to launch such attacks, for now they have little incentive. And the ones with the most sophisticated weapons also have the most sophisticated infrastructure and plenty to lose, which is why most activity is at the level of espionage rather than war.

However, there's no reason why this should remain the case forever. When countries spend billions on building up a stockpile of weapons, there is always the increased risk of confrontation, especially when the rules of engagement are still in flux.

But even now a new and even more dangerous battlefield is being built. As we connect more devices — especially the ones in our homes — to the Web, cyberwar is poised to become much more personal.

As thermostats, fridges and cars become part of the Internet of things, their usefulness to us may increase, but so does the risk of them being attacked. Just as more industrial systems are being connected up, we're doing the same to our homes.

The Internet of things, and even wearable tech, bring with them great potential, but unless these systems are incredibly well-secured, they could be easy targets to compromise.

Cyberwarfare might seem like a remote, fanciful threat, but digital weapons could create the most personal attacks possible. Sure, it's hard to see much horror lurking in a denial of service attack against your internet-enabled toothbrush (or the fabled Internet fridge) but the idea of an attack that turned your gadgets, your car or even your home against you is one we need to be aware of.

We tend to think of our use of technology as insulating us from risk, but in future that may no longer be the case. If cyberwar ever becomes a reality, the home front could become an unexpected battlefield.

9.6 Key Terms and Review Questions

1. Technical Terms

hacker	黑客	9.1
cybersecurity	计算机安全	9.1
code injection	代码注入	9.1
malpractice	玩忽职守	9.1
Internet of Things	物联网	9.1
confidentiality	机密性	9.1
integrity	完整性	9.1
availibity	可用性	9.1
authentication	认证	9.1
authorization	授权	9.1
nonrepudiation	非否认	9.1
privacy	隐私	9.1
unauthorized	未授权的，非法的	9.1
denial of service	拒绝服务攻击	9.1

security attack	安全攻击	9.1
security mechanism	安全机制	9.1
security service	安全服务	9.1
Back Door attacks	后门攻击	9.1
masquerading	伪装，冒充	9.1
IP Spoofing	IP 地址欺骗	9.1
DNS Spoofing	域名欺骗	9.1
Man-in-the-middle attack	中间人攻击	9.1
Replay attack	重放攻击	9.1
TCP/IP Hijacking	TCP/IP 动持	9.1
attacker	攻击者	9.1
encryption	加密	9.1
decryption	解密	9.1
brute force attack	暴力破解攻击	9.1
substitution	代替	9.1
permutation	置换	9.1
Dictionary password attack	字典攻击	9.1
intruder	入侵者	9.1
scramble	混乱	9.1
opponent	对手	9.1
trusted third party	可信第三方	9.1
vulerability	弱点	9.1
countermeasure	对抗措施	9.2
principle of least privilege	最小特权原则	9.2
defense in depth	纵深防御	9.2
audit trails	审计跟踪	9.2
accountability	可说明性	9.2
reproducible	可再生的，可复制的	9.2
risk assessment	风险评估	9.2
Intrusion Detection System	入侵检测系统	9.2
post-attack forensic	攻击后取证	9.2
big data	大数据	9.2
scanner	扫描器	9.2
two-factor authentication	双因素认证	9.2
dongle	电子狗、加密狗	9.2
social engineering attack	社会工程攻击	9.2
patches and updates	补丁和更新	9.2
Evaluation Assurance Levels	评估保证等级	9.2
security breaches	安全漏洞	9.2

jurisdiction	管辖	9.2
anonymizing	匿名的	9.2
cryptology	密码术	9.3
adversary	敌对者	9.3
rotor cipher machines	转子密码机	9.3
cipher	密码	9.3
key	密钥	9.3
plaintext	明文	9.3
ciphertext	密文	9.3
cryptosystem	密码体制	9.3
symmetric system	对称（密码）系统	9.3
asymmetric system	非对称（密码）系统	9.3
cryptanalysis	密码分析术	9.3
cryptolinguistics	密码语言学	9.3
Caesar cipher	凯撒密码	9.3
Steganography	隐写术	9.3
frequency analysis	频率分析	9.3
polyalphabetic cipher	多表置换密码	9.3
scytale	密码棒	9.3
integer factorization	整数因子分解	9.3
discrete logarithm	离散对数	9.3
one-time pad	一次性密码本	9.3
quantum computing	量子计算	9.3
triple-DES	三重 DES	9.3
block ciphers	块密码，分组密码	9.3
stream ciphers	流密码	9.3
key management	密钥管理	9.3
hash function	散列函数	9.3
digital signature	数字签名	9.3
authenticity	真实性，可靠性	9.3
public key	公开密钥	9.3
private key	私有密钥	9.3
modular multiplication	模数乘法	9.3
exponentiation	乘方	9.3
cyberespionage	网络间谍行为	9.4
under false pretenses	假借名义	9.4
fallout	附带后果	9.4
cyber warfare	网络战	9.5
digital weapons	数字武器	9.5

arms race	军备竞赛	9.5
Stuxnet worm	震网病毒	9.5
bespoke	定制的	9.5
levelled at	瞄准	9.5
target intelligence	目标情报	9.5
zero day flaws	零日漏洞	9.5

2. Translation Exercises

(9-1) Authentication and authorization go hand in hand. Users must be authenticated before carrying out the activity they are authorized to perform. Security is strong when the means of authentication cannot later be refuted — the user cannot later deny that he or she performed the activity. This is known as nonrepudiation.

(9-2) On systems which rely solely on a login name and password the security of the entire system is only as strong as the passwords chosen by the users. The best way to ensure passwords are not cracked is to avoid the use of simple words or phrases which can be found in a dictionary. This needs to be balanced with making the passwords easy enough to remember so that users do not write them on pieces of paper and stick them on their laptops or monitors for others to find.

(9-3) A firewall can be defined as a way of filtering network data between a host or a network and another network, such as the Internet, and can be implemented as software running on the machine, hooking into the network stack to provide real time filtering and blocking. Another implementation is a so-called "physical firewall", which consists of a separate machine filtering network traffic.

(9-4) Symmetric-key cryptosystems use the same key for encryption and decryption of a message, though a message or group of messages may have a different key than others. A significant disadvantage of symmetric ciphers is the key management necessary to use them securely. Each distinct pair of communicating parties must, ideally, share a different key, and perhaps each ciphertext exchanged as well. The number of keys required increases as the square of the number of network members, which very quickly requires complex key management schemes to keep them all consistent and secret.

(9-5) Cryptographic hash functions are a third type of cryptographic algorithm. They take a message of any length as input, and output a short, fixed length hash, which can be used in (for example) a digital signature. For good hash functions, an attacker cannot find two messages that produce the same hash.

(9-6) Public-key algorithms are most often based on the computational complexity of "hard" problems, often from number theory. For example, the hardness of RSA is related to the integer factorization problem, while Diffie–Hellman and DSA are related to the discrete logarithm problem. Because of the difficulty of the underlying problems, most public-key algorithms involve operations such as modular multiplication and exponentiation, which are much more computationally expensive than the techniques used in most block ciphers, especially with typical key sizes.

References

[1] Stallings. W NETWORK SECURITY ESSENTIALS: APPLICATIONS AND STANDARDS FOURTH EDITION. Prentice Hall,2011.

[2] Nalin A, Steve L, Michael S. Designing a Mobile Game to Teach Conceptual Knowledge of Avoiding "Phishing Attacks"[J]. International Journal for e-Learning Security, 2012,2 (1): 127-132.

[3] Michael S, Gheorghita G, Nalin A. Assessing the Role of Conceptual Knowledge in an Anti-Phishing Educational Game[C]. Proceedings of the 14th IEEE International Conference on Advanced Learning Technologies, IEEE, 2014: 218.

[4] Cryptography[DB/OL]. [2017-07-09]. https://en.wikipedia.org/wiki/Cryptography.

[5] Phillip B M., Introduction to Modern Cryptography[M]. 2005:10.

Chapter 10 Latest Progresses in Computer Science

10.1 Quantum Information Science

Quantum information science is an area of study based on the idea that information science depends on quantum effects in physics. It includes theoretical issues in computational models as well as more experimental topics in quantum physics including what can and cannot be done with quantum information. The term "quantum information theory" is sometimes used, but it fails to encompass experimental research in the area.

10.1.1 Quantum Computing

1. A Brief Introduction

The massive amount of processing power generated by computer manufacturers has not yet been able to quench our thirst for speed and computing capacity. In 1947, American computer engineer Howard Aiken said that just six electronic digital computers would satisfy the computing needs of the United States. Others have made similar errant predictions about the amount of computing power that would support our growing technological needs. Of course, Aiken didn't count on the large amounts of data generated by scientific research, the proliferation of personal computers or the emergence of the Internet, which have only fueled our need for more, more and more computing power.

Will we ever have the amount of computing power we need or want? (10-1)If, as Moore's Law states, the number of transistors on a microprocessor continues to double every 18 months, the year 2020 or 2030 will find the circuits on a microprocessor measured on an atomic scale. And the logical next step will be to create quantum computers, which will harness the power of atoms and molecules to perform memory and processing tasks. Quantum computers have the potential to perform certain calculations significantly faster than any silicon-based computer.

Scientists have already built basic quantum computers that can perform certain calculations; but a practical quantum computer is still years away.

You don't have to go back too far to find the origins of quantum computing. While computers have been around for the majority of the 20th century, quantum computing was first theorized less than 30 years ago, by a physicist at the Argonne National Laboratory. Paul Benioff is credited with first applying quantum theory to computers in 1981. Benioff theorized about creating a quantum Turing machine.

Today's computers, like a Turing machine, work by manipulating bits that exist in one of two states: a 0 or a 1. Quantum computers aren't limited to two states; they encode information as quantum bits, or **qubits**, which can exist in **superposition**. Qubits represent atoms, **ions**, **photons** or electrons and their respective control devices that are working together to act as computer memory and a processor. Because a quantum computer can contain these multiple states simultaneously, it has the potential to be millions of times more powerful than today's most powerful supercomputers.

The superposition of qubits is what gives quantum computers their inherent parallelism. According to physicist David Deutsch, this parallelism allows a quantum computer to work on a million computations at once, while your desktop PC works on one. A 30-qubit quantum computer would equal the processing power of a conventional computer that could run at 10 **teraflops** (trillions of floating-point operations per second). Today's typical desktop computers run at speeds measured in gigaflops (billions of floating-point operations per second).

Quantum computers also utilize another aspect of quantum mechanics known as **entanglement**. One problem with the idea of quantum computers is that if you try to look at the subatomic particles, you could bump them, and thereby change their value. If you look at a qubit in superposition to determine its value, the qubit will assume the value of either 0 or 1, but not both (effectively turning your spiffy quantum computer into a mundane digital computer). To make a practical quantum computer, scientists have to devise ways of making measurements indirectly to preserve the system's integrity. Entanglement provides a potential answer. In quantum physics, if you apply an outside force to two atoms, it can cause them to become entangled and the second atom can take on the properties of the first atom. So if left alone, an atom will spin in all directions. The instant it is disturbed it chooses one spin, or one value; and at the same time, the second entangled atom will choose an opposite spin, or value. This allows scientists to know the value of the qubits without actually looking at them.

Quantum computers could one day replace silicon chips, just like the transistor once replaced the vacuum tube. But for now, the technology required to develop such a quantum computer is beyond our reach. Most research in quantum computing is still very theoretical.

The most advanced quantum computers have not gone beyond manipulating more than 16 qubits, meaning that they are a far cry from practical application. However, the potential remains that quantum computers one day could perform, quickly and easily, calculations that are incredibly time-consuming on conventional computers. Several key advancements have been made in quantum computing in the last few years.

2. The Potential and Power of Quantum Computing

In a traditional computer, information is encoded in a series of bits, and these bits are manipulated via Boolean logic gates arranged in succession to produce an end result. Similarly, a quantum computer manipulates qubits by executing a series of quantum gates, each a **unitary transformation** acting on a single qubit or pair of qubits. In applying these gates in succession, a quantum computer can perform a complicated unitary transformation to a set of qubits in some initial state. The qubits can then be measured, with this measurement serving as the final computational result. This similarity in calculation between a classical and quantum computer affords that in theory, a classical computer can accurately simulate a quantum computer. In other words, a classical computer would be able to do anything a quantum computer can. So why bother with quantum computers? Although a classical computer can theoretically simulate a quantum computer, it is incredibly inefficient, so much so that a classical computer is effectively incapable of performing many tasks that a quantum computer could perform with ease. The simulation of a quantum computer on a classical one is a computationally hard problem because the correlations among quantum bits are qualitatively different from correlations among classical bits, as first explained by John Bell. Take for example a system of only a few hundred qubits, this exists in a Hilbert space of dimension $\sim10^{90}$ that in simulation would require a classical computer to work with exponentially large matrices (to perform calculations on each individual state, which is also represented as a matrix), meaning it would take an exponentially longer time than even a primitive quantum computer.

Richard Feynman was among the first to recognize the potential in quantum superposition for solving such problems much much faster. For example, a system of 500 qubits, which is impossible to simulate classically, represents a quantum superposition of as many as 2^{500} states. Each state would be classically equivalent to a single list of 500 1's and 0's. Any quantum operation on that system — a particular pulse of radio waves, for instance, whose action might be to execute a controlled — NOT operation on the 100th and 101st qubits — would simultaneously operate on all 2^{500} states. Hence with one fell swoop, one tick of the computer clock, a quantum operation could compute not just on one machine state, as serial computers do, but on 2^{500} machine states at once! Eventually, however, observing the system would cause it to collapse into a single quantum state corresponding to a single answer, a single list of 500 1's and 0's, as dictated by the measurement axiom of quantum mechanics. The reason this is an exciting result is because this answer, derived from the massive **quantum parallelism** achieved through superposition, is the equivalent of performing the same operation on a classical super computer with $\sim10^{150}$ separate processors (which is of course impossible)!

Early investigators in this field were naturally excited by the potential of such immense computing power, and soon after realizing its potential, the hunt was on to find something interesting for a quantum computer to do. Peter Shor, a research and computer scientist at AT&T's Bell Laboratories in New Jersey, provided such an application by devising the first quantum computer algorithm. Shor's algorithm harnesses the power of quantum superposition to rapidly factor very large numbers (on the order $\sim10^{200}$ digits and greater) in a matter of seconds. The premier application of a quantum computer capable of implementing this algorithm lies in the field of encryption, where one common (and best) encryption code, known as RSA, relies heavily on the difficulty of factoring very large composite numbers into their primes. A computer which can do this easily is naturally of great interest to numerous government agencies that use RSA — previously considered to be "uncrackable" — and anyone interested in electronic and financial privacy.

Encryption, however, is only one application of a quantum computer. In addition, Shor has put together a toolbox of mathematical operations that can only be performed on a quantum computer, many of which he used in his factorization algorithm. Furthermore, Feynman asserted that a quantum computer could function as a kind of simulator for quantum physics, potentially opening the doors to many discoveries in the field. Currently the power and capability of a quantum computer is primarily theoretical speculation; the advent of the first fully functional quantum computer will undoubtedly bring many new and exciting applications.

3. Obstacles and Research

The field of quantum information processing has made numerous promising advancements since its conception, including the building of two- and three- qubit quantum computers capable of some simple arithmetic and data sorting. However, a few potentially large obstacles still remain that prevent us from "just building one," or more precisely, building a quantum computer that can rival today's modern digital computer. Among these difficulties, error correction, **decoherence**, and hardware architecture are probably the most formidable. Error correction is rather self-explanatory, but what errors need correction? The answer is primarily those errors that arise as a direct result of decoherence, or the tendency of a quantum computer to decay from a given quantum state into an incoherent state as it interacts, or entangles, with the state of the environment. These interactions between the environment and qubits are unavoidable, and induce the breakdown of information stored in the quantum computer, and thus errors in computation. Before any quantum computer will be capable of solving hard problems, research must devise a way to maintain decoherence and other potential sources of error at an acceptable level. Thanks to the theory (and now reality) of quantum error correction, first proposed in 1995 and continually developed since, small scale quantum computers have been built and the prospects of large quantum computers are looking up. Probably the most important idea in this field is the application of error correction in **phase coherence** as a means to extract information and reduce error in a quantum system without actually measuring that system. In 1998, researches at Los Alamos National Laboratory and MIT led by Raymond Laflamme managed to spread a single bit of quantum information (qubit) across three nuclear spins in each molecule of a liquid solution of **alanine** or **trichloroethylene** molecules. They accomplished this using the techniques of nuclear magnetic resonance (NMR). This experiment is significant because spreading out the information actually made it harder to corrupt. Quantum mechanics tells us that directly measuring the state of a qubit invariably destroys the superposition of states in which it exists, forcing it to become either a 0 or 1. The technique of spreading out the information allows researchers to utilize the property of entanglement to study the interactions between states as an indirect method for analyzing the quantum information. Rather than a direct measurement, the group compared the spins to see if any new differences arose between them without learning the information itself. This technique gave them the ability to detect and fix errors in a qubit's phase coherence, and thus maintain a higher level of coherence in the quantum system. This milestone has provided argument against skeptics, and hope for believers.

At this point, only a few of the benefits of quantum computation and quantum computers are readily obvious, but before more possibilities are uncovered theory must be put to the test. In order to do this, devices capable of quantum computation must be constructed. Quantum computing hardware is, however, still in its infancy. As a result of several significant experiments, nuclear magnetic resonance (NMR) has become the most popular component in quantum hardware architecture. Only within 1999 a group from Los Alamos National Laboratory and MIT constructed the first experimental demonstrations of a quantum computer using nuclear magnetic resonance (NMR) technology. Currently, research is underway to discover methods for battling the destructive effects of decoherence, to develop an optimal hardware architecture for designing and building a quantum computer, and to further uncover quantum algorithms to utilize the immense computing power available in these devices. Naturally this pursuit is intimately related to quantum error correction codes and quantum algorithms, so a number of groups are doing simultaneous research in a number of these fields. To date, designs have involved **ion traps**, cavity quantum electrodynamics (QED), and NMR. Though these devices have had mild success in performing interesting experiments, the technologies each have serious limitations. Ion trap computers are limited in speed by the **vibration** frequency of the modes in the trap. NMR devices have an exponential attenuation of signal to noise as the number of qubits in a system increases. Cavity QED is slightly more promising; however, it still has only been demonstrated with a few qubits. The future of quantum computer hardware architecture is likely to be very different from what we know today; however, the current research has helped to provide insight as to what obstacles the future will hold for these devices.

10.1.2 Quantum Cryptography

Quantum cryptography is the science of exploiting quantum mechanical properties to perform cryptographic tasks. The best known example of quantum cryptography is quantum key distribution which offers an information-theoretically secure solution to the key exchange problem. Currently used popular public-key encryption and signature schemes (e.g., RSA and ElGamal) can be broken by quantum adversaries. The advantage of quantum cryptography lies in the fact that it allows the completion of various cryptographic tasks that are proven or conjectured to be impossible using only classical (i.e. non-quantum) communication. For example, it is impossible to copy data encoded in a quantum state and the very act of reading data encoded in a quantum state changes the state. This is used to detect eavesdropping in quantum key distribution.

Quantum cryptography uses **Heisenberg's uncertainty principle** formulated in 1927 and the **No-cloning theorem** first articulated by Wootters and Zurek and Dieks in 1982. Werner

Heisenberg discovered one of the fundamental principles of quantum mechanics: "At the instant at which the position of the electron is known, its momentum therefore can be known only up to magnitudes which correspond to that discontinuous change; thus, the more precisely the position is determined, the less precisely the momentum is known, and conversely" This simply means that observation of quanta changes its behavior. By measuring the velocity of quanta we would affect it, and thereby change its position; if we want to find a quant's position, we are forced to change its velocity. Therefore, we cannot measure a quantum system's characteristics without changing it and we cannot record all characteristics of a quantum system before those characteristics are measured. The No-cloning theorem demonstrates that it is impossible to create a copy of an arbitrary unknown quantum state. This makes unobserved eavesdropping impossible because it will be quickly detected, thus greatly improving assurance that the communicated data remains private.

The most well-known and developed application of quantum cryptography is quantum key distribution, which is the process of using quantum communication to establish a shared key between two parties (Alice and Bob, for example) without a third party (Eve) learning anything about that key, even if Eve can eavesdrop on all communication between Alice and Bob. If Eve tries to learn information about the key being established, key establishment will fail causing Alice and Bob to notice. Once the key is established, it is then typically used for encrypted communication using classical techniques. For instance, the exchanged key could be used as for symmetric cryptography.

Following the discovery of quantum key distribution and its unconditional security, researchers tried to achieve other cryptographic tasks with unconditional security. One such task was **commitment**. A commitment scheme allows a party Alice to fix a certain value (to "commit") in such a way that Alice cannot change that value while at the same time ensuring that the recipient Bob cannot learn anything about that value until Alice reveals it. Such commitment schemes are commonly used in cryptographic protocols. In the quantum setting, they would be particularly useful: Crépeau and Kilian showed that from a commitment and a quantum channel, one can construct an unconditionally secure protocol for performing so-called **oblivious transfer**. Oblivious transfer, on the other hand, had been shown by Kilian to allow implementation of almost any distributed computation in a secure way (so-called secure multi-party computation). (Notice that here we are a bit imprecise: The results by Crépeau and Kilian together do not directly imply that given a commitment and a quantum channel one can perform secure multi-party computation. This is because the results do not guarantee "composability", that is, when plugging them together, one might lose security.

Unfortunately, early quantum commitment protocols were shown to be flawed. In fact, Mayers showed that (unconditionally secure) quantum commitment is impossible: a computationally unlimited attacker can break any quantum commitment protocol.

Yet, the result by Mayers does not preclude the possibility of constructing quantum commitment protocols (and thus secure multi-party computation protocols) under assumptions that they are much weaker than the assumptions needed for commitment protocols that do not use quantum communication. The bounded quantum storage model described below is an example for a setting in which quantum communication can be used to construct commitment protocols. A breakthrough in November 2013 offers "unconditional" security of information by harnessing quantum theory and relativity, which has been successfully demonstrated on a global scale for the first time.

The security of quantum key distribution can be proven mathematically without imposing any restrictions on the abilities of an eavesdropper, something not possible with classical key distribution. This is usually described as "unconditional security", although there are some minimal assumptions required, including that the laws of quantum mechanics apply and that Alice and Bob are able to authenticate each other, i.e. Eve should not be able to impersonate Alice or Bob as otherwise a **man-in-the-middle attack** would be possible.

One aspect of quantum key distribution is that it is secure against quantum computers. Its strength does not depend on mathematical complexity, like post-quantum cryptography, but on physical principles.

The goal of position-based quantum cryptography is to use the geographical location of a player as its (only) **credential**. For example, one wants to send a message to a player at a specified position with the guarantee that it can only be read if the receiving party is located at that particular position. In the basic task of position-verification, a player, Alice, wants to convince the (honest) verifiers that she is located at a particular point. It has been shown by Chandran et al. that position-verification using classical protocols is impossible against **colluding adversaries** (who control all positions except the prover's claimed position). Under various restrictions on the adversaries, schemes are possible.

Under the name of "quantum tagging", the first position-based quantum schemes have been investigated in 2002 by Kent. A US-patent was granted in 2006, but the results only appeared in the scientific literature in 2010. After several other quantum protocols for position verification have been suggested in 2010, Buhrman et al. were able to show a general impossibility result: using an enormous amount of quantum entanglement (they use a doubly exponential number of EPR pairs, in the number of qubits the honest player operates on), colluding adversaries are always able to make it look to the verifiers as if they were at the claimed position. However, this result does not exclude the possibility of practical schemes in the bounded- or noisy-quantum-storage model (see above). Later Beigi and König improved the amount of EPR pairs needed in the general attack against position-verification protocols to exponential. They also showed that a particular protocol remains secure against adversaries who controls only a linear amount of EPR pairs.

Quantum computers may become a technological reality; it is therefore important to study cryptographic schemes used against adversaries with access to a quantum computer. The study of such schemes is often referred to as post-quantum cryptography. The need for post-quantum cryptography arises from the fact that many popular encryption and signature schemes (such as

RSA and its variants, and schemes based on elliptic curves) can be broken using Shor's algorithm for factoring and computing discrete logarithms on a quantum computer.

There is also research into how existing cryptographic techniques have to be modified to be able to cope with quantum adversaries. For example, when trying to develop **zero-knowledge proof systems** that are secure against quantum adversaries, new techniques need to be used: In a classical setting, the analysis of a zero-knowledge proof system usually involves "rewinding", a technique that makes it necessary to copy the internal state of the adversary. In a quantum setting, copying a state is not always possible (no-cloning theorem); a variant of the rewinding technique has to be used.

Post quantum algorithms are also called "**quantum resistant**", because it is not known or provable that there will not be potential future quantum attacks against them. Even though they are not vulnerable to Shor's algorithm, the NSA is announcing plans to transition to quantum resistant algorithms. The National Institute of Security and Technology (NIST) believes that it is time to think of quantum-safe primitives.

10.2 Deep Learning①

10.2.1 Introduction

In the early days of artificial intelligence, the field rapidly tackled and solved problems that are intellectually difficult for human beings but relatively straightforward for computers — problems that can be described by a list of formal, mathematical rules. The true challenge to artificial intelligence proved to be solving the tasks that are easy for people to perform but hard for people to describe formally — problems that we solve intuitively, that feel automatic, like recognizing spoken words or faces in images.

This solution is to allow computers to learn from experience and understand the world in terms of a hierarchy of concepts, with each concept defined in terms of its relation to simpler concepts. By gathering knowledge from experience, this approach avoids the need for human operators to formally specify all of the knowledge that the computer needs. The hierarchy of concepts allows the computer to learn complicated concepts by building them out of simpler ones. If we draw a graph showing how these concepts are built on top of each other, the graph is deep, with many layers. For this reason, we call this approach to AI deep learning.

Many of the early successes of AI took place in relatively sterile and formal environments and did not require computers to have much knowledge about the world. For example, IBM's Deep Blue chess-playing system defeated world champion Garry Kasparov in 1997 (Hsu, 2002). Chess is of course a very simple world, containing only sixty-four locations and thirty-two pieces that can move in only rigidly circumscribed ways. Devising a successful chess strategy is a

① Yoshua Bengio, *et al*. Deep Learning. MIT Press, 2016.

tremendous accomplishment, but the challenge is not due to the difficulty of describing the relevant concepts to the computer. Chess can be completely described by a very brief list of completely formal rules, easily provided ahead of time by the programmer.

Ironically, abstract and formal tasks that are among the most difficult mental undertakings for a human being are among the easiest for a computer. Computers have long been able to defeat even the best human chess player, but are only recently matching some of the abilities of average human beings to recognize objects or speech. A person's everyday life requires an immense amount of knowledge about the world, and much of this knowledge is subjective and intuitive, and therefore difficult to articulate in a formal way. Computers need to capture this same knowledge in order to behave in an intelligent way. One of the key challenges in artificial intelligence is how to get this informal knowledge into a computer.

Several artificial intelligence projects have sought to hard-code knowledge about the world in formal languages. A computer can reason about statements in these formal languages automatically using logical inference rules. This is known as the knowledge base approach to artificial intelligence. None of these projects has lead to a major success. One of the most famous such projects is Cyc (Lenat and Guha, 1989). Cyc is an inference engine and a database of statements in a language called CycL. These statements are entered by a staff of human supervisors. It is an unwieldy process. People struggle to devise formal rules with enough complexity to accurately describe the world. For example, Cyc failed to understand a story about a person named Fred shaving in the morning (Linde, 1992). Its inference engine detected an inconsistency in the story: it knew that people do not have electrical parts, but because Fred was holding an electric razor, it believed the entity "FredWhileShaving" contained electrical parts. It therefore asked whether Fred was still a person while he was shaving.

The difficulties faced by systems relying on hard-coded knowledge suggest that AI systems need the ability to acquire their own knowledge, by extracting patterns from raw data. This capability is known as machine learning. The introduction of machine learning allowed computers to tackle problems involving knowledge of the real world and make decisions that appear subjective. A simple machine learning algorithm called logistic regression can determine whether to recommend cesarean delivery (Mor-Yosef et al., 1990). A simple machine learning algorithm called naive Bayes can separate legitimate e-mail from spam e-mail.

The performance of these simple machine learning algorithms depends heavily on the representation of the data they are given. For example, when logistic regression is used to recommend cesarean delivery, the AI system does not examine the patient directly. Instead, the doctor tells the system several pieces of relevant information, such as the presence or absence of a uterine scar. Each piece of information included in the representation of the patient is known as a feature. Logistic regression learns how each of these features of the patient correlates with various outcomes. However, it cannot influence the way that the features are defined in any way. If logistic regression was given a 3-D MRI image of the patient, rather than the doctor's formalized report, it would not be able to make useful predictions. Individual voxels[①] in an MRI scan have negligible correlation with any complications that might occur during delivery.

This dependence on representations is a general phenomenon that appears throughout computer science and even daily life. In computer science, operations such as searching a collection of data can proceed exponentially faster if the collection is structured and indexed intelligently. People can easily perform arithmetic on Arabic numerals, but find arithmetic on Roman numerals much more time consuming. It is not surprising that the choice of representation has an enormous effect on the performance of machine learning algorithms.

Many artificial intelligence tasks can be solved by designing the right set of features to extract for that task, then providing these features to a simple machine learning algorithm. For example, a useful feature for speaker identification from sound is the pitch. The pitch can be formally specified — it is the lowest frequency major peak of the spectrogram. It is useful for speaker identification because it is determined by the size of the vocal tract, and therefore gives a strong clue as to whether the speaker is a man, woman, or child.

However, for many tasks, it is difficult to know what features should be extracted. For example, suppose that we would like to write a program to detect cars in photographs. We know that cars have wheels, so we might like to use the presence of a wheel as a feature. Unfortunately, it is difficult to describe exactly what a wheel looks like in terms of pixel values. A wheel has a simple geometric shape but its image may be complicated by shadows falling on the wheel, the sun glaring off the metal parts of the wheel, the fender of the car or an object in the foreground obscuring part of the wheel, and so on.

One solution to this problem is to use machine learning to discover not only the mapping from representation to output but also the representation itself.

This approach is known as representation learning. Learned representations often result in much better performance than can be obtained with hand-designed representations. They also allow AI systems to rapidly adapt to new tasks, with minimal human intervention. A representation learning algorithm can discover a good set of features for a simple task in minutes, or a complex task in hours to months. Manually designing features for a complex task requires a great deal of human time and effort; it can take decades for an entire community of researchers.

The quintessential example of a representation learning algorithm is the autoencoder.

An autoencoder is the combination of an encoder function that converts the input data into a different representation, and a decoder function that converts the new representation back into the original format. Autoencoders are trained to preserve as much information as possible when an input is run through the encoder and then the decoder, but are also trained to make the new representation have various nice properties. Different kinds of autoencoders aim to achieve different kinds of properties.

---

① A voxel is the value at a single point in a 3-D scan, much as a pixel as the value at a single point in an image.

When designing features or algorithms for learning features, our goal is usually to separate the factors of variation that explain the observed data. In this context, we use the word "factors" simply to refer to separate sources of influence; the factors are usually not combined by multiplication. Such factors are often not quantities that are directly observed but they may exist either as unobserved objects or forces in the physical world that affect observable quantities, or they are constructs in the human mind that provide useful simplifying explanations or inferred causes of the observed data. They can be thought of as concepts or abstractions that help us make sense of the rich variability in the data. When analyzing a speech recording, the factors of variation include the speaker's age and sex, their accent, and the words that they are speaking. When analyzing an image of a car, the factors of variation include the position of the car, its color, and the angle and brightness of the sun.

(10-2) A major source of difficulty in many real-world artificial intelligence applications is that many of the factors of variation influence every single piece of data we are able to observe. The individual pixels in an image of a red car might be very close to black at night. The shape of the car's silhouette depends on the viewing angle. Most applications require us to disentangle the factors of variation and discard the ones that we do not care about.

Of course, it can be very difficult to extract such high-level, abstract features from raw data. Many of these factors of variation, such as a speaker's accent, can only be identified using sophisticated, nearly human-level understanding of the data. When it is nearly as difficult to obtain a representation as to solve the original problem, representation learning does not, at first glance, seem to help us.

Deep learning solves this central problem in representation learning by introducing representations that are expressed in terms of other, simpler representations. Deep learning allows the computer to build complex concepts out of simpler concepts.

The quintessential example of a deep learning model is the **multilayer perceptron** (MLP). A multilayer perceptron is just a mathematical function mapping some set of input values to output values. The function is formed by composing many simpler functions. We can think of each application of a different mathematical function as providing a new representation of the input.

The idea of learning the right representation for the data provides one perspective on deep learning. Another perspective on deep learning is that it allows the computer to learn a multi-step computer program. Each layer of the representation can be thought of as the state of the computer's memory after executing another set of instructions in parallel. Networks with greater depth can execute more instructions in sequence. Being able to execute instructions sequentially offers great power because later instructions can refer back to the results of earlier instructions. According to this view of deep learning, not all of the information in a layer's representation of the input necessarily encodes factors of variation that explain the input. The representation is also used to store state information that helps to execute a program that can make sense of the input. This state information could be analogous to a counter or pointer in a traditional computer program. It has nothing to do with the content of the input specifically, but it helps the model to organize its processing.

There are two main ways of measuring the depth of a model.

The first view is based on the number of sequential instructions that must be executed to evaluate the architecture. We can think of this as the length longest path through a flow chart that describes how to compute each of the model's outputs given its inputs. Just as two equivalent computer programs will have different lengths depending on which language the program is written in, the same function may be drawn as a **flow chart** with different depths depending on which functions we allow to be used as individual steps in the flow chart.

Another approach used by **deep probabilistic models**, examples not the depth of the computational graph but the depth of the graph describing how concepts are related to each other. (10-3) In this case, the depth of the flow chart of the computations needed to compute the representation of each concept may be much deeper than the graph of the concepts themselves. This is because the system's understanding of the simpler concepts can be refined given information about the more complex concepts. For example, an AI system observing an image of a face with one eye in shadow may initially only see one eye. After detecting that a face is present, it can then infer that a second eye is probably present as well. In this case, the graph of concepts only includes two layers — a layer for eyes and a layer for faces — but the graph of computations includes $2n$ layers if we refine our estimate of each concept given the other n times.

Because it is not always clear which of these two views — the depth of the computational graph, or the depth of the probabilistic modeling graph — is most relevant, and because different people choose different sets of smallest elements from which to construct their graphs, there is no single correct value for the depth of an architecture, just as there is no single correct value for length of a computer program. Nor is there a consensus about how much depth a model requires to qualify as "deep". However, deep learning can safely be regarded as the study of models that either involve a greater amount of composition of learned functions or learned concepts than traditional machine learning does.

To summarize, deep learning, is an approach to AI. Specifically, it is a type of machine learning, a technique that allows computer systems to improve with experience and data. Machine learning is considered by most computer scientists to be the only viable approach to build AI systems that can operate in complicated, real-world environments. Deep learning is a particular kind of machine learning that achieves great power and flexibility by learning to represent the world as a nested hierarchy of concepts and representations, with each concept defined in relation to simpler concepts, and more abstract representations computed in terms of less abstract ones.

10.2.2 Historical Trends in Deep Learning

It is easiest to understand deep learning with some historical context. Rather than providing a detailed history of deep learning, we identify a few key trends:

(1) Deep learning has had a long and rich history, but has gone by many names reflecting different philosophical viewpoints, and has waxed and waned in popularity.

(2) Deep learning has become more useful as the amount of available training data has increased.

(3) Deep learning models have grown in size over time as computer hardware and software infrastructure for deep learning has improved.

(4) Deep learning has solved increasingly complicated applications with increasing accuracy over time.

We expect that many readers have heard of deep learning as an exciting new technology, and are surprised to see a mention of "history" in a book about an emerging field. In fact, deep learning has a long and rich history.

Deep learning only appears to be new, because it was relatively unpopular for several years preceding its current popularity, and because it has gone through many different names. While the term "deep learning" is relatively new, the field dates back to the 1950s. The field has been rebranded many times, reflecting the influence of different researchers and different perspectives.

A comprehensive history of deep learning is beyond the scope of this book. However, some basic context is useful for understanding deep learning. Broadly speaking, there have been three waves of development of deep learning: deep learning known as **cybernetics** in the 1940s—1960s, deep learning known as **connectionism** in the 1980s—1990s, and the current resurgence under the name deep learning beginning in 2006.

Some of the earliest learning algorithms we recognize today were intended to be computational models of biological learning, i.e. models of how learning happens or could happen in the brain. As a result, one of the names that deep learning has gone by is artificial neural networks (ANNs). The corresponding perspective on deep learning models is that they are engineered systems inspired by the biological brain (whether the human brain or the brain of another animal). The neural perspective on deep learning is motivated by two main ideas. One idea is that the brain provides a proof by example that intelligent behavior is possible, and a conceptually straightforward path to building intelligence is to **reverse engineering** the computational principles behind the brain and duplicate its functionality. Another perspective is that it would be deeply interesting to understand the brain and the principles that underlie human intelligence, so machine learning models that shed light on these basic scientific questions are useful apart from their ability to solve engineering applications.

The modern term "deep learning" goes beyond the **neuroscientific** perspective on the current breed of machine learning models. It appeals to a more general principle of learning multiple levels of composition, which can be applied in machine learning frameworks that are not necessarily neurally inspired.

Today, neuroscience is regarded as an important source of inspiration for deep learning researchers, but it is no longer the predominant guide for the field. The main reason for the diminished role of neuroscience in deep learning research today is that we simply do not have enough information about the brain to use it as a guide. To obtain a deep understanding of the actual algorithms used by the brain, we would need to be able to monitor the activity of (at the very least) thousands of interconnected neurons simultaneously. Because we are not able to do this, we are far from understanding even some of the most simple and well-studied parts of the brain.

Neuroscience has given us a reason to hope that a single deep learning algorithm can solve many different tasks. Neuroscientists have found that ferrets can learn to "see" with the **auditory processing region** of their brain if their brains are rewired to send visual signals to that area. This suggests that much of the mammalian brain might use a single algorithm to solve most of the different tasks that the brain solves. Before this hypothesis, machine learning research was more fragmented, with different communities of researchers studying natural language processing, vision, motion planning, and speech recognition.

Today, these application communities are still separate, but it is common for deep learning research groups to study many or even all of these application areas simultaneously.

Media accounts often emphasize the similarity of deep learning to the brain. While it is true that deep learning researchers are more likely to cite the brain as an influence than researchers working in other machine learning fields such as kernel machines or Bayesian statistics, one should not view deep learning as an attempt to simulate the brain. Modern deep learning draws inspiration from many fields, especially applied math fundamentals like linear algebra, probability, information theory, and numerical optimization. While some deep learning researchers cite neuroscience as an important influence, others are not concerned with neuroscience at all.

It is worth noting that the effort to understand how the brain works on an algorithmic level is alive and well. This endeavor is primarily known as "computational neuroscience" and is a separate field of study from deep learning. It is common for researchers to move back and forth between both fields. The field of deep learning is primarily concerned with how to build computer systems that are able to successfully solve tasks requiring intelligence, while the field of computational neuroscience is primarily concerned with building more accurate models of how the brain actually works.

In the 1980s, the second wave of neural network research emerged in great part via a movement called connectionism or parallel distributed processing. Connectionism arose in the context of cognitive science. Cognitive science is an interdisciplinary approach to understanding the mind, combining multiple different levels of analysis. During the early 1980s, most cognitive scientists studied models of symbolic reasoning. Despite their popularity, symbolic models were difficult to explain in terms of how the brain could actually implement them using neurons. The connectionists began to study models of cognition that could actually be grounded in neural implementations, reviving many ideas dating back to the work of psychologist Donald Hebb in the 1940s.

The central idea in connectionism is that a large number of simple computational units can achieve intelligent behavior when networked together. This insight applies equally to neurons in biological nervous systems and to hidden units in computational models.

Several key concepts arose during the connectionism movement of the 1980s that remain central to today's deep learning.

One of these concepts is that of **distributed representation**. This is the idea that each input to a system should be represented by many features, and each feature should be involved in the representation of many possible inputs. For example, suppose we have a vision system that can recognize cars, trucks, and birds and these objects can each be red, green, or blue. One way of representing these inputs would be to have a separate neuron or hidden unit that activates for each of the nine possible combinations: red truck, red car, red bird, green truck, and so on.

This requires nine different neurons, and each neuron must independently learn the concept of color and object identity. One way to improve on this situation is to use a distributed representation, with three neurons describing the color and three neurons describing the object identity. This requires only six neurons total instead of nine, and the neuron describing redness is able to learn about redness from images of cars, trucks, and birds, not only from images of one specific category of objects.

Another major accomplishment of the connectionist movement was the successful use of back-propagation to train deep neural networks with internal representations and the popularization of the **back-propagation** algorithm. This algorithm has waxed and waned in popularity but is currently the dominant approach to training deep models.

The second wave of neural networks research lasted until the mid-1990s. At that point, the popularity of neural networks declined again. This was in part due to a negative reaction to the failure of neural networks to fulfill excessive promises made by a variety of people seeking investment in neural network-based ventures, but also due to improvements in other fields of machine learning: **kernel machines** and graphical models.

Kernel machines enjoy many nice theoretical guarantees. In particular, training a kernel machine is a **convex optimization problem** which means that the training process can be guaranteed to find the optimal model efficiently. This made kernel machines very amenable to software implementations that "just work" without much need for the human operator to understand the underlying ideas.

During this time, neural networks continued to obtain impressive performance on some tasks. The Canadian Institute for Advanced Research (CIFAR) helped to keep neural networks research alive via its Neural Computation and Adaptive Perception research initiative. At this point in time, deep networks were generally believed to be very difficult to train. We now know that algorithms that have existed since the 1980s work quite well, but this was not apparent circa 2006. The issue is perhaps simply that these algorithms were too computationally costly to allow much experimentation with the hardware available at the time.

The third wave of neural networks research began with a breakthrough in 2006. Geoffrey

Hinton showed that a kind of neural network called a **deep belief network** could be efficiently trained using a strategy called greedy layer-wise pretraining. This wave of neural networks research popularized the use of the term deep learning to emphasize that researchers were now able to train deeper neural networks than had been possible before, and to emphasize the theoretical importance of depth. Deep neural networks displaced kernel machines with manually designed features for several important application areas during this time — in part because the time and memory cost of training a kernel machine is quadratic in the size of the dataset, and datasets grew to be large enough for this cost to outweigh the benefits of convex optimization.

This third wave of popularity of neural networks continues, though the focus of deep learning research has changed dramatically within the time of this wave. The third wave began with a focus on new unsupervised learning techniques and the ability of deep models to generalize well from small datasets, but today there is more interest in much older supervised learning algorithms and the ability of deep models to leverage large labeled datasets.

10.3 Cloud Computing①

In 1969, Leonard Kleinrock, one of the chief scientists of the original Advanced Research Projects Agency Network (ARPANET), which seeded the Internet, said:

As of now, computer networks are still in their infancy, but as they grow up and become sophisticated, we will probably see the spread of "computer utilities" which, like present electric and telephone utilities, will service individual homes and offices across the country.

This vision of computing utilities based on a service-provisioning model anticipated the massive transformation of the entire computing industry in the 21st century, whereby computing services will be readily available on demand, just as other utility services such as water, electricity, telephone, and gas are available in today's society. Similarly, users (consumers) need to pay providers only when they access the computing services. In addition, consumers no longer need to invest heavily or encounter difficulties in building and maintaining complex IT infrastructure.

In such a model, users access services based on their requirements without regard to where the services are hosted. This model has been referred to as utility computing or, recently (since 2007), as **cloud computing**. The latter term often denotes the infrastructure as a "cloud" from which businesses and users can access applications as services from anywhere in the world and on demand.

Hence, cloud computing can be classified as a new paradigm for the dynamic provisioning of computing services supported by state-of-the-art data centers employing virtualization technologies for consolidation and effective utilization of resources.

Cloud computing allows renting infrastructure, runtime environments, and services on a

① Excerpted from *Mastering Cloud Computing*

pay-per-use basis. This principle finds several practical applications and then gives different images of cloud computing to different people. Chief information and technology officers of large enterprises see opportunities for scaling their infrastructure on demand and sizing it according to their business needs. End users leveraging cloud computing services can access their documents and data anytime, anywhere, and from any device connected to the Internet. Many other points of view exist. One of the most diffuse views of cloud computing can be summarized as follows:

I don't care where my servers are, who manages them, where my documents are stored, or where my applications are hosted. I just want them always available and access them from any device connected through Internet. And I am willing to pay for this service for as a long as I need it.

The concept expressed above has strong similarities to the way we use other services, such as water and electricity. In other words, cloud computing turns IT services into utilities. Such a delivery model is made possible by the effective composition of several technologies, which have reached the appropriate maturity level. Web 2.0 technologies play a central role in making cloud computing an attractive opportunity for building computing systems. They have transformed the Internet into a rich application and service delivery platform, mature enough to serve complex needs. Service orientation allows cloud computing to deliver its capabilities with familiar abstractions, while virtualization confers on cloud computing the necessary degree of customization, control, and flexibility for building production and enterprise systems.

Besides being an extremely flexible environment for building new systems and applications, cloud computing also provides an opportunity for integrating additional capacity or new features into existing systems. The use of dynamically provisioned IT resources constitutes a more attractive opportunity than buying additional infrastructure and software, the sizing of which can be difficult to estimate and the needs of which are limited in time. This is one of the most important advantages of cloud computing, which has made it a popular phenomenon. With the wide deployment of cloud computing systems, the foundation technologies and systems enabling them are becoming consolidated and standardized. This is a fundamental step in the realization of the long-term vision for cloud computing, which provides an open environment where computing, storage, and other services are traded as computing utilities.

10.3.1 The Vision of Cloud Computing

Cloud computing allows anyone with a credit card to provision **virtual hardware**, runtime environments, and services. These are used for as long as needed, with no up-front commitments required.

The entire stack of a computing system is transformed into a collection of utilities, which can be provisioned and composed together to deploy systems in hours rather than days and with virtually no maintenance costs. This opportunity, initially met with skepticism, has now become a practice across several application domains and business sectors. The demand has fasttracked technical development and enriched the set of services offered, which have also become more sophisticated and cheaper.

Despite its evolution, the use of cloud computing is often limited to a single service at a time or, more commonly, a set of related services offered by the same vendor. Previously, the lack of effective standardization efforts made it difficult to move hosted services from one vendor to another. The long-term vision of cloud computing is that IT services are traded as utilities in an open market, without technological and legal barriers. In this cloud marketplace, cloud service providers and consumers, trading cloud services as utilities, play a central role.

Many of the technological elements contributing to this vision already exist. Different stakeholders leverage clouds for a variety of services. The need for ubiquitous storage and compute power on demand is the most common reason to consider cloud computing. A scalable runtime for applications is an attractive option for application and system developers that do not have infrastructure or cannot afford any further expansion of existing infrastructure. The capability for Web based access to documents and their processing using sophisticated applications is one of the appealing factors for end users.

In all these cases, the discovery of such services is mostly done by human intervention: a person (or a team of people) looks over the Internet to identify offerings that meet his or her needs. We imagine that in the near future it will be possible to find the solution that matches our needs by simply entering our request in a global digital market that trades cloud computing services. The existence of such a market will enable the automation of the discovery process and its integration into existing software systems, thus allowing users to transparently leverage cloud resources in their applications and systems. The existence of a global platform for trading cloud services will also help service providers become more visible and therefore potentially increase their revenue. A global cloud market also reduces the barriers between service consumers and providers: it is no longer necessary to belong to only one of these two categories. For example, a cloud provider might become a consumer of a competitor service in order to fulfill its own promises to customers.

These are all possibilities that are introduced with the establishment of a global cloud computing marketplace and by defining effective standards for the unified representation of cloud services as well as the interaction among different cloud technologies. A considerable shift toward cloud computing has already been registered, and its rapid adoption facilitates its consolidation.

Moreover, by concentrating the core capabilities of cloud computing into large **datacenters**, it is possible to reduce or remove the need for any technical infrastructure on the service consumer side.

This approach provides opportunities for optimizing datacenter facilities and fully utilizing their capabilities to serve multiple users. This consolidation model will reduce the waste of energy and carbon emissions, thus contributing to a greener IT on one end and increasing revenue on the other end.

10.3.2 Defining a Cloud

Cloud computing has become a popular buzzword; it has been widely used to refer to different technologies, services, and concepts. It is often associated with virtualized infrastructure or hardware on demand, utility computing, **IT outsourcing**, platform and software as a service, and many other things that now are the focus of the IT industry.

The term cloud has historically been used in the telecommunications industry as an abstraction of the network in system diagrams. It then became the symbol of the most popular computer network: the Internet. This meaning also applies to cloud computing, which refers to an Internet-centric way of computing. The Internet plays a fundamental role in cloud computing, since it represents either the medium or the platform through which many cloud computing services are delivered and made accessible. This aspect is also reflected in the definition given by Armbrust et al. :

Cloud computing refers to both the applications delivered as services over the Internet and the hardware and system software in the datacenters that provide those services.

This definition describes cloud computing as a phenomenon touching on the entire stack: from the underlying hardware to the high-level software services and applications. It introduces the concept of everything as a service, mostly referred as XaaS (standing for X-as-a-Service), where the different components of a system — IT infrastructure, development platforms, databases, and so on — can be delivered, measured, and consequently priced as a service. This new approach significantly influences not only the way that we build software but also the way we deploy it, make it accessible, and design our IT infrastructure, and even the way companies allocate the costs for IT needs. The approach fostered by cloud computing is global: it covers both the needs of a single user hosting documents in the cloud and the ones of a CIO (Chief Information Officer) deciding to deploy part of or the entire corporate IT infrastructure in the public cloud. This notion of multiple parties using a shared cloud computing environment is highlighted in a definition proposed by the U.S. National Institute of Standards and Technology (NIST):

(10-4) Cloud computing is a model for enabling ubiquitous, convenient, on-demand network access to a shared pool of configurable computing resources (e.g., networks, servers, storage, applications, and services) that can be rapidly provisioned and released with minimal management effort or service provider interaction.

Another important aspect of cloud computing is its **utility-oriented** approach. More than any other trend in distributed computing, cloud computing focuses on delivering services with a given pricing model, in most cases a "**pay-per-use**" strategy. It makes it possible to access online storage, rent virtual hardware, or use development platforms and pay only for their effective usage, with no or minimal up-front costs. All these operations can be performed and billed simply by entering the credit card details and accessing the exposed services through a Web browser. This helps us provide a different and more practical characterization of cloud computing. According to Reese[①], we can define three criteria to discriminate whether a service is delivered in the cloud computing style:

(1) The service is accessible via a Web browser (nonproprietary) or a Web services application programming interface (API).

(2) Zero capital expenditure is necessary to get.

(3) You pay only for what you use as you use it.

Even though many cloud computing services are freely available for single users, enterpriseclass services are delivered according a specific pricing scheme. In this case users subscribe to the service and establish with the service provider a service-level agreement (SLA) defining the quality-of-service parameters under which the service is delivered. The utility-oriented nature of cloud computing is clearly expressed by Buyya et al.

A cloud is a type of parallel and distributed system consisting of a collection of interconnected and virtualized computers that are dynamically provisioned and presented as one or more unified computing resources based on service-level agreements established through negotiation between the service provider and consumers.

10.3.3 A Closer Look

Cloud computing is helping enterprises, governments, public and private institutions, and research organizations shape more effective and demand-driven computing systems. Access to, as well as integration of, cloud computing resources and systems is now as easy as performing a credit card transaction over the Internet. Practical examples of such systems exist across all market segments.

(1) Large enterprises can **offload** some of their activities to cloud-based systems. Recently, the **New York Times** has converted its digital library of past editions into a Web-friendly format. This required a considerable amount of computing power for a short period of time. By renting Amazon EC2 and S3 Cloud resources, the **Times** performed this task in 36 hours and relinquished these resources, with no additional costs.

(2) Small enterprises and **start-ups** can afford to translate their ideas into business results more quickly, without excessive **up-front** costs. Animoto is a company that creates videos out of images, music, and video fragments submitted by users. The process involves a considerable amount of storage and backend processing required for producing the video, which is finally made available to the user. Animoto does not own a single server and bases its computing infrastructure entirely on Amazon **Web Services**, which are sized on demand according to the overall workload to be processed. Such workload can vary a lot and require instant scalability. 3Up-front investment is clearly not an effective solution for many companies, and cloud computing systems become an appropriate alternative.

① Reese G. Cloud application architectures: building applications and infrastructure in the cloud. Sebastopol, CA, USA: O'Reilly Media Inc.; 2009.

(3) System developers can concentrate on the **business logic** rather than dealing with the complexity of infrastructure management and scalability. Little Fluffy Toys is a company in London that has developed a **widget** providing users with information about nearby bicycle rental services. The company has managed to back the widget's computing needs on Google AppEngine and be on the market in only one week.

(4) End users can have their documents accessible from everywhere and any device. Apple iCloud is a service that allows users to have their documents stored in the Cloud and access them from any device users connect to it. This makes it possible to take a picture while traveling with a smartphone, go back home and edit the same picture on your laptop, and have it show as updated on your tablet computer. This process is completely transparent to the user, who does not have to set up cables and connect these devices with each other.

How is all of this made possible? The same concept of IT services **on demand** — whether computing power, storage, or runtime environments for applications — on a pay-as-you-go basis accommodates these four different scenarios. Cloud computing does not only contribute with the opportunity of easily accessing IT services on demand, it also introduces a new way of thinking about IT services and resources: as utilities.

The three major models for deploying and accessing cloud computing environments are **public clouds, private/enterprise clouds**, and **hybrid clouds**.

Public clouds are the most common deployment models in which necessary IT infrastructure (e.g., virtualized datacenters) is established by a third-party service provider that makes it available to any consumer on a subscription basis. Such clouds are appealing to users because they allow users to quickly leverage compute, storage, and application services. In this environment, users' data and applications are deployed on cloud datacenters on the vendor's premises.

Large organizations that own massive computing infrastructures can still benefit from cloud computing by replicating the cloud IT service delivery model in-house. This idea has given birth to the concept of private clouds as opposed to public clouds. In 2010, for example, the U.S. federal government, one of the world's largest consumers of IT spending (around $76 billion on more than 10,000 systems) started a cloud computing initiative aimed at providing government agencies with a more efficient use of their computing facilities. The use of cloud-based in-house solutions is also driven by the need to keep confidential information within an organization's premises. Institutions such as governments and banks that have high security, privacy, and regulatory concerns prefer to build and use their own private or enterprise clouds.

(10-5) Whenever private cloud resources are unable to meet users' quality-of-service requirements, hybrid computing systems, partially composed of public cloud resources and privately owned infrastructures, are created to serve the organization's needs. These are often referred as hybrid clouds, which are becoming a common way for many stakeholders to start exploring the possibilities offered by cloud computing.

10.3.4 The Cloud Computing Reference Model

A fundamental characteristic of cloud computing is the capability to deliver, on demand, a variety of IT services that are quite diverse from each other. This variety creates different perceptions of what cloud computing is among users. Despite this lack of uniformity, it is possible to classify cloud computing services offerings into three major categories: **Infrastructure-as-a-Service** (IaaS), Platform-as-a-Service (PaaS), and Software-as-a-Service (SaaS). These categories are related to each other as described in Fig.10-1, which provides an organic view of cloud computing. We refer to this diagram as the Cloud Computing Reference Model.

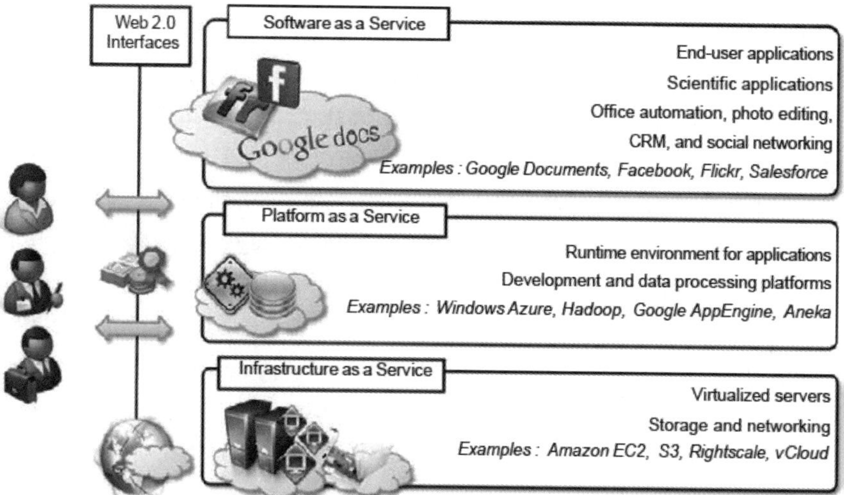

Fig.10-1 The Cloud Computing Reference Model

The model organizes the wide range of cloud computing services into a layered view that walks the computing stack from bottom to top.

At the base of the stack, Infrastructure-as-a-Service solutions deliver infrastructure on demand in the form of virtual hardware, storage, and networking. Virtual hardware is utilized to provide compute on demand in the form of virtual machine instances. These are created at users' request on the provider's infrastructure, and users are given tools and interfaces to configure the software stack installed in the virtual machine. The pricing model is usually defined in terms of dollars per hour, where the hourly cost is influenced by the characteristics of the virtual hardware. Virtual storage is delivered in the form of raw disk space or object store. The former complements a virtual hardware offering that requires persistent storage. The latter is a more high-level abstraction for storing entities rather than files. Virtual networking identifies the collection of services that manage the networking among virtual instances and their connectivity to the Internet or private networks.

Platform-as-a-Service solutions are the next step in the stack. They deliver scalable and elastic runtime environments on demand and host the execution of applications. These services are backed by a core middleware platform that is responsible for creating the abstract environment where applications are deployed and executed. It is the responsibility of the service provider to provide scalability and to manage fault tolerance, while users are requested to focus on the logic of the application developed by leveraging the provider's APIs and libraries. This approach increases the level of abstraction at which cloud computing is leveraged but also constrains the user in a more controlled environment.

At the top of the stack, Software-as-a-Service solutions provide applications and services on demand. Most of the common functionalities of desktop applications — such as office automation, document management, photo editing, and customer relationship management (CRM) software — are replicated on the provider's infrastructure and made more scalable and accessible through a browser on demand. These applications are shared across multiple users whose interaction is isolated from the other users. The SaaS layer is also the area of social networking Websites, which leverage cloud-based infrastructures to sustain the load generated by their popularity.

Each layer provides a different service to users. IaaS solutions are sought by users who want to leverage cloud computing from building dynamically scalable computing systems requiring a specific software stack. IaaS services are therefore used to develop scalable Websites or for background processing. PaaS solutions provide scalable programming platforms for developing applications and are more appropriate when new systems have to be developed. SaaS solutions target mostly end users who want to benefit from the elastic scalability of the cloud without doing any software development, installation, configuration, and maintenance. This solution is appropriate when there are existing SaaS services that fit users needs (such as email, document management, CRM, etc.) and a minimum level of customization is needed.

10.4 Big Data

10.4.1 Let the Data Speak

The fruits of the information society are easy to see, with a cellphone in every pocket, a computer in every backpack, and big information technology systems in back offices everywhere. But less noticeable is the information itself. Half a century after computers entered mainstream society, the data has begun to accumulate to the point where something new and special is taking place. Not only is the world awash with more information than ever before, but that information is growing faster. The change of scale has led to a change of state. The quantitative change has led to a qualitative one. The sciences like astronomy and genomics, which first experienced the explosion in the 2000s, coined the term "**big data**". The concept is now migrating to all areas of human endeavor.

There is no rigorous definition of big data. Initially the idea was that the volume of information had grown so large that the quantity being examined no longer fit into the memory that computers use for processing, so engineers needed to **revamp** the tools they used for analyzing it all. That is the origin of new processing technologies like Google's MapReduce[①] and its open-source equivalent, Hadoop, which came out of Yahoo. These let one manage far larger quantities of data than before, and the data — importantly — need not be placed in tidy rows or classic database tables. Other **data-crunching technologies** that dispense with the rigid hierarchies and homogeneity of yore are also on the horizon. At the same time, because Internet companies could collect vast troves of data and had a burning financial incentive to make sense of them, they became the leading users of the latest processing technologies, superseding offline companies that had, in some cases, decades more experience.

One way to think about the issue today is this: big data refers to things one can do at a large scale that cannot be done at a smaller one, to extract new insights or create new forms of value, in ways that change markets, organizations, the relationship between citizens and governments, and more.

But this is just the start. The era of big data challenges the way we live and interact with the world. Most strikingly, society will need to shed some of its obsession for causality in exchange for simple correlations: not knowing why but only what. This overturns centuries of established practices and challenges our most basic understanding of how to make decisions and comprehend reality.

Big data marks the beginning of a major transformation. Like so many new technologies, big data will surely become a victim of **Silicon Valley's**[②] notorious **hype cycle**: after being feted on the cover of magazines and at industry conferences, the trend will be dismissed and many of the data-smitten startups will flounder. But both the infatuation and the damnation profoundly misunderstand the importance of what is taking place. Just as the telescope enabled us to comprehend the universe and the microscope allowed us to understand germs, the new techniques for collecting and analyzing huge bodies of data will help us make sense of our world in ways we are just starting to appreciate.

To appreciate the degree to which an information revolution is already under way, consider trends from across the spectrum of society. Our digital universe is constantly expanding. Take astronomy. When the **Sloan Digital Sky Survey** began in 2000, its telescope in New Mexico collected more data in its first few weeks than had been amassed in the entire history of astronomy. By 2010 the survey's archive teemed with a whopping 140 **Terabytes** of information. But a successor, the **Large Synoptic Survey Telescope** in Chile, due to come on stream in 2016, will acquire that quantity of data every five days.

Such astronomical quantities are found closer to home as well. When scientists first

① MapReduce is a programming model and an associated implementation for processing and generating big data sets with a parallel, distributed algorithm on a cluster.

② Silicon Valley is a nickname for the southern portion of the San Francisco Bay Area, where is now the home to many of the world's largest high-tech corporations.

decoded the human genome in 2003, it took them a decade of intensive work to sequence the three billion base pairs. Now, a decade later, a single facility can sequence that much DNA in a day. In finance, about seven billion shares change hands every day on U.S. equity markets, of which around two-thirds is traded by computer algorithms based on mathematical models that crunch mountains of data to predict gains while trying to reduce risk.

Internet companies have been particularly swamped. Google processes more than 24 **petabytes** of data per day, a volume that is thousands of times the quantity of all printed material in the U.S. Library of Congress. Facebook, a company that didn't exist a decade ago, gets more than 10 million new photos uploaded every hour. Facebook members click a "like" button or leave a comment nearly three billion times per day, creating a digital trail that the company can mine to learn about users' preferences. Meanwhile, the 800 million monthly users of Google's YouTube service upload over an hour of video every second. The number of messages on Twitter grows at around 200 percent a year and by 2012 had exceeded 400 million tweets a day.

From the sciences to healthcare, from banking to the Internet, the sectors may be diverse yet together they tell a similar story: the amount of data in the world is growing fast, outstripping not just our machines but our imaginations.

Many people have tried to put an actual figure on the quantity of information that surrounds us and to calculate how fast it grows. They've had varying degrees of success because they've measured different things. One of the more comprehensive studies was done by Martin Hilbert of the University of Southern California's Annenberg School for Communication and Journalism. He has striven to put a figure on everything that has been produced, stored, and communicated. That would include not only books, paintings, emails, photographs, music, and video (analog and digital), but video games, phone calls, even car navigation systems and letters sent through the mail. He also included broadcast media like television and radio, based on audience reach.

By Hilbert's reckoning, more than 300 **exabytes** of stored data existed in 2007. To understand what this means in slightly more human terms, think of it like this. A full-length feature film in digital form can be compressed into a one gigabyte file. An **exabyte** is one billion gigabytes. In short, it's a lot. Interestingly, in 2007 only about 7 percent of the data was analog (paper, books, photographic prints, and so on). The rest was digital. But not long ago the picture looked very different. Though the ideas of the "information revolution" and "digital age" have been around since the 1960s, they have only just become a reality by some measures. As recently as the year 2000, only a quarter of the stored information in the world was digital. The other three-quarters were on paper, film, vinyl LP records, magnetic cassette tapes, and the like.

Things really are speeding up. The amount of stored information grows four times faster than the world economy, while the processing power of computers grows nine times faster. Little wonder that people complain of information overload. Everyone is whiplashed by the changes.

Consider an analogy from nanotechnology — where things get smaller, not bigger. The principle behind nanotechnology is that when you get to the molecular level, the physical properties can change. Knowing those new characteristics means you can devise materials to do things that could not be done before. At the nanoscale, for example, more flexible metals and stretchable ceramics are possible. Conversely, when we increase the scale of the data that we work with, we can do new things that weren't possible when we just worked with smaller amounts.

Sometimes the constraints that we live with, and presume are the same for everything, are really only functions of the scale in which we operate. Take another analogy, again from the sciences. For humans, the single most important physical law is gravity: it reigns over all that we do. But for tiny insects, gravity is mostly immaterial. For some, like water striders, the operative law of the physical universe is surface tension, which allows them to walk across a pond without falling in.

With information, as with physics, size matters. Hence, Google is able to identify the prevalence of the flu just about as well as official data based on actual patient visits to the doctor. It can do this by combing through hundreds of billions of search terms — and it can produce an answer in near real time, far faster than official sources. Likewise, Etzioni's Farecast can predict the price volatility of an airplane ticket and thus shift substantial economic power into the hands of consumers. But both can do so well only by analyzing hundreds of billions of data points.

(10-6) These two examples show the scientific and societal importance of big data as well as the degree to which big data can become a source of economic value. They mark two ways in which the world of big data is poised to shake up everything from businesses and the sciences to healthcare, government, education, economics, the humanities, and every other aspect of society.

10.4.2 Definition and Characteristic of Big Data

The term Big Data has been in use since the 1990s, with some giving credit to John Mashey for coining or at least making it popular. Big data usually includes data sets with sizes beyond the ability of commonly used software tools to capture, curate, manage, and process data within a tolerable elapsed time. Big Data philosophy encompasses unstructured, semi-structured and structured data, however the main focus is on unstructured data. Big data "size" is a constantly moving target, as of 2012 ranging from a few dozen terabytes to many petabytes of data. Big data requires a set of techniques and technologies with new forms of integration to reveal insights from datasets that are diverse, complex, and of a massive scale.

In a 2001 research report and related lectures, META Group (now Gartner) defined data growth challenges and opportunities as being three-dimensional, i.e. increasing volume (amount of data), velocity (speed of data in and out), and variety (range of data types and sources). Gartner, and now much of the industry, continue to use this "3Vs" model for describing big data. In 2012, Gartner updated its definition as follows: "Big Data is high-volume, high-velocity and/or high-variety information assets that demand cost-effective, innovative forms of information processing that enable enhanced insight, decision making, and process automation." Gartner's definition of the 3Vs is still widely used, and in agreement with a consensual definition that states that "Big Data represents the Information assets characterized by such a High Volume,

Velocity and Variety to require specific Technology and Analytical Methods for its transformation into Value". Additionally, a new V "Veracity" is added by some organizations to describe it, revisionism challenged by some industry authorities.

Big data can be described by the following characteristics.

Volume: the quantity of generated and stored data. The size of the data determines the value and potential insight- and whether it can actually be considered big data or not.

Variety: the type and nature of the data. This helps people who analyze it to effectively use the resulting insight.

Velocity: in this context, the speed at which the data is generated and processed to meet the demands and challenges that lie in the path of growth and development.

Variability: inconsistency of the data set can hamper processes to handle and manage it.

Veracity: the quality of captured data can vary greatly, affecting the accurate analysis.

10.4.3 Value of Big Data

With big data, the value of data is changing. In the digital age, data shed its role of supporting transactions and often became the good itself that was traded. In a big data world, things change again. Data's value shifts from its primary use to its potential future uses. This has profound consequences. It affects how businesses value the data they hold and who they let access it. It enables, and may force, companies to change their business models. It alters how organizations think about data and how they use it.

Information has always been essential for market transactions. Data enables price discovery, for instance, which is a signal for how much to produce. This dimension of data is well understood. Certain types of information have long been traded on markets. Content found in books, articles, music, and movies is an example, as is financial information like stock prices. These have been joined in the past few decades by personal data. Specialized data brokers in the United States such as Acxiom, Experian, and Equifax charge handsomely for comprehensive **dossiers** of personal information on hundreds of millions of consumers. With Facebook, Twitter, LinkedIn, and other social media platforms, our personal connections, opinions, preferences, and patterns of everyday living have joined the pool of personal information already available about us.

In short, although data has long been valuable, it was either seen as ancillary to the core operations of running a business, or limited to relatively narrow categories such as intellectual property or personal information. In contrast, in the age of big data, all data will be regarded as valuable.

When we say, "All data," we mean even the rawest, most seemingly mundane bits of information. Think of readings from a heat sensor on a factory machine. Or the real-time stream of GPS coordinates, accelerometer readings, and fuel levels from a delivery vehicle — or a fleet of 60,000 of them. Or think of billions of old search queries, or the price of nearly every seat on every commercial airline flight in the United States going back years.

Until recently there were no easy ways to collect, store, and analyze such data, which severely limited the opportunities to extract its potential value. In Adam Smith's celebrated example of the pin maker, with which he discussed the division of labor in the eighteenth century, it would have required observers watching all the workers not just for one particular study, but at all times everyday, taking detailed measurements, and counting the output on thick paper with feathery quill pens. When classical economists considered the factors of production (land, labor, and capital), the idea of harnessing data was largely absent. Though the cost of gathering and using data has declined over the past two centuries, until fairly recently it remained relatively expensive.

What makes our era different is that many of the inherent limitations on the collection of data no longer exist. Technology has reached a point where vast amounts of information often can be captured and recorded cheaply. Data can frequently be collected passively, without much effort or even awareness on the part of those being recorded. And because the cost of storage has fallen so much, it is easier to justify keeping data than discarding it. All this makes much more data available at lower cost than ever before. Over the past half-century, the cost of digital storage has been roughly cut in half every two years, while storage density has increased 50 million-fold. In light of informational firms like Farecast or Google — where raw facts go in at one end of a digital assembly line and processed information comes out at the other — data is starting to look like a new resource or factor of production.

The immediate value of most data is evident to those who collect it. In fact, they probably gather it with a specific purpose in mind. Stores collect sales data for proper financial accounting. Factories monitor their output to ensure it conforms to quality standards. Websites log every click users make — sometimes even where the mouse-cursor moves — for analyzing and optimizing the content the sites present to visitors. These primary uses of the data justify its collection and processing. When Amazon records not only the books that customers buy but the web pages they merely look at, it knows it will use the data to offer **personalized recommendations**. Similarly, Facebook tracks users' "status updates" and "likes" to determine the most suitable ads to display on its website to earn revenue.

Unlike material things — the food we eat, a candle that burns — data's value does not diminish when it is used; it can be processed again and again. Information is what economists call a "non-rivalrous" good: one person's use of it does not impede another's. And information doesn't wear out with use the way material goods do. Hence Amazon can use data from past transactions when making recommendations to its customers — and use it repeatedly, not only for the customer who generated the data but for many others as well.

Just as data can be used many times for the same purpose, more importantly, it can be harnessed for multiple purposes as well. This point is important as we try to understand how much information will be worth to us in the era of big data. We've seen some of this potential realized already, as when Walmart searched its database of old sales receipts and spotted the lucrative correlation between hurricanes and Pop-Tarts sales.

All this suggests that data's full value is much greater than the value extracted from its first use. It also means that companies can exploit data effectively even if the first or each subsequent use only brings a tiny amount of value, so long as they utilize the data many times over.

10.4.4 Risk of Big Data

With big data promising valuable insights to those who analyze it, all signs seem to point to a further surge in others' gathering, storing, and reusing our personal data. The size and scale of data collections will increase by leaps and bounds as storage costs continue to plummet and analytic tools become ever more powerful. If the Internet age threatened privacy, does big data endanger it even more? Is that the dark side of big data?

Yes, and it is not the only one. Here, too, the essential point about big data is that a change of scale leads to a change of state. As we'll explain, this transformation not only makes protecting privacy much harder, but also presents an entirely new **menace**: penalties based on **propensities**. That is the possibility of using big-data predictions about people to judge and punish them even before they've acted. Doing this negates ideas of fairness, justice, and free will.

In addition to privacy and propensity, there is a third danger. We risk falling victim to a dictatorship of data, whereby we fetishize the information, the output of our analyses, and end up misusing it. Handled responsibly, big data is a useful tool of rational decision-making. Wielded unwisely, it can become an instrument of the powerful, who may turn it into a source of repression, either by simply frustrating customers and employees or, worse, by harming citizens.

Still, much of the data that's now being generated does include personal information. And companies have a welter of incentives to capture more, keep it longer, and reuse it often. The data may not even explicitly seem like personal information, but with big-data processes it can easily be traced back to the individual it refers to. Or intimate details about a person's life can be deduced.

The important question, however, is not whether big data increases the risk to privacy (it does), but whether it changes the character of the risk. If the threat is simply larger, then the laws and rules that protect privacy may still work in the big-data age; all we need to do is redouble our existing efforts. On the other hand, if the problem changes, we may need new solutions. Unfortunately, the problem has been transformed. With big data, the value of information no longer resides solely in its primary purpose. As we've argued, it is now in secondary uses.

This change undermines the central role assigned to individuals in current privacy laws. Today they are told at the time of collection which information is being gathered and for what purpose; then they have an opportunity to agree, so that collection can commence. While this concept of "notice and consent" is not the only lawful way to gather and process personal data, according to Fred Cate, a privacy expert at Indiana University, it has been **transmogrified** into a cornerstone of privacy principles around the world.

Strikingly, in a big-data age, most innovative secondary uses haven't been imagined when the data is first collected. How can companies provide notice for a purpose that has yet to exist? How can individuals give informed consent to an unknown? Yet in the absence of consent, any big-data analysis containing personal information might require going back to every person and asking permission for each reuse. Can you imagine Google trying to contact hundreds of millions of users for approval to use their old search queries to predict the flu? No company would shoulder the cost, even if the task were technically feasible.

The alternative, asking users to agree to any possible future use of their data at the time of collection, isn't helpful either. Such a wholesale permission emasculates the very notion of informed consent. In the context of big data, the tried and trusted concept of notice and consent is often either too restrictive to unearth data's latent value or too empty to protect individuals' privacy.

Other ways of protecting privacy fail as well. If everyone's information is in a dataset, even choosing to "opt out" may leave a trace. Take Google's Street View, its cars collected images of roads and houses in many countries. In Germany, Google faced widespread public and media protests. People feared that pictures of their homes and gardens could aid gangs of burglars in selecting lucrative targets. Under regulatory pressure, Google agreed to let homeowners opt out by blurring their houses in the image. But the opt-out is visible on Street View — you notice the obfuscated houses — and burglars may interpret this as a signal that they are especially good targets.

A technical approach to protecting privacy — anonymization — also doesn't work effectively in many cases. **Anonymization** refers to stripping out from datasets any personal identifiers, such as name, address, credit card number, date of birth, or Social Security number. The resulting data can then be analyzed and shared without compromising anyone's privacy. That works in a world of small data. But big data, with its increase in the quantity and variety of information, facilitates re-identification. Consider the cases of seemingly unidentifiable web searches and movie ratings.

In the era of big data, the three core strategies long used to ensure privacy — individual notice and consent, opting out, and anonymization — have lost much of their effectiveness. Already today many users feel their privacy is being violated. Just wait until big-data practices become more commonplace.

10.5 Key Terms and Review Questions

1. Technical Terms

quantum	量子	10.1
qubit	量子位	10.1
superposition	迭加	10.1
ions	离子	10.1
photons	光子	10.1

teraflop	每秒万亿次浮点运算	10.1
gigaflop	每秒百万次浮点运算	10.1
entanglement	纠缠	10.1
unitary transformation	酉变换	10.1
quantum parallelism	量子并行	10.1
decoherence	消相干	10.1
alanine	丙氨酸	10.1
trichloroethylene	三氯乙烯	10.1
phase coherence	相位相干	10.1
ion traps	离子阱	10.1
vibration	振动，振荡	10.1
quantum cryptography	量子密码学	10.1
Heisenberg's uncertainty principle	海森堡不确定性原理	10.1
no-cloning theorem	不可克隆原理	10.1
commitment	承诺，保证	10.1
oblivious transfer	不经意传输	10.1
man-in-the-middle attack	中间人攻击	10.1
credential	证书，凭据	10.1
colluding adversaries	共谋的敌对者	10.1
post-quantum cryptography	后量子密码学	10.1
zero-knowledge proof systems	零知识证明系统	10.1
quantum resistant	量子抵抗	10.1
deep learning	深度学习	10.2
circumscribed	受限制的	10.2
logistic regression	逻辑回归	10.2
naive Bayes	朴素贝叶斯	10.2
voxel	体元，立体像素	10.2
pitch	音高	10.2
vocal tract	声道，音腔	10.2
autoencoder	自编码	10.2
silhouette	轮廓，剪影	10.2
viewing angle	视角	10.2
disentangle	解脱，清理	10.2
multilayer perceptron	多层感知器	10.2
flow chart	流程图	10.2
deep probabilistic models	深度概率模型	10.2
cybernetics	控制论	10.2
connectionism	连接主义	10.2
artifical neural network	人工神经网络	10.2

reverse engineering	逆向工程	10.2
neuroscientific	神经科学的	10.2
auditory processing region	听觉处理区	10.2
hypothesis	假设	10.2
symbolic reasoning	符号推理	10.2
distributed representation	分散式表达	10.2
back-propagation	反向传播	10.2
kernel machines	核函数机	10.2
deep belief network	深度信度网络	10.2
cloud computing	云计算	10.3
virtual hardware	虚拟硬件	10.3
datacenters	数据中心	10.3
IT outsourcing	IT 外包	10.3
on-demand	按需；请求式	10.3
utility-oriented	面向应用的	10.3
pay-per-use	按次付费	10.3
offload	分流；卸下	10.3
start-ups	初创公司	10.3
up-front	预付的	10.3
Web Services	网络服务	10.3
business logic	业务逻辑	10.3
widget	桌面小程序	10.3
on demand	按需	10.3
pay-as-you-go	现购现付	10.3
public clouds	公有云	10.3
private/enterprise clouds	私有云/企业云	10.3
hybrid clouds	混合云	10.3
Infrastructure-as-a-Service	基础设施即服务	10.3
Platform-as-a-Service	平台即服务	10.3
Software-as-a-Service	软件即服务	10.3
big data	大数据	10.4
revamp	改进	10.4
data-crunching technology	数据处理技术	10.4
Silicon Valley	硅谷	10.4
hype cycle	技术成熟曲线	10.4
Sloan Digital Sky Survey	斯隆数字巡天	10.4
Terabyte	2^{40}～10^{12} 字节	10.4
Large Synoptic Survey Telescope	大型巡天望远镜	10.4

petabyte	2^{50}～10^{15}字节	10.4
exabyte	2^{60}～10^{18}字节	10.4
Spambot	垃圾邮件程序	10.4
inboxe	收件箱	10.4
registrant	注册人	10.4
decipher	破译	10.4
dossier	档案	10.4
personalized recommendation	个性化推荐	10.4
menace	威胁	10.4
propensity	偏好	10.4
transmogrify	变形	10.4
anonymization	匿名化	10.4
veracity	真实性，可靠性	10.4
delineate	描绘	10.4
nonlinear relationship	非线性关系	10.4
causal effect	因果效应	10.4

2. Translation Exercises

(10-1) If, as Moore's Law states, the number of transistors on a microprocessor continues to double every 18 months, the year 2020 or 2030 will find the circuits on a microprocessor measured on an atomic scale. And the logical next step will be to create quantum computers, which will harness the power of atoms and molecules to perform memory and processing tasks. Quantum computers have the potential to perform certain calculations significantly faster than any silicon-based computer.

(10-2) A major source of difficulty in many real-world artificial intelligence applications is that many of the factors of variation influence every single piece of data we are able to observe. The individual pixels in an image of a red car might be very close to black at night. The shape of the car's silhouette depends on the viewing angle. Most applications require us to disentangle the factors of variation and discard the ones that we do not care about.

(10-3) In this case, the depth of the flow chart of the computations needed to compute the representation of each concept may be much deeper than the graph of the concepts themselves. This is because the system's understanding of the simpler concepts can be refined given information about the more complex concepts. For example, an AI system observing an image of a face with one eye in shadow may initially only see one eye. After detecting that a face is present, it can then infer that a second eye is probably present as well. In this case, the graph of concepts only includes two layers — a layer for eyes and a layer for faces — but the graph of computations includes $2n$ layers if we refine our estimate of each concept given the other n times.

(10-4) Cloud computing is a model for enabling ubiquitous, convenient, on-demand network access to a shared pool of configurable computing resources (e.g., networks, servers, storage, applications, and services) that can be rapidly provisioned and released with minimal management effort or service provider interaction.

(10-5) Whenever private cloud resources are unable to meet users' quality-of-service requirements, hybrid computing systems, partially composed of public cloud resources and privately owned infrastructures, are created to serve the organization's needs. These are often referred as hybrid clouds, which are becoming a common way for many stakeholders to start exploring the possibilities offered by cloud computing.

(10-6) These two examples show the scientific and societal importance of big data as well as the degree to which big data can become a source of economic value. They mark two ways in which the world of big data is poised to shake up everything from businesses and the sciences to healthcare, government, education, economics, the humanities, and every other aspect of society.

References

[1] Quantum computing[DB/OL]. [2017-07-08].https://en.wikipedia.org/wiki/Quantum_computing.

[2] Bonsor K, Strickland J. How Quantum Computers Work[Z/OL]. [2017-07-08].http://computer.howstuffworks.com/quantum-computer.htm.

[3] Bengio Y, et al. Deep Learning[M], MIT Press, 2016.

[4] Deep learning[DB/OL]. [2017-07-09] .https://en.wikipedia.org/wiki/Deep_learning#Fundamental_concepts.

[5] Buyya R, Vecchiola C, Selvi S T. Mastering Cloud Computing: Foundations and Applications Programming[M]. Morgan Kaufmann Publishers,2013.

[6] Mayer-Schönberger V, Cukier K. Big Data: A Revolution That Will Transform How We Live, Work, and Think[M]. Eamon Dolan/Houghton Mifflin Harcourt, 2013.

图书资源支持

感谢您一直以来对清华版图书的支持和爱护。为了配合本书的使用，本书提供配套的资源，有需求的读者请扫描下方的“书圈”微信公众号二维码，在图书专区下载，也可以拨打电话或发送电子邮件咨询。

如果您在使用本书的过程中遇到了什么问题，或者有相关图书出版计划，也请您发邮件告诉我们，以便我们更好地为您服务。

我们的联系方式：

地　　址：北京市海淀区双清路学研大厦 A 座 701

邮　　编：100084

电　　话：010－62770175－4608

资源下载：http://www.tup.com.cn

客服邮箱：tupjsj@vip.163.com

QQ：2301891038（请写明您的单位和姓名）

资源下载、样书申请

书圈

扫一扫，获取最新目录

用微信扫一扫右边的二维码，即可关注清华大学出版社公众号“书圈”。